Thermal Food Engineering Operations

Scrivener Publishing
100 Cummings Center, Suite 541J
Beverly, MA 01915-6106

Bioprocessing in Food Science

Series Editor: Anil Panghal, PhD

Scope: Bioprocessing in Food Science will comprise a series of volumes covering the entirety of food science, unit operations in food processing, nutrition, food chemistry, microbiology, biotechnology, physics and engineering during harvesting, processing, packaging, food safety, and storage and supply chain of food. The main objectives of this series are to disseminate knowledge pertaining to recent technologies developed in the field of food science and food process engineering to students, researchers and industry people. This will enable them to make crucial decisions regarding adoption, implementation, economics and constraints of the different technologies.

As the demand of healthy food is increasing in the current global scenario, so manufacturers are searching for new possibilities for occupying a major share in a rapidly changing food market. Compiled reports and knowledge on bioprocessing and food products is a must for industry people. In the current scenario, academia, researchers and food industries are working in a scattered manner and different technologies developed at each level are not implemented for the benefits of different stake holders. However, the advancements in bioprocesses are required at all levels for betterment of food industries and consumers.

The volumes in this series will be comprehensive compilations of all the research that has been carried out so far, their practical applications and the future scope of research and development in the food bioprocessing industry. The novel technologies employed for processing different types of foods, encompassing the background, principles, classification, applications, equipment, effect on foods, legislative issue, technology implementation, constraints, and food and human safety concerns will be covered in this series in an orderly fashion. These volumes will comprehensively meet the knowledge requirements for the curriculum of undergraduate, postgraduate and research students for learning the concepts of bioprocessing in food engineering. Undergraduate, post graduate students and academicians, researchers in academics and in the industry, large- and small-scale manufacturers, national research laboratories, all working in the field of food science, agri-processing and food biotechnology will benefit.

Publishers at Scrivener
Martin Scrivener (martin@scrivenerpublishing.com)
Phillip Carmical (pcarmical@scrivenerpublishing.com)

Thermal Food Engineering Operations

Edited by

Nitin Kumar
Anil Panghal
and
M. K. Garg

Scrivener Publishing

WILEY

This edition first published 2022 by John Wiley & Sons, Inc., 111 River Street, Hoboken, NJ 07030, USA and Scrivener Publishing LLC, 100 Cummings Center, Suite 541J, Beverly, MA 01915, USA

For more information about Scrivener publications please visit www.scrivenerpublishing.com.

Wiley Global Headquarters
111 River Street, Hoboken, NJ 07030, USA

For details of our global editorial offices, customer services, and more information about Wiley products visit us at www.wiley.com.

Library of Congress Cataloging-in-Publication Data

ISBN 9781119775591

Cover image: Wikimedia Commons and the editors
Cover design by Russell Richardson

Set in size of 11pt and Minion Pro by Manila Typesetting Company, Makati, Philippines

Printed in the USA

10 9 8 7 6 5 4 3 2 1

Contents

Preface

Thermal processing is a significant component of the undergraduate and postgraduate degrees in agriculture engineering, food engineering and food science technology throughout the world. Thermal food engineering operations are considered one of the core competencies for these programs and in industries as well. Researchers will be able to use the information as a guide in establishing the direction of future research on thermophysical properties and food processing. The audience for this volume will be the student preparing for a career as a food engineer, practicing engineers in the food and related industries, and scientists and technologists seeking information about processes and the information needed in design and development of thermal food engineering processes and operations. Simultaneously, improving food quality and food safety are continue to be critical issues during thermal processing. So, quality, food safety and role of technology in food industry are discussed to cover these areas of food industry.

A great variety of topics is covered, with the emphasis on the most recent development in thermal operations in food industry. The chapters presented in this volume throw light on a number of research subjects that have provided critical information on different thermal processes, their impact on different food components, and their feasibility in food industry. Each chapter also provides background information of the changes in different thermal operations which changed drastically over the years. The authors emphasis on newer thermal technologies which are making a great impact on the industry and the resulting finished products. The adoption of modern technology has increased efficiency and productivity within the factory. Most importantly, utilizing the newer thermal operations has greatly improved product quality. All chapters are supported with a wealth of useful references that should prove to be an invaluable source for the reader. Self-explanatory illustrations and tables have been incorporated in each chapter complimentary to the main text.

Thanks are due to all authors for contributing their knowledgeable chapters in this volume and helping us to complete the book. We also thank the authorities of Chaudhary Charan Singh Haryana Agricultural University, Hisar for their help and support. Finally, we also express indebtedness and thankfulness to Scrivener Publishing and Wiley team for their unfailing guidance and helpful assistance.

Nitin Kumar
Anil Panghal
M.K. Garg

1

Novel Thermal Technologies: Trends and Prospects

Amrita Preetam[1]*, Vipasha[1], Sushree Titikshya[1], Vivek Kumar[1], K.K. Pant[2] and S. N. Naik[1]

[1]Centre for Rural Development and Technology, IIT Delhi, Delhi, India
[2]Department of Chemical Engineering, IIT Delhi, Delhi, India

Abstract
Heating is possibly the most traditional way of processing foods. The technologies involved in heating have been continuously developing for the past many years as per consumer need, satisfaction and demand. Techniques such as dielectric heating, ohmic heating, and infrared heating are evolving and can substitute for the conventional heating methods for improving quality and shelf life, and providing a faster production rate. The conventional technologies are primarily based on convective, conductive, and radiative heat transfer. But the new novel thermal methods are mainly relying on the electromagnetic field or electrical conductivity and are having cleaner environmental impacts such as energy saving, water savings, improved efficiencies, fewer emissions, and eventually decreasing dependency on non-renewable resources. The chapter discusses novel thermal technologies. Definitions, basic principles, environmental impacts, current trends, and future perspectives are described along with the mechanism and advantages of the novel thermal technologies. The novel thermal technologies are continuously emerging and evolving as per consumer requirements and need.

Keywords: Novel thermal technologies, infrared heating, ohmic heating, microwave heating, radiofrequency heating

1.1 Introduction

The primary goal for food processors is quality and safety assurance. To ensure microbiological food safety, the use of heat by thermal operation

**Corresponding author*: amritapreetam92@gmail.com

Nitin Kumar, Anil Panghal and M. K. Garg (eds.) Thermal Food Engineering Operations, (1–44)

involving drying, sterilization, evaporation, and other methods are common practices. The conventional heating methods rely on principles such as convection, radiation, and conduction [36] that primarily rely on heat generation exterior of the product to be warmed up. But there are limitations attached to it. These conventional ways of processing, due to the decrease in efficiency of heat transfer, by excessive heating because of time reach the thermal center of foods for conducting sufficient heat or losses because of the heat on the surface of equipment and installation. Some of these problems can be resolved by technical solutions such as heat recycling or advanced designing and installation methods but at high expense.

Therefore research has been made for raising the quality and safety and economic aspects of food through technological development. The novel thermal technologies in which the main processing factor is temperature change as the main parameter responsible for food processing can be considered as the promising alternative in food processing as compared to the traditional process. Unlike traditional technologies, novel thermal technologies are based on electromagnet field (EMI) or electric conductivity. Novel thermal technologies are based on the heat generations directly inside the food. The novel thermal technologies have successfully helped in enhancing the effectiveness of heat processing along with ensuring food safety and maintaining nutritional food properties. Infrared heating has also evolved for the processing of food. The thermal technologies involve the equipment plotted to heat the food to process it, whereas in non-thermal techniques the food is virtually processed without the involvement of food. The general definition of common technologies involved in novel thermal techniques and their basic differences are discussed below.

Ohmic heating is also called Joule heating, electrical residence heating. It is a method of heating the food by the passage of an electric current, so heat is generated due to the electrical current. It is a direct method, as the heat energy is directly dissipated into the food. It is primarily used to preserve food. Electric energy is dissipated into heat, which results in quick and uniform heating followed by maintaining the nutritional value and color. The key variable in electrical conductivity is designing of an effective ohmic meter. Ohmic heating uses a normal electrical supply frequency which is of 50-60Hz. Ohmic heating instantly penetrates directly into the food. The applications of ohmic heating include UHT sterilization, pasteurization, and others.

Dielectric heating is another novel process that provides volumetric heating, for uniform sterilization or preserving of food. It is also a direct method and is based on the process of heating the material by causing dielectric motion in its molecule using alternating electric fields, microwave electromagnetic radiation, or radio wave. The intensity of the electric field and the dielectric properties of the product regulate the volumetric power and absorption and the rate of heat generation. Both microwave heating and radiofrequency belong to this category and follow the principle of dielectric heating. The depth of penetration is directly related to frequency in the case of dielectric heating. The thermal conductivity is not so important in dielectric heating. The few application of microwave and radiofrequency are in freeze-drying, baking, sterilization, rendering, frying, and many others.

Infrared heating is mainly utilized to modify the eating characteristics of food by varying its color, texture, flavor, and odor. Radiant heat is less managed and has a broader range of frequencies. The thermal conductivity is a limiting factor in infrared heating. It acts as an indirect method of heating. Infrared is simply absorbed and converted into heat. It has limited penetration depth in food. It has several advantages over conventional methods such as decreased heating time, reduce quality loss, and uniform heating, versatility, easy to operate and compact equipment, and many others. It also has a vast area of application includes drying, frying, baking, cooking, freeze-drying, pasteurization, sterilization, blanching, and many others.

The other technique is non-thermal heating technologies which are based on pulsed light, pulsed electric fields, ultrasound, and gamma radiation, and others, where the temperature may change also but is not the prime parameter for food processing. The purpose of this chapter is to deliver a general outlook of novel thermal technologies in the food processing sector along with their environmental impact, current trends, and future perspective.

1.2 Novel Thermal Technologies: Current Status and Trends

The most common approach for food processing in the last 50 years is thermal processing because a huge amount of microorganisms are removed at elevated temperatures by killing them. Thermal processing protects food by pasteurization, hot air drying, and others, induces variations to improve food quality by baking, blanching, roasting, frying, and cooking. Time and temperature used are the key variable depending on the application used. In the case of thermal processing, sometimes the high temperature may lead to loss of nutrients or bioactive compounds which results in low-processed food and low-grade food.

So in such a situation, novel thermal techniques or with the combination of traditional technologies are used to modify the quality and shelf life while decreasing the change in sensory properties. The food industry is continuously developing in order to fulfill customer demand for food nutrition, natural flavors, food quality, and taste. Innovation and research are continuously growing all over the world to maintain and improve standards. Currently, consumers demand food with the least or no chemical additives and should be minimally processed [37]. These developing technologies are called 'novel' technologies because they are successfully fulfilling the needs of consumers and are an improvement of conventional technologies. Depending upon the principle used, it can be thermal or non-thermal. Techniques such as microwave, ultrasound, and pulsed electric field can be an alternative proved by many researchers to develop nutritious and safe food [10, 15, 16]. Such techniques are being used broadly by many innovative food companies [6]. As compared to traditional technologies these new emerging technologies have many benefits over traditional techniques such as more heat and mass transfer, improved product quality, short process and residence time, better functionality, enhanced preservation, and others. The processing of the food is important for taste, nutritional content, texture, and appearance [36]. The benefits of novel processing technologies over traditional techniques are improved functional characteristics and retention of sensory attributes by using the promising next-generation food [62]. The development, research, and large-scale set-up of these novel technologies are taking place internationally. It is evident from the number of publications on the benefits of novel thermal technologies in food processing in various food and agriculture processing research journals [51].

Microwave: The most popular and extensive technology studied worldwide both domestically and industrially is microwave processing due to its various advantages such as easy operation, lower maintenance requirement, and cleaner environment [77]. But despite all the advantages, microwave is facing two main hurdles, i.e., irregular distribution of temperature within the food product and high cost of energy regarding this technique [6]. Furthermore, the set-up operated at 2450 MHz may give rise to serious boundary and surface overheating of the food to outstretch the desired elevated temperature in cold spots. For those cases, continuous microwave systems have been used to provide uniform temperatures for the heating of foods. Some authors have suggested fusion with water as the heating medium, pulsed microwave [24, 51]. The most common technology is microwave-assisted thermal sterilization system (MATS™) based on

915MHz single-mode cavities using a shallow bed with water food immersion; it penetrates deeper in food and water offers to reduce the edge heating. It got approval in 2009 by the Food and Drug Administration (FDA) [72].

Infrared heating: Proved by many researchers, Infrared heating (IRH) is an efficient process for the purification of pathogenic microorganisms in food. Many operational variables such as food temperature, size and kind of food materials, IR power intensity, IR power intensity, and others are necessary for microbial inactivation. At a commercial scale, IRH had been found as the replacement or substitute to decrease non-uniform temperature distribution which occurs in microwave heating [6]. Internationally, IRH is used for blanching, drying, baking, roasting, and peeling. At the industrial level IRH has considerable advantages such as a large heat delivery rate, no medium required, high energy efficiency, low environmental footprint, and others [39]. But because of less penetration depth, this technology is not successful at the commercial level; for example, it cannot be utilized in in-packaging food processing. The major successful large-scale (commercial) applications of IRH is drying of low-moisture foods (grains, pasta, tea, etc.), also the applications in baking (e.g., pizzas, biscuits, and others) and in the oven for roasting of cereals, coffee, etc.

Radiofrequency: Another thermal technology is Radiofrequency heating (RFH), used from the 1940s. Earlier applications were to warm bread, dry up and blanch vegetables and others. RFH has a greater industrial interest because of its unique properties such as deeper penetration due to its lesser frequencies, uniform electric field distribution, and longer wavelength. Major applications are in the food-drying sectors for pasta, snacks, and crackers and sterilization or pasteurization process, treatment of seeds and disinfection of product [19]. As compared to microwave heating, RF has the potential to reduce surface overheating and can also give better results at a commercial scale [81]. On a commercial scale, such as for treating bulk materials, sterilization of packaged foods is successful because they are simple to construct, have a more uniform heating pattern, and have greater penetration depth. Drawbacks of this technology include, at industrial scale, the design equipment is complicated, there is a high investment cost and technical issues such as dielectric failure and thermal runaway heating that can damage package and product [1]. Another common thermal technique is ohmic heating (OH) where internal heat generation takes place by passing a current into the materials.

Ohmic heating (OH): Compared to other technologies, ohmic heating has advantages such as larger temperature in particles than in liquid, decreased fouling, energy-efficient, uniform heating (achieved by thermal, physical, and rheological properties), and lower cost [64]. The drawbacks include the requirement of aseptic packing after OH heating, the possibility of corrosion, direct exposure of the electrode with food.

Major utilization of OH are blanching, sterilization, evaporation, dehydration, extraction, and evaporation. The basic procedure involved in OH of microbial inactivation is thermal harm and in some cases by electroporation. In comparison to traditional heating, OH heating can attain lesser heating times, can keep away from hot surfaces, and can decrease the temperature gradients.

Since the 1990s, OH is now utilized in developing countries and all over the world. Almost a hundred processing plants have been placed all over the world. The market is in the developing stage and evolving constantly. OH equipment is installed all over the world such as in Italy, France, Spain, Greece, and Mexico [54]. The application of OH is not much commercialized for solid food products. For liquids, viscous liquids, and pumpable multiphase products, the installed set-ups perform the sterilization and pasteurization of numerous food products with great characteristics with main applications in vegetables and fruit areas.

Overall, the major issue involved in commercialization of electromagnetic techniques for numerous food applications is the lack of heat uniformity, which has a major impact on key variables of food processing and safety. To avoid this downside, hybrid systems are proposed, i.e., the combination of traditional and volumetric heating [54, 63]. The hybrid system offers advantages such as safety, improved process efficiency, and product properties. Successful hybrid techniques are IR-convective drying, a combination of IRH, IR-heat pump drying, and microwave heating, and many others are still in progress because of the magnified energy throughput.

1.2.1 Environmental Impact of Novel Thermal Technologies

The emergence of novel thermal technologies and non-thermal processes in food processing industries is capable of producing high-quality and standardized products. Both of them are environmentally sound and efficient in nature as compared to conventional technologies. Here we will consider more on the environmental footprints of novel thermal technologies. The primary objective in the food industry is food safety which requires high energy consumption, but novel thermal technologies are successfully able to balance energy saving and energy consumption.

The high value of hygiene and safety of food requires large use of water in both hot and cold cycles in production which consequently increases the environmental footprint. Processes such as cooking, sterilization, drying, and pasteurization require various types of energy. Novel thermal technologies are promising, attractive, and efficient in nature. They are capable of providing improved quality and reduced environmental effects which will eventually reduce environmental footprints. Novel thermal technologies can reduce processing costs followed by improving and maintaining the value-added products. Overall the primary types of energy used based on conventional thermal processing techniques are fossil fuel and electricity, majorly utilized in refrigeration and mechanical power in pumps. A heat exchanger is commonly used in the pasteurization of beverages where the pathogens are killed when heated to a particular residence time. During thermal treatment, convection and conduction play a major role to transfer heat to the products. For viscous fluids, directing heating process is applied, e.g., steam injection and steam infusion are utilized for thermal treatments. In the food and beverages industry, regarding the distribution of energy in 2002, Denmark suggests that total consumption of energy (TJ/Year) is 135,200 including the amount of heating and power. Adapted from [58].

This concludes that major heat is used in frying, evaporation, drying, and heating for thermal processes. Until the present moment, this trend is still functioning. Novel thermal technologies such as radio frequency, ohmic heating, microwave, etc., for food processing being continuously evolving. These novel thermal technologies have reduced emissions, reliability, improved productivity, high product quality, energy saving, water saving and consequently have less impact on the environment; [45] investigated that for Orange juice and cookies manufacturing, radio frequency drying (RF) can range up to 0 to 73.8 TJ per year in terms of primary energy saving. The major kinds of gas emissions from food industries are linked to power and heat production particulate matter and gases such as SO_2, CO_2, NO, from combustion processes. The particulate matter and volatile organic compounds (VOCs) and other chemical emissions are from methods such as size reduction, heating, refrigeration system, and cooking methods.

Conventionally 33% of the overall energy consumed in food processing corresponds to the production of steam. The steam is commonly used in drying, concentrating liquids, cooking, sterilizing, etc., in the processing of food processes. The generation of steam used in food industries involves the utilization of boilers. To remove the dissolved solid from the boiler system a large quantity of water is periodically drained from the bottom, which is called a blow-down. Inadequate blow-down may lead to the gathering of dirt which reduces the heat transfer rates and increases the loss of energy. Irregular boiler maintenance can decrease the efficiency of the boiler up to 20-30%. The efficiency of a boiler is affected by losses of heat by convection and radiation [58]. Improper boiler maintenance can also emit large emissions of CO_2 and loss of energy. [58] also mentions the losses that occurred of a boiler or steam generation system composed of: gases from the combustion of air, or incomplete combustion, radiation losses, boiler blow-down water, heat convection, and fouling of heat transfer surfaces from hot boiler surface. Many attempts have been made to evolve a sustainable sector for lowering the emission of gases, e.g., CO_2 and enhance the energy efficiency of devices and methods using renewable energy is now the main concern for every method. Therefore, using electricity in food powering systems may show an environmental benefit as compared to conventional techniques used. Overall novel technologies are considered sustainable, once they reduce the consumption of boilers or steam generation systems and eventually decrease the waste-water, heat loss and increase energy-saving and water-saving as well. Furthermore, the electricity is produced by an eco-friendly source of renewable energy; after that these methods will efficiently contribute to decreasing the pollution, assisting them to protect the environment. [82] shows the balancing by ohmic heating decreased the extent of solid leaching irrespective of the dimensions of the product. It is concluded by [65], OH blanching offers benefit in aspects such as water-saving by maintaining the quality of the processed products. Novel thermal can efficiently accelerate the drying processes when related to traditionally heat pre-treated samples allowing exact control of the process temperature and eventually it can decrease costs of energy, reduce the gas consumption and lower combustion-related emissions [53].

So we can conclude that novel thermal technologies are one of the most novel techniques in food preservation processes. Novel thermal technologies are quite efficient in all aspects such as the efficiency of energy, saving of water, and reduced emissions. Most of the processes involved in novel thermal technologies are green and hence more environmentally friendly, having the least environmental impact as compared to conventional technologies.

1.2.2 The Objective of Thermal Processing

The major purpose of thermal processing is to maintain certain quality standards, to reduce enzymatic activities, reduce microbial activities to enhance its shelf life, increase digestibility, and maintain certain physical and chemical variations to ensure its characteristic and safety of food. The objective also includes adding values such as maintaining its texture, flavor, color, etc., and make varieties of new products; it should also be needed by the specific section of the population. Over the past several years, consumer demands have improved standard, convenient and varied food which required the modification and development in existing traditional process for the new food preservation technologies. For that, the new novel thermal technologies evolved. Novel thermal technologies are better not only in terms of their quality improvements of food and heating efficiency but also in other important aspects such as water-saving, energy-saving, and reduced emissions. Most of these technologies are green and have less environmental impact and improve the added value of foods.

1.2.3 Preservation Process

The basic definition of food indicates that food is the materials, formulated or processed which are consumed orally by living organisms for development, pleasure, needs, and fulfillment. The chemical composition of food includes majorly water, fats, lipids, and carbohydrates with less amount of minerals and compounds containing organics. The different categories of food are perishable, synthetic, non-perishable, fresh, medical food, harvested, manufactured, preserved, and others. The preservation of food majorly depends on the type of food required to be produced and formulated. Preservation of food is defined as maintaining its properties at the desired level for long as possible. Safety with sustainability and innovation are the major aspects and priorities to ensure the preservation of food. In the modern era, the preservation of and processing of food not only includes the safety of the foods but also maintains sustainable innovation, economic feasibility, customer satisfaction, nutritional aspects, absence of chemical preservative, andshould be environmental sound [62].

Food preservation is necessary for ensuring desired quality level, consumer satisfaction, to maintain preservation length and also to focus on the group for whom the products are to be preserved [10]. The reason for preservation also includes to form the value-added products, provide modifications in diet, and most importantly to overcome the improper planning in agricultural sectors. Preservation loss not only results in minor deterioration of food but also results in the transformation of the food to a severely toxic state.

After a certain period of time, the quality and characteristics of food may get deteriorated and become undesirable for consumption so it becomes the prime factor to study the rate of variations of quality attributes which indicates its shelf life; this is a very important parameter to consider. The quality of products can rely on appearance, yield, eating characteristics, microbial characteristics, and the consumer's overall experience. The deterioration of food depends upon mechanical, chemical, physical, and microbial reactions. The quality of the product was maintained at every stage of food production and overall processing chains such as manufacture, storage, distribution, and sale. Additionally, the need for preservation must depend on its purpose and use, and consider the population for whom the preservation is to be done as the nutritional requirement and food restriction apply differently to different sections of groups.

There are many measures for food preservation; inhibition, inactivation, and avoiding recontamination are the common ones. Each method contains several processes of preservation such as inhibition, which includes a decrease of oxygen, adding preservatives, control of pH, freezing, drying, surface coating, gas removal, fermentation, and many others. Inactivation includes irradiation, sterilization, extrusion, and others; avoiding contamination involves packaging, hygienic processing, aseptic processing, and others.

Thermal technology has been the backbone of food production and preservation for many years. In this technology, the temperature is assumed to be the major parameter for preservation and processing mechanism to make food commercially sterile, i.e., to get rid of pathogens and microorganisms which usually grow in the normal shelf life of the food product. Thermally processing the food provides real importance to the food by increasing and preserving its shelf life longer than the chilled food processing technologies.

In novel thermal technologies, preservation is done by the use of electricity. Various forms of electrical energy are utilized for food preservation such as ohmic heating, high intensity pulsed electric field, high-voltage arc discharge, microwave heating, and low electric field stimulation. Ohmic heating is the most common and is based on volumetric heating which prevents the overheating of food, provides uniform and quick heating; it depends on the principle that generation of heat in the food is an outcome of electrical residence when an electric current is moved through the food product. Furthermore, ohmic heating prevents thermal damage and promotes the efficiency of energy. Similarly, microwave heating is also very common and utilized in almost every household and the food industry but its low penetration depth of microwave into solid provides thermal non-uniformity. The other available methods utilizing electric energy are also very versatile, useful, and efficient for the preservation of food.

The current electro heating can be used to produce to form new and up-to-date products with diversified functionality.

1.3 Types of Thermal Technologies

Thermal processings are perhaps most essential in the food sector which has been used for the past many years; it has been discovered to increase the quality and shelf life of food with heat treatments. Thermal processing is the heating of foods at a particular temperature for a specified period of time. There are several techniques available in thermal technologies such as radiofrequency heating, ohmic heating, blanching, drying, frying, chilling, infrared heating, freezing and microwave heating, and extrusion. Combined high-pressure thermal treatment of food is also a very efficient prospect for the processing and preservation of food. The most common among them are microwave heating, ohmic heating, combined microwave vacuum-drying, radiofrequency processing, and new hybrid drying technologies.

Hybrid technologies are the recent development in engineering in the operation and design of the dryers to attain dried products with desired characteristics. In hybrid technologies, the drying technologies are combined with the new drying techniques to achieve a new age drying process to reduce energy consumption and enhance product quality. New age drying technologies would be very helpful for the bioproducts in agricultural sectors for all economic, environmental, and product quality aspects. Instant infusion is another new process for the heat treatment of food, depending upon the product requirement providing mild pasteurization and sterilization. For effective and efficient pasteurization and sterilization, the following are the needs: rapid and small heating time, accurate, and small residence time at sterilizing temperature, and rapid cooling time.

For examining the food when it goes under thermal process, nuclear magnetic resonance (NMR) and magnetic resonance imaging (MRI) can be used as they possess some unique properties for the same. Both of them can be used to investigate variation in the food during processing. Both of them are non-invasive and able to detect water mobility. NMR is the most versatile analytical technique used in modern times. It is capable of revealing complicated multivariate information inside the optically opaque and complex food matrix, also the thermal transformation in liquid, suspension, and gels regarding food samples. NMR and MRI are based on the magnetic properties of atomic nuclei and many elements have isotopes with such properties. Both of them are superior to any other instrumental methods because both are non-invasive, non-destructive, both measure volumes instead of surfaces, and are able to extract both physical and chemical information. By both techniques, it can extract knowledge about diffusion, flow, water distribution, and others. Furthermore, processes such as heating, freezing, hydration, dehydration, and salting can be detected and monitored non-invasively.

1.3.1 Infrared Heating

1.3.1.1 Principal and Mechanism

Infrared was discovered by William Herschel in the 1800s. Infrared depicted below red (infra: below); red is the longest wavelength of visible light. IR heating is the transmission of thermal energy in the form of electromagnetic waves. Wavelength between 0.7 and 1000 micrometer, wavelength larger than visible light but smaller than those of radio waves are the infrared waves. Three major types of infrared waves are [39]:

1. Short waves 0.76-2 μm (near IR waves), temperature above 1000°C
2. Medium waves: 2-4 μm (medium IR waves), when the temperature is above 400 - 1000°C
3. Long waves: 4-1000 μm (far IR waves), when the temperature is below 400°C

The working principle of infrared waves includes: IR energy is electromagnetic radiation emitted by hot objects (quartz lamb, quartz tubes, or metal body) by vibrations and rotation of molecules. When it is absorbed, the radiation provides up its energy to heat materials. An object is a "blackbody", if it absorbs (or emits) 100% of incident IR radiation. The quantity of heat emitted from a perfect radiator (blackbody) is expressed by Stefan-Boltzmann law equation:

$$Q = \sigma_{SB} .A.T^4$$

Where Q is defined as the rate of heat emission, σ_{SB} is the Stefan-Boltzmann constant, T is defined as the absolute temperature, and A is defined as the surface area.

When radiant heaters and food products are not perfect absorbers, the Stefan-Boltzmann equation was modified and the concept of "grey body" was found:

$$Q = \sigma_{SB} .\varepsilon.A.T^4$$

Where ε defined as the emissivity of the grey body (ranging from 0 to 1).

This property changes with the wavelength of emitted radiation and temperature of the grey body.

The heating level depends on the absorbed energy, which then rely on the composition of food and the radiation frequency.

Mathematically, the transfer of heat rate to food is expressed as,

$$Q = \sigma_{SB}.\varepsilon.A.(T^4{}_1 - T^4)_2$$

Where T_1 depicts the temperature of the emitter and T_2 depicts the temperature of the absorber. Heat transfer rate relies on:

1. Surface temperatures of heating and receiving bodies or materials,
2. Surface characteristics of both bodies or materials,
3. Shapes of the emitting and receiving materials.

Quantities indicate infrared radiations are the perfect source of energy for heating purposes. They indicate the factors such as larger heat transfer capacity, heat penetration directly into the product, no heating of surrounding air, and fast process control. A perfect balance required for optimal heating between the body and surface heating is attained with IR. The parameters that are important to control to achieve optimal heating results are radiator temperature, infrared penetration characteristics, radiator efficiency, and infrared reflection or absorption properties.

1.3.1.2 Advantages of IR Heating

There are several advantages of IR over traditional heating techniques: [1].

Instant heat: Electric IR system forms heat instantly so there is no need for heat build-up.

Reduced operating costs: The energy can reach 50%, depending upon the insulation, types of construction, and other factors. Furthermore, operations maintenance is limited to the cleaning of reflectors and heat source changing.

Clean and safe: Operating IR is a low-risk task and there is no production of by-products.

Zone control: The IR energy is absorbed only where it is directed and does not propagate. **Other advantages of IR are as follows:**

1. Quick heating rate
2. Shorter residence time

3. Uniform drying temperature
4. A high degree of process control
5. Higher thermal efficiency
6. Cleaner work environment
7. Alternate source of energy

The utilization of IR technique in the food sector is still developing; many attempts are continuously growing for the development of IR technologies and the future research is focusing on process control and equipment design development, expanding the areas of applications of IR heating and understanding the interaction between heating process and product characteristics.

1.3.1.3 *Applications of IR Heating*

There are wide applications for IR (infrared radiation) which include medical, paper industries dye, automobile, and others. Table 1.1 discusses the several application of IR heating [1]. In industrial applications, medium to long-range wavelengths seem to be beneficial, for all materials to be heated or dried give the largest absorption in the 3-10 mm region. Moreover, the applications in short waves are continuously evolving. Applications of IR are mostly within the area of food for drying and many other processes during the period from the 1950s to the 1970s from the Soviet Union, the United States, and the Eastern European countries. During the 1970s, much research was performed about industrial frying or meat products cooking and the utilization of near-infrared (NIR) techniques is initiated [28, 43]. In the 1970s and 1980s, several types of research were carried out to apply this technique in the sector of food, mostly at [Swedish Institute of Food and Biotechnology] and gained a set of knowledge. Recent work is experimental in nature and performed in Japan, Taiwan, and several other countries. Applications are mostly from areas such as dehydration, drying of vegetables, fish, rice, roasting of coffee, cocoa, and cereals, heating of floor, frying of meat, baking of pizza, biscuits, and bread, enzymes, and pathogens inactivation. Also, for thawing, blanching, sterilization, pasteurization of packing materials, and surface pasteurization, these techniques have been used.

The major effects on food involve the quick heating of food surfaces sealed in moisture and aroma compounds. Variation to components of food surfaces is equivalent to those that happen during baking.

1.3.2 Microwave Heating

1.3.2.1 *Principal and Mechanism*

With the increasing demand for healthy foods, there is a repeated effort given to enhance and optimize different processing techniques in food,

Table 1.1 Applications of Infrared heating: [1].

Industry	**Methods**
Agriculture	Incubation and warming
Bottling	Drying
Glass	Drying, curing the varnish or paint on back—mirrors and tempering layers
Medical-applications	Incubation and warming
Environmental chambers	Heating
Food	Toasting, cooking, food warming, drying, broiling, and melting
Pharmaceutical	Drying water from powder—tablets
Metal treatment	Preheating—aluminum; steel
Paper	Laminating Calendaring—rolls Adhesive—labels Drying water from—towels
Paint	Primer, topcoat alkyd, acrylic—steel panels, Drying—bicycles, vehicles bodies, aluminium bodies
Textiles	Moisture elimination—carpets Latex and PVC backing Moisture elimination from dyes
Plastics	Laminating Annealing Forming Embossing

to meet the expectations of consumers. With the advancement of emerging technologies, microwave energy has become an indispensable part of every household system. The use of microwave has expanded from heating and defrosting to thawing, blanching, sterilization, drying, etc., in food industries [20, 69]. Microwave is electromagnetic waves with a frequency which ranges from 300 Mhz to 300 GHz. Frequency of microwave used for domestic purposes is 2.45 GHz, whereas the frequency for industrial purposes is 915 MHz [8].

Microwave is a varying magnetic field that generates heat on interaction with, and absorption by, certain dielectric materials, and with the positioning of the direction of the electric field, the native thermal motion of the polarity molecule changes [26]. Water, the dominant polar molecule, consists of separated molecules of an oxygen atom with a negative charge and hydrogen atom with a positive charge which combinedly structure into an electric dipole. When these dipoles fluctuate swiftly back and forth from positive to negative in the direction of the electric field numerous times per second, these express reversals produce frictional heat. This implies that the polar molecules in food play a vital role in the heating performance of food in the microwave system. Due to this frictional rise in the temperature of the water, food components get heated up by convection and conduction. The loss factor dielectric constant of the food determines the depth of penetration of both microwaves and RF energy [49]. This also relies on the varying temperature and moisture concentration of the sample material plus the frequency of the electric field. Overall, with lesser frequency and loss factors, we get more depth ofpenetration. Energy distribution varies with food samples which also governs the depth of penetration of the microwave inside the food. As the food material to be heated in the microwave matches with the wavelength of the material, it becomes difficult to manage the heat uniformity of microwave heating which can be taken as a crucial constraint for industrial application of microwave heating. Thus, a central obligation for microwave energy application and microwave equipment in the food industry is the potential to accurately regulate heating uniformity. Microorganisms are not affected as a result of microwave radiation but are susceptible to the heat generated because of the radiation. Microwaves are likely to be a channel through ceramics, thermoplastics, and glass whereas they are absorbed by carbon and water; reflected by metals but conceivably transmitted using metal hollow tubes and on transiting amongst diverse materials get refracted like the visible light. Microwaves can also be focused on a beam [77].

The microwave energy is transferred to food through the contactless transmission of the wave. This system ensures the uniform heating of food samples during the operation. The equipment comprises a magnetron which is the generator, guide waves which are the aluminium tubes, and for a continuous operation, it has a tunnel attached with a conveyor or a metal compartment for batch operation. These chambers and tunnels are sealed by absorbers or traps to prevent the microwave from escaping and causing injury to the operator [11].

In the microwave system, the two oscillating perpendicular fields, i.e., electric and magnetic, act directly on the heating material, converting the part of absorbed energy to thermal energy. The interaction of the microwave radiation with chemically bound water present in the food material generates high pressure and temperature due to the absorption of the characteristic photonic energy of electromagnetic waves. This process causes moisture evaporation, resulting in pressure exertion on the plant material to cellular and subcellular level leading to swell up and rupture eventually [44].

Microwaves are defined by two mechanisms:

(a) Ionic polarisation: ions present in the solution, when suspended to the electric field, orient themselves, experiencing acceleration and an upsurged kinetic energy. When ions collide with each other it gets converted into heat. This frequent collision increases the density or the concentration of the solution which is also known as the ionic polarization effect [3], whereas in gases the collision becomes difficult due to the spacing between the molecules. In food material, cations are generated by the presence of salts of sodium, potassium, or calcium whilst chlorine produces anions.
(b) Dipole rotation: When the polar molecule strives to situate itself into the fluctuating electric field caused by the microwave, the dipole rotation is created, where the oscillation of the dipolar species leads to the collision with the surrounding producing heat [80]. With the increase in temperature, the dipole movement decreases, whereas ionic conduction increases hence, food samples with both the compounds when heated by the microwave, first governed by the dipole rotation and then with the increase in temperature governed by ionic conduction. The comparative involvement of these methods of heating hinges on the concentration and flexibility of sample ions, plus on the sample's relaxation time [36].

1.3.2.2 Advantages of Microwave in Food Industry

Microwaves have the capacity of penetrating deep into the food materials which offers a remarkable advantage of the reduction of the processing time for varied different processes like sterilization, drying, etc. (Table 1.2). Microwave heating also provides unvarying temperature gradients, avoiding charring of surfaces of the food; and when utilized for drying, the probability of giving case hardening is less during microwave heating [4]. It could operate both in the continuous and batch process because of the integration of the microwave energy with the convection process. Microwave aids in the general energy conservation which is created from

Table 1.2 List of some techniques combined with microwave technology.

Mode	Applications	Benefits	Drawbacks	Reference
Ultrasound	Enzyme activity Drying Extraction	Precise electronic control. Competent energy savings.	-	[38]
Cold Plasma	Sterilization of microorganisms	Maintained product quality.	Inappropriate for impenetrable peels. Probability of getting slenderize.	[35]
Infrared Heating	Tempering Baking Drying	Refining rehydration Properties and quality. Reduced processing time by 95%.	Probability of escalating compactness and slenderness standards. Manufacturing constrained by the equipment size and operating cost.	[55]
Freeze Drying	Dehydration	Quick energy dissipation. Energy savings up to 40%. Advanced volatiles retention level.	Tough to control quality at high MRP. Lengthier drying time at low power.	[30]
Convective Drying	Dehydration	Reduce overheating.	Induces slight discoloration reactions.	[68]

the entire plant volume. Microwave heating is also the predominant pertinent to food processing since it has the aptitude to evade the charring of temperature-sensitive resources. A key shortcoming can be the equipment's capital rate and consequently most expected to be utilized for top-quality products [47].

1.3.2.3 Application of Microwave in Food Processing Technologies

The food processing industry utilizes microwaving immensely for different purposes like cooking, preservation, drying, sterilization, and heating of foods [26]. These particular applications have several advantages, such as microwave drying offers lower bulk density and lower shrinkage along with overhead rehydration ratio and saves power when compared to customary drying [27]. Similarly, the antioxidant activity and bioactive compounds, as well as the striking colors of different fruits and vegetables cooked with or without water, could also be maintained through microwave cooking or heating. It can also minimize antinutritional aspects, temporarily upsurge in digestibility of in-vitro protein. And when it comes to microwave sterilization, it ensures not only food safety, but also reduces the potential risk of any microbes' attack on the food, inactivating enzymes to preserve the nourishment of food. This section reviews various reports on different applications of the microwave, their advantages, and effects onthe quality parameter of food materials.

Microwave Drying

Drying is a complex volumetric heating process that involves heat and mass transfer [13]. The strong microwave radiation when penetrates inside the food item generates vapor and a pressure gradient that heats the food from the inside and outside at the same time with a simultaneous increase of temperature.

Microwave drying improves the quality of some food products with minimum drying time. A microwave uses high-frequency electromagnetic energy and converts it into heat. Wet products manage the energy absorption strength which carefully heats the interior parts of food samples selectively. The moisture present in the food vigorously evaporates and travels towards the surface without affecting the exterior parts of the sample [41]. The microwave drying process goes through two successive stages, i.e., liquid evaporation [26], and three phases of drying include heating, constant rate, and falling rate [5]. Limiting diffusion rate during the falling rate drying period results in shrinkage of the structure of the food. Nevertheless, drying in the microwave generates vapor inside and develops a core pressure gradient outside the product, prevents the shrinkage to food material, and therefore, the drying in the falling rate period is appraised to be very beneficial in microwave drying. Microwave drying when combined with various other methods for example microwave-convection, hot air microwave, vacuum-microwave, and microwave-freeze, microwave-infrared gives more efficient results in terms of quality of the food products which is not achieved by only microwave drying and other conventional methods [8].

Microwave-assisted Freeze Drying

Heat sensitive foods like tomatoes or berries undergo the freeze-drying (FD) method for moisture removal which promotes easy rehydration and prevents chemical decomposition. However, FD takes longer drying time as well as being expensive, which ultimately leads to excessive energy cost and lower productivity [18, 29]. Therefore, combining FD with radiation significantly eases the limitations of FD with shorter processing time, higher energy saving plus efficient drying in the falling rate period as compared to the convention freezing process [18]. Dehydration of Fuji apple was stated by [41] using FD merged with Microwave-Vacuum, the study reported that time for drying is reduced by 40% with nil nutritional change using this double-step technique.

Microwave-assisted Vacuum Drying

In recent years, with the rapid accepted growth and popularity, this method comes with the combination of volumetric heating and vacuum drying. The advantages this combination provides are express moisture evaporation and minimum structural and chemical changes of the final dried products [7]. The final results revealed that a combination of both the techniques at 90°C restored the anthocyanins and augmented antioxidant activity when related to supplementary approaches enlisted [84].

Microwave-assisted Infrared Drying

IR drying has been utilized in the past years for a varied range of agricultural products due to its acceptance as an alternative technique. However, because of its low penetrating power into the food material, it is combined with microwave energy, and their synergistic effect was observed and correlated by [57] on the drying attributes of kiwifruit and banana. For both the samples, they reported a good amount of moisture loss with reduced drying time up to 98% when related to traditional drying. Similarly, an alternative study [66] also reported the standard quality of raspberries with the drying kinetics which showed the superior class final product at varied vacuum pressures and power levels yielding 17.55% better anthocyanin retention, 2.4 times exclusive crispness value, 21.21% advanced radical-scavenging action, and 25.63% higher rehydration properties than infrared drying (IRD) at finest settings.

Microwave Heating

Microwave heating relies on volumetric heating of the food material instantaneously and can also be combined with the convective and radiant heating process [33]. The electric field induces the dipole rotation, generating friction between molecules inside the microwave which assists in heating the food materials [2]. The penetration depth of the microwave is dependent on the food composition and its accompanying changes related tothe chemical composition of the food, i.e., cook loss, bioactive components, antioxidant activity, and anti-nutritional factors, comprising phytic acid, trypsin inhibitor, tannins, and saponins [40]. The chicken streak rigidity was lowered after microwave heating which was not significant when cooked with grilling or boiling reported by [9]. However, it was observed differently in the case of beef burgundy which perceived the tougher texture by microwaving than the convection oven [32]. [52] reported a remarkable increase in cooking loss with augmenting core temperature and time of bovine muscle during microwave heating.

Microwave-assisted Infrared Heating

In the food sector infrared (IR) radiation offers a wide range of advantages such as an express regulation response, rapid heating with minimum changes in product quality. However, as discussed earlier, its weak penetration power makes this technique only used for surface heating. Besides, there are also chances of unwanted fraction and swelling of the material due to prolonged exposure to IR radiation. Nevertheless, merging microwave with IR heating becomes profitable and could help to tackle all these drawbacks which occur during the process on the surface and inside of food [60].

Microwave-assisted Infrared Baking

Microwave heating combined with IR increases the manufacture of confectioneries which was restricted before due to two major aspects, i.e., uneven dispersal of moisture inside the food and poor penetration power. Merging this energy becomes a possible approach to overcome the challenge as the microwave helps to reclaim the time of processing and administrating an effective dispersal of temperature within the material, whereas IR heating remarkably subscribes to crust formation and browning [20]. [55] observed that the legume cakes hold a better characteristic texture with increased volume and desirable surface color when treated in a combination of infrared and microwave as compared to the only conventional oven. Similarly, a gluten-free bread was prepared in an MW-IR oven using different concentration of flaxseed and gums which was kept at a frozen temperature at 20°C for 10 days before baking [56]. The results turned out to be much better as compared to the conventional oven with a darker color, softer texture, and higher volume.

Microwave-assisted Infrared Roasting

Roasting is considered to be an appropriate method for flavor, texture, and color enhancement which can be done with limited investment giving to a high production quantity. However, there are not many reports with the explored application of MW-IR grouped for roasting. [76] studied the roasting characteristics of hazelnut using and compared with conventionally roasted ones. In the MW-IR oven, the optimum roasting time was 2.5 min at the power level of 90% with lower and upper halogen lamps powered at 20% and 60%, respectively, whereas conventionally it took 20 min at 150°C. When it comes to quality, the hazelnuts roasted in both techniques showed similar characteristics attributes concerning moisture content, color, and fatty acid composition. However, the above results confirm the reduction of the roasting time significantly which is therefore also recommended for other food materials.

Microwave Cooking

The foremost usage of microwave is cooking. This section reports various studies of microwave and effect on the various cooking parameters such as color retention, quality, and taste for different food materials. [70] studied the chemical changes associated with skipjack tuna (*Katsuwonus pelamis*) during the process of boiling at 100°C, frying with sunflower oil at 180°C, then put through canning and at the end microwave heating for 10, 15 and 20 s. It was found that the health beneficial PUFA loss was minimum with boiling, 70–85% during frying, 100% with the canning, and 20–55% during microwave heating. Cholesterol content slightly increased in microwaving with no increase while cooking whereas the highest content was observed with canning and it got lowered during the frying process which could be leaching cholesterol from tuna while frying into the oil. Thus, taking into account all the methods more fatty acids can be preserved with microwave heating [70].

Blanching is usually utilized for the inactivation of enzyme and color retention, for varied fruits and vegetables by immersing it in hot boiling water or steaming with acids or salts solution. Microwave blanching offered the extreme retention of color, chlorophyll, and ascorbic acid contents with better preservation of quality parameters. [17] reported microwave blanching as a better technique than the traditional blanching for peanuts in terms of time and energy saving. However, processing at a high temperature in microwave blanching results in ashy off-flavor.

Microwave-assisted Ultrasonication

Ultrasound plays a major role in the food industry and has been applied to vary processing techniques like extraction, drying, sterilization, and freezing with various advantages like maintaining the food quality parameters, augmented food preservation, and also assists in thermal treatments. On the other hand, it also reduces the cost of production by eliminating some of the purification steps [73]. However, ultrasonication does affect the physiochemical parameters, degraded the quality, exhibiting off-flavor in the food material [12]. Therefore, the fusion of microwave and ultrasound making it the microwave-assisted ultra-sonification technique renders a collaborative effect eliminating the drawbacks attached to the individual techniques [13], and therefore, the collective skill has been extensively premeditated for the food sector. The ultrasonication technique with microwave assistance has been verified as an innovative process for rapid and effectual extraction. The most unique feature which it has is the exceptional achievement of weakening the hydrogen bonds and subsequently augmenting the penetration rate of solvent into the matrix by amplifying the dipole rotationhence enabling systematic solvation [42].

Microwave Sterilization

The intention of sterilization or pasteurization is usually done to kill or make inactive all the microorganisms present in the food, ultimately strengthening the safety and encompassing the serviceable life of the food. A study was conducted by [17] where the different effect of sterilization was observed without fluctuation of the microwave conditions (temperature and power). The log cycle 5.12 reduction of *Salmonella typhimurium* on jalapeno pepper was observed using a microwave with water-assistance at 950 W to reach a temperature of 63°C for 25 s. 4.45 log reduction was also studied on coriander foliage at 3×10^8 CFU/g at 63°C for 10 s. A similar study was conducted for Salmonella enteritidis in potato omelette which was treated under varied microwaving conditions to test the inactivation rate of 6.30 log CFU/g. It is noted that the inactivation of microorganisms quicker with the increase in power level during microwave sterilization [78].

Microwave-powered Cold Plasma

As the name suggests it is a non-thermal technique, especially engaged in the food sector for microbial decontamination. DNA in the chromosomes inactivate its microbes which are later destroyed through plasma [74]. [83] examined the effectiveness of the microwave treatment combining with cold plasma to prevent the growth of *Penicillium italicumandim* and observe the storage stability of the mandarin at 4 and 25°C. The outcome presented the highest inhibition of *Penicillium italicumandim* i.e., 84% reduction in disease incidence for 10 mins at a power level of 900 W, combined with nitrogen (N_2). Besides, no visible differences in titratable acidity, soluble solids, or weight loss were marked. Likewise, [67] verified that the population of *Salmonella typhimurium and Escherichia coli* O157:H7 reduced drastically up to 2.8 log CFU/g on lettuce using microwave-cold plasma treatment with nitrogen gas at 400 W. it showed the bactericidal effect with no damage to the quality and sensory parameter of lettuce.

1.3.3 Radiofrequency (RF) Heating

1.3.3.1 Principal and Mechanism

Radiofrequency (RF) ranges from 300 kHz to 300 MHz in the electromagnetic spectrum [54]. It comes under dielectric heating in which direct, in-depth penetration happens inside the food in the range of 1–100MHz EM waves [31]. Figure 1.1 Illustrates the schematic diagram for radiofrequency heating [25]. The lower frequency is applied and is more appropriate for processing the large volume materials; hence it is observed as a fast and volumetric heating method [14]. The frequencies used in industries for heating applications are 13.56 and 27.12 MHz. The food placed between electrodes is heated using transmitted electromagnetic energy. The RF energy is transferred over the free space and through the not resistant packaged materials. High field strengths generated provide sufficiently higher heating rates in foods. Dielectric heating is useful in colder temperature range or even less than the freezing point of foods [54]. The product mainly targeted RF heating but not in the surrounding environment. The heat is generated inside the food material through ionic conductance and dipole rotation. In the treatment, the moisture got equalized within the product without any over-drying or heating of the material. Following the environment-friendly perspective, the more efficient use of dielectric techniques plays a vital role in thefood processing sector.

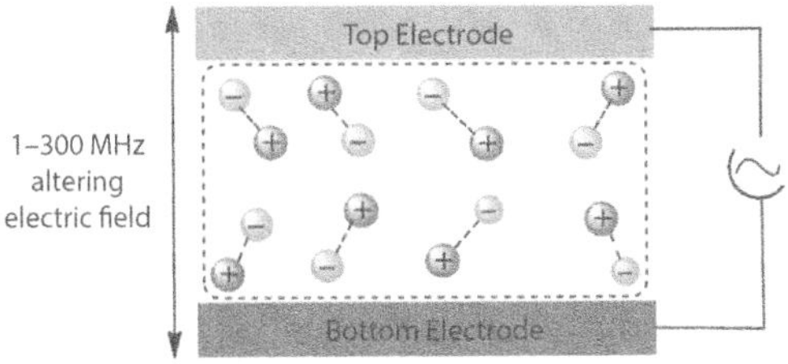

Figure 1.1 Illustration of radiofrequency heating [25].

As stated in studies, an excellent depletion in several microbes and pests achieved by radiofrequency heating in various food products such as eggs and its products, poultry, meat and its products, fish and shellfish, fruit juice and jam, canned fruit, starch, soy milk, molasses, pea protein concentrates, ready to cook meals, milk, and milk products, sweet desserts, cereals, and bakery products, spices, etc.

RF energy combining with other different thermal methods shows synergistic effects, notably making the RF pasteurization efficient particularly for agricultural materials having less moisture content.

RF uses a uniform and non-ionizing form of electromagnetic energy. RF system consists of two electrode plates made of metals in which the conducting materials are kept, generating the alternate electromagnetic field inside.

Electrodes are designed to provide an invariable electric field for different food shapes. Foods containing high moistures use the conventional layout of electrodes and rod-type electrodes used for dry products, which provides stray fields on the material on a conveyor belt.

RF is suitable for food materials in bulk with high ionic conductance [79].

According to the federal commission, the assigned RF frequencies mostly used are 27.12, 13.56, and 40.68 MHz for usage in industries, science, and medicine fields [25].

The heating systems used in the industry or R&D field are mainly free-running oscillators and the 50 ohms RF system.

Food placed between power generated oscillating circuit, consisting of a coil, condenser plates, a source of energy, and an amplifier [54]. An alternating electrical field generated between the electrodes causes the materials to reorient themselves towards the electrode poles of opposite charges.

While the food is heated, the given frequency is consistently monitored and maintained [54]. The generation of heat is dependent on various measures which are frequency, twice the value of the voltage applied, product proportions, and the dielectric loss factor of the material.

The 50 ohm RF system adjusts its impedance to 50 Ω, which should be the same as the generator impedance, thus delivering a stable heating process [25].

The high-frequency heating principle described as:

$$\frac{dT}{dt} = 0.239 \times 10^{-6} \frac{P}{c\rho}$$

$$P = 2\pi f E^2 \varepsilon_0 \varepsilon''$$

The equation yielded to;

$$P = 55.61\, fE^2\varepsilon'' \times 10^{-12} \text{ (since } \varepsilon_0 = 8.854\times10^{-12}\ \mathbf{F{\cdot}m^{-1}})$$

In the above equation, the terms are as follows: dT/dt (°C/s), P (Watt/m3), c gives heat capacity of the dielectric material (J/kg), ρ is density in kg/m3, f is the frequency in Hz, dielectric loss factor E (V/m), and ε" is the unreal value of the complex relative permittivity, ε* = ε"- jε".

Higher the value of ε", higher energy absorbance would be there at a particular voltage and frequency. RF heating rate observed to be in direct proportion with the value of ε" and twice the electrical strength, but found to have an inverse relation with the heat capacity and density of the product.

In the circuit, the triode valve assists the oscillations produced, and the potentiometer adjusts the output power. The galvanometer shows the supply of incident power as well as reflected power in the generator [79].

1.3.3.2 *Advantages and Disadvantages*

Advantages:

a. Suitable for large and thick foods
b. The investment cost is less or Low maintenance costs
c. Easy to control and understand [59]
d. Improves the level of the moisture
e. RF is energy efficient
f. Over dehydration and heating is prevented on the surfaces
g. The shortcomings faced in energy or heat transfer controlled using
h. shorter process lines [79]
i. Increased throughput
j. Simpler construction [54]

Disadvantages:

a. Risk of arching
b. The frequency band is narrow
c. Occupy large floor space [59]
d. The high initial capital cost of equipment and operating cost
e. Fluctuations in electrical costs
f. Different designs and applicators required to meet product-specific requirements [79]
g. Reduced power density [54]

Using a good quality of applicator's design with specific fine-tuning, technologically advanced tools and highly skilled technicians sums up to be an efficient RF system.

An exemplary RF system would have greater penetration depth, the volumetric heating advantage of dielectric heating than other techniques, along the hybrid systems produce effectual, swift, and graded results [79].

1.3.3.3 *Applications*

RF dielectric has various applications in lumber, textile, and food industries, etc.

RF treatment is giving after the biscuits baked in cereals, to dry foods (herbs, spices, vegetables, potato products, pasta products), thawing, etc. The efficient role of RF is seen in moisture removal during the baking process.

It is also used for the thawing of meats and fish in large amounts, and its volumetric nature enhances the acceleration in the process without affecting the distribution of temperature throughout the product [54].

i) RF application in Fresh food processing:
In Fresh fruits: A significant reduction in brown rot was seen when dielectric treatment (27.12MHz, 15kW) was given with hot water. Infection in peaches being reduced using RF heating and brown rots controlled in nectarines.
In Fresh Vegetables: Vacuum-packaged Caixin (a green leafy vegetable) with an adequate microbial count resulted in a significant decrease in the number using ohmic heating with a 20 mm gap in electrode without affecting any other property of the material [26].

ii) RF heating on fresh meats showed results when treated while meat processing, as reduced *E. coli* count, shelf life extension wherein it has higher heating uniformity and had a shorter cooking time [31, 75].
iii) Significant reduction of *A. parasiticus* in corns seen using a combination of RF heating with hot air [25, 26].
iv) RF heating system in food industries used to inactivate bacteria in wheat flour, caixin, peach, pepper spice, apple juice, stone fruit, almonds, ham, etc., using free-running or 50Ω type system at a frequency of 27.12MHz [31].
v) It also used for disinfestation of rice, lentils, legumes, walnuts, etc., using a free-running oscillator at a frequency of 27.12MHz [31].
vi) Also used in cooking purposes in beef, egg white, meat emulsion, pork ham, turkey, meat, etc., using free running or 50 ohms at a frequency of 27.12MHz.
vii) Used in the thawing of pork, tuna fish, beef using 50 ohms at a frequency of 13.56-27.12MHz [31].
viii) For enzyme inactivation in miso paste, apple, myrosinase using 50 ohms at 27.12MHz frequency.
ix) Application in roasting and drying of nuts as well for improving energy efficiency and product quality [31].
x) Dry-blanching with RF was observed to be an efficient method to treat products in which a 95% reduction in enzyme activity (POD) was seen in 3-7 min. time with a gap of 8-8.6 cm; rehydration enhancement, retention of the textural properties were also recorded [24].
xi) Application in wood drying (10-30MHz), agricultural product drying (27MHz), food drying, etc. [79].
xii) Production of Biodiesel from moist microalgae: Cell disruption of Algae done using RF heater at a frequency of 27.12MHz and maximum power output of 6kW and the esterification/transesterification reactions were also promoted [48].

1.3.4 Ohmic Heating

1.3.4.1 Principal and Mechanism

Heating technology has observed some recent advances with the following technology development, viz Dielectric heating (Radiofrequency and microwaves), induction and Ohmic heating. Ohmic Heating or joule heating is electro-heating, or electroconductive heating or direct electric resistance heating in which heat generated within the food material is due to the generation of alternating electric current passed through the food material. In 1900, the Ohmic heating technique was utilized for the Pasteurization of infant milk. In the early twentieth century, the electro-pure process was relevant commercially, but later with time, it became prevalent during the 1980s.

The food itself acts conductor in electric resistance heating as the voltages are applied, the current between the electrode and ground helps in determining the resistivity of the food as per Ohm's law. Increasing the current or voltage, thereby increase in temperature increases the conductivityof foods [54].

In Ohmic heating, unlike other thermal methods, the electrode is in contact with food; less frequency is applied compared to the frequency of radio or microwaves, and the waveform is usually a sine wave. Resistance heating systems help with the production of products with high storage stability through proper maintenance of food in terms of color and nutritional value [34]. Figure 1.2 depicts the circuit diagram of static (batch type) resistance heating process [46].

Ohmic heating is defined as the amount of heat generated in which electrical current passes through the food and current resisting the flow of electricity. Its principle is based on the direct application of Ohm's Law, wherein, the current through the conductor between the two points is directly proportional to the voltage.

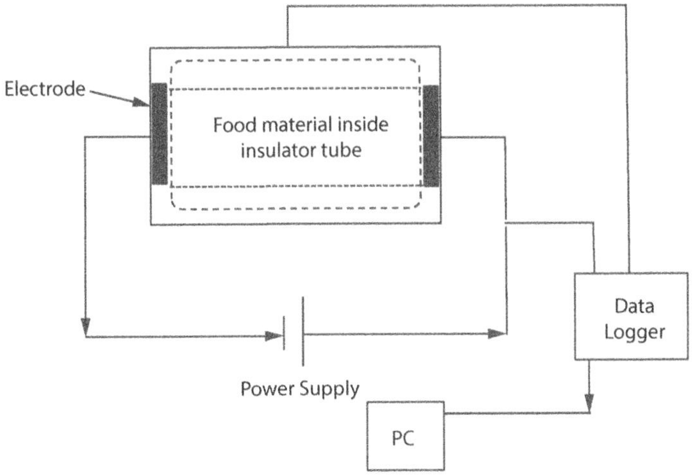

Figure 1.2 Circuit diagram of static (batch type) resistance heating process [46].

Heat is generated internally due to the alternating current applied to the food, and the OH ratio is directly dependent on the electrical conductivity of food.

Voltage by the alternating current applied at both ends of the electrode, which interacts directly with the supplied food. Food treated in the ohmic heating chamber should come in contact with the electrodes. Since food contains the right amount of ionic salts and water molecules, electrical energy is converted into thermal energy to generate heat internally, and almost all energy inputs convert to heat without a loss factor. The heat generated can be used for other food processes such as Pasteurization, sterilization, and blanching related to OH applications. Resistance heating devices instrumentation includes an electrode (electrodes may be of stainless steel, aluminum, titanium, or platinum-coated titanium), power supply, an insulating tube or container or a heating chamber trapping the food sample inside it, data logger system, a current sensor, a thermocouple, and a Personal computer. The significant parameters involved are electrode configuration (flow of current across or parallel to the product flow direction), the gap between the two electrodes, heater shape, AC frequency, the voltage applied, the speed of the product, a supply of electricity, amount of charge per unit time. Some other factors considered are the type of product and its properties in particular specific conductance and rate of heating; total solids in food material, the heat capacity, viscousness of the material, density, the size and shape of the particles, and inclination of an electric field. The electric field intensity, E and σ, the electrical conductance, and Q are the heating rate that is directly proportional to the field strength and the square of the electrical conductivity. The increased temperature, conductance determine the efficiency of OH and the values of temperature, voltage, current, the time displayed in the data acquisition system [61].

The electric field in the homogeneous medium determined by:

E0 = ∇U, in which U is the electrostatic potential

So, generation of heat per unit volume would be;

$$G_0 = \sigma_L (E_0)^2,$$

where σ_L defined as the electrical conductivity of dispersed or liquid materials

Changes in the conductance with the temperature for solid and liquid foods mostly taken in a linear form, described as follows:

$$\sigma = \sigma in\ [1 + m\ (T - Tin)]$$

wherein, σ_{in} described as the electrical conductivity at the start point of temperature Tin and m is described as a constant of proportionality (°C^{-1})

The resistance in the ohmic heating defined as:

$$R = (Rs * x / RA)$$

where R is total heat resistance in ohms, Rs (ohms per meter) is specific product resistance, x(m) is the gap between the electrodes, and A(m^2) is the electrode area.

1.3.4.2 Advantages and Disadvantages

Advantages

a. Particulate foods less than one cubic inch considered appropriate in the resistance heating process; solids content range of 20-70% is deemed to be significant in the liquid-particle mixture flow reaching the plug flow.
b. Mixtures of particles and liquids are evenly heated in some situations because the properties of fluids and the particles are similar such as viscosity, conductance, heat capacity, flow rate, etc.
c. The required temperatures can be achieved rapidly in the case of ultra-high temperature processes as well.
d. There is no hot surface for heat transfer, so the risk from flaming or overprocessing of the product is low.
e. High-energy conversion efficiency.
f. Capital cost relatively low.
g. Colour and nutritional value of food retained.
h. System is environment-friendly.
i. Process control is better than the conventional heating method and more straightforward as well.
j. Reduced fouling compared to that seen in conventional heating.
k. It is an exemplary process for shear-sensitive products since the flow rate is low.
l. Fast and uniform processing of liquid and solid phases which reduces any harm due to heat and retention of

nutrients is there [34]. Products treated in ohmic heating have better textural properties than traditional heating procedures.

m. Rapidly heated and heating is volumetric in nature.
n. Easily controllable.
o. Rapid inactivation of microbes and enzymes [46].
p. Storage and distribution in ambient temperature combined with the aseptic filling system.

Disadvantages

a. Installation cost and operation costs are high for ohmic heating systems compared to conventional processing methods.
b. Fat globules present in food are not appropriately heated or heated slowly in ohmic systems because of no conductance (since no salt and water presence) [34].
c. Longer processing time may be required in heating systems to achieve appropriable alterations in food materials, e.g., process of gelatinization of starch.
d. Nutritional and physical attributes of the material bear alterations in the rapid ohmic heating process.
e. Ohmic electrodes may get corroded after the cooking process, and hence final product may also get affected.
f. In continuous ohmic textural softening in beans; changes in the uniformity of product (volume enlargement) were seen, low quality of duct plugging (because of the large size of soybeans) [22, 23].
g. Change in the cell membrane of raw produce due to heat transfer by electro-thermal effect resistance heating, particularly at a low frequency.
h. Loss in Nutritional values and increase in BOD of effluents, e.g., cell-based material coming out from the heating medium is inappropriate for some products or processes
i. Capital funding and safety issues [21–23].
j. Inert electrode material.
k. No proper control in the electrical conductance of all the food constituents because then different heating rates would be there for various components [54].
l. Unavailability of data on essential parameters affects the heating, such as dwell time, direction, level of loading.

m. Inappropriate temperature conformance technology to identify cold/hot locations spot [54].

1.3.4.3 Applications

Ohmic heating has several uses in the food processing industry. It has successful applications in various food products (solids or liquid/fluids or a mixture of both); like juices, fruits and vegetables, stews, meats, sauces, pasta, soups, and seafood.

In the initial years, the application of ohmic heating in beetroots resulted in improvement in the diffusion of betanin from the beetroot tissue, and its concentration and electric field strength were studied. Then studies carried out on radish in which heating rate was analyzed concerning frequency, which stated an increase in AC frequency leads to an increase in heating rate. Also, electrically heated foods are influenced by their mass transfer properties.

With time, its potential applications discovered in a huge amount such as in blanching, dehydration, fermentation, evaporation, extraction, etc.

The electric resistance heating process provides a chance for the production of new, high value-based, shelf-stable food products of quality that could not be realized before with current sterilization technology [54].

1. Blanching: Ohmic heating can enhance process efficiency in blanching, e.g, Ohm heating has rapid and uniform heating; hence there is no need to dice vegetables due to an effective method of ohmic heating, which has a fast turnaround time and less surface area. Blanching of mushrooms with the application of ohmic heating resulted in shrinking at lower temperatures with less water use compared with conventional ones.
2. Evaporation: Using electric resistance heating results in a threefold increase in the evaporation rate and hence improved product quality.
3. Dehydration: Improved dehydration rates of vegetable tissues, for example, Treating (with ohmic heating) sweet potatoes before dehydration significantly speeds up hot air drying compared to the traditionally processed samples. It also has seen the acceleration in vacuum drying rate of sweet with minimal ohmic pretreatment.
4. Fermentation: Enhanced extraction of components from different food products. For example, Ohmically pretreated

apple tissue showed an increase in the yield of mechanical juice extraction. Also, ohmically treated rice bran showed an increased extraction of RBO (rice bran oil) (particularly at lower frequencies (say, 1Hz)) using ohmic heating.

5. Microbial Inactivation: Resistance heating is fatal to almost all microorganisms. Lower D-value and z-values were recorded in studies for microbial inactivation. A study by [71] resulted in a higher microbial death rate deactivating viable milk aerobes and S. Thermophilus using ohmic heating. Death of microbes in ohmic heating is due to the higher temperatures or increased electric field strength [34].
6. Enzymes Inactivation: Electric resistance heating generates an electric field which is found to be efficient for rapidly inactivating the enzymes such as lipoxygenase, polyphenol oxidase, pectin esterase than traditional heating systems [34]. For example,
   (i) Dynamics of polyphenol oxidase deactivation during ohm heating of grape juice studied, and ohmic heat treatment at various potential gradients applied to grape juice at specific temperatures. Changes in PPO enzymatic activity were noted.
   (ii) Declination in the activity of exogenic pectin methylesterase enzyme present in cloudberry jam and apple juice using ohmic heating and it follows rate law.
   (iii) Resistance heating with continuous alternating current resulted in successful deactivation of Bacillus subtilis spores in orange juice in a short duration of time, using a pressurized electric sterilization system; also reduction in peculiar smell and loss of ascorbic acid observed in the studies.
7. Extraction rate increase: Plant tissues pretreated ohmically at moderate temperatures and maximum efficiency in the extraction of juices by pressing was seen, e.g., increased efficiency of sucrose extraction from sugar beet; improvement in the dispersion of soymilk from soybeans. Ohmically treated products have higher retention value of color and nutrition, lower processing times, and higher extraction yields [34].

Other Applications of Ohmic Heating: (Ohmically treated food materials) [21]

a. Peeling and tissue softening of Ohmically treated and cooked Red beet, carrot, golden carrot, radish, turnip, potato, cabbage, apple, etc.
b. Resistance heating improved the folic acid content of apple due to electropermeabilization.
c. Inactivation of enzymes and reduction in textural changes during storage in pineapple observed using resistance heating.
d. Thermal and non-thermal effects of OH in the meat thawing process and beef cuts.
e. Application in producing products (e.g., ham and bologna emulsions, frankfurters, burger patties, etc.) with properties like soft, less cohesive, less elastic, mushier, less resilient, chewier, and good textural products in a short span of time.
f. Improved firmness in pound cakes.
g. Thermal denaturation of Tofu: Better process control observed and increased in apparent breaking strength and young's modulus.

8. Rapid heating rates in different types of pickles (1-5°C per second), full-cream milk (7-50°C per second), cut pieces of pork and beef; Low heating and conductivity in fruits, margarine, etc. [61].
9. Time reduction in Bread dough proofing to achieve higher expansion because of high heating rates.
10. The heat source for desalination processes and reduction in processing time in ethanol distillation; better process control in ohmic assisted hydrodistillation.
11. Food Package for long-duration space missions and military; producing shelf-stable food products with better process control, e.g., Pulsed OH treated retort pouches for space missions [46].
12. Reduction of PPO in ohmically treated sugarcane juice, the complete deactivation of PPO in apples, in watermelon juice, in grape juice at a particular temperature-time combination.
13. POD inactivation in broccoli, potato, carrot, sugarcane juice, pumpkin, and artichoke heads in shorter duration and retention of nutrients [46].

1.4 Future Perspective of Novel Thermal Technologies

Microwave ovens are the most common heating techniques currently used in everyday life for reheating or preparing new meals. The advantages of the microwave for future application include time saver, volume heating, high product quality, and many others. The combination of the microwave with the conventional, infrared heating system may have the potential to build a completely new cooking system in future perspectives. Despite the high potential, a microwave is less common in industrial applications; maybe the reason behind this is the high energy cost, which can be balanced by the low research budget and quality of products formed.

The future utilization of RF heating is based on the consumer's food preferences and need. The current increasing demand for hygienic and safe food will continue to increase demand for RF heating. Increasing the demand for pre-packaged food will result in the growth of RF heating technology. IR can be very useful for continuous baking, drying, and surface pasteurization. The utilization and application of IR in food industries are limited; equipment available for these techniques are not fully optimized and need more research and exploration. With the development of science and technology, the IR techniques might evolve and show their full potential. The ohmic heating has the promising way of providing food with produce high-grade value-added, safe, and better shelf-life products.

Ohmic heating can be efficiently utilized to heat high-acid food products, and for quick heating of liquid products. In future applications, ohmic heating technologies can be utilized in blanching, fermentation, evaporation, dehydration, and extraction. The considerable factors to note for future perspective in the future of ohmic heating are technologies are, considering different factors such as residence time, levels of loading, etc. Lack of temperature for locating a hot and cold spot and lack of models account for differences in electrical conductivity between solid and liquid phase, the response of two-phase with change in temperature [75].

1.5 Conclusion

Safety of food and maintaining its quality is the main concern in any heating process. Fulfilling consumer demand and needs is the main concern for any food processing industry. It has been found that ohmic heating is considered to be the most common and green process in food processing industries [50]. Infrared heating is still evolving and widely applied in drying especially low moisture food. Radiofrequency heating is less widespread as compared to other novel thermal heating technologies. Microwave heating is the most flexible and explored technology worldwide. It has a broad range of applications in hybrid systems. Microwave has great capabilities in sterilization and pasteurization. Novel thermal technologies have great potential and applications in producing safe and quality products with great efficiencies. A few drawbacks attached with these technologies are high investment cost and less control over all variables, which may delay its industrial implementation but its technical solutions are continuously evolving and are underway. Novel thermal technologies have many advantages over their drawbacks such as providing quality value-added products, energy-efficient, and environmentally sound, which proves it's promising and novel as compared to conventional technologies.

References

1. Antonio Vicente and Ines Alexandra Castro. (2007). Novel thermal Processing Technologies. In *Advances in Thermal and Non Thermal Food Preservation* (pp. 99–130).
2. Aguilar-Reynosa, A., Romani, A., Rodriguez-Jasso, R. M., Aguilar, C. N., Garrote, G., & Ruiz, H. A. (2017). Microwave heating processing as alternative of pretreatment in second-generation biorefinery: An overview. *Energy Conversion and Management, 136*, 50-65.
3. Álvarez, A., Fayos-Fernández, J., Monzó-Cabrera, J., Cocero, M. J., & Mato, R. B. (2017). Measurement and correlation of the dielectric properties of a grape pomace extraction media. Effect of temperature and composition. *Journal of Food Engineering, 197*, 98-106.
4. Aydogdu, A., Sumnu, G., & Sahin, S. (2015). Effects of microwave-infrared combination drying on quality of eggplants. *Food and Bioprocess Technology, 8*(6), 1198-1210.
5. Bal, L. M., Kar, A., Satya, S., & Naik, S. N. (2010). Drying kinetics and effective moisture diffusivity of bamboo shoot slices undergoing microwave drying. *International Journal of Food Science & Technology, 45*(11), 2321-2328.
6. Barba, F. J., Zhu, Z., Koubaa, M., Sant'Ana, A. S., & Orlien, V. (2016). Green alternative methods for the extraction of antioxidant bioactive compounds from winery wastes and by- products: A review. *Trends in Food Science and Technology, 49*, 96–109. https://doi.org/10.1016/j.tifs.2016.01.006.
7. Bórquez, R., Melo, D., & Saavedra, C. (2015). Microwave–vacuum drying of strawberries with automatic temperature control. *Food and Bioprocess Technology, 8*(2), 266-276.
8. Chandrasekaran, S., Ramanathan, S., & Basak, T. (2013). Microwave food processing—A review. *Food Research International, 52*(1), 243-261.

9. Choi, Y. S., Hwang, K. E., Jeong, T. J., Kim, Y. B., Jeon, K. H., Kim, E. M., ... & Kim, C.J. (2016). Comparative study on the effects of boiling, steaming, grilling, microwaving and superheated steaming on quality characteristics of marinated chicken steak. *Korean Journal for Food Science of Animal Resources, 36*(1), 1.
10. Chemat, F., Rombaut, N., Meullemiestre, A., Turk, M., Perino, S., Fabiano-Tixier, A. S., & Abert-Vian, M. (2017). Review of Green Food Processing techniques. Preservation, transformation, and extraction. *Innovative Food Science and Emerging Technologies, 41*, 357–377. https://doi.org/10.1016/j.ifset.2017.04.016
11. Chemat, F., & Cravotto, G. (Eds.). (2012). *Microwave-assisted extraction for bioactive compounds: theory and practice* (Vol. 4). Springer Science & Business Media.
12. Chemat, F., & Khan, M. K. (2011). Applications of ultrasound in food technology: processing, preservation and extraction. *Ultrasonics sonochemistry, 18*(4), 813-835.
13. Chen, F., Zhang, X., Zhang, Q., Du, X., Yang, L., Zu, Y., & Yang, F. (2016). Simultaneous synergistic microwave–ultrasonic extraction and hydrolysis for preparation of trans-resveratrol in tree peony seed oil-extracted residues using imidazolium-based ionic liquid. *Industrial Crops and Products, 94*, 266-280.
14. Chen, Z., Li, Y., Wang, L., Liu, S., Wang, K., Sun, J., & Xu, B. (2017). Evaluation of the possible non-thermal effect of microwave radiation on the inactivation of wheat germlipase. *Journal of Food Process Engineering, 40*(4), e12506.
15. Clodoveo, M. L. (2013). An overview of emerging techniques in virgin olive oil extraction process: strategies in the development of innovative plants. *Journal of Agricultural Engineering, 44*(2s), 297–305. https://doi.org/10.4081/jae.2013.s2.e60
16. Clodoveo, M. L., Dipalmo, T., Rizzello, C. G., Corbo, F., & Crupi, P. (2016). Emerging technology to develop novel red winemaking practices: An overview. *Innovative Food Science and Emerging Technologies, 38*, 41–56. https://doi.org/10.1016/j.ifset.2016.08.020
17. De La Vega-Miranda, B., Santiesteban-Lopez, N. A., Lopez-Malo, A., & Sosa-Morales,
18. M. E. (2012). Inactivation of *Salmonella Typhimurium* in fresh vegetables using water- assisted microwave heating. *Food Control, 26*(1), 19-22.
19. Duan, X., Liu, W. C., Ren, G. Y., Liu, L. L., & Liu, Y. H. (2016). Browning behavior of button mushrooms during microwave freeze-drying. *Drying Technology, 34*(11), 1373-1379.
20. E.E. Tănase, A.C. Miteluţ, M.E. Popa, G. A. Ştefănoiu and M. D. (2015). Radio frequency heating for food safety and preservation - State of the art. https://www.researchgate.net/publication/284323081_Radio_frequency_heating_for_food_safety_and_preservation_-_State_of_the_art

21. Ekezie, F. G. C., Sun, D. W., Han, Z., & Cheng, J. H. (2017). Microwave-assisted food processing technologies for enhancing product quality and process efficiency: A review of recent developments. *Trends in Food Science & Technology, 67*, 58-69.
22. Gavahian, M., Sastry, S., Farhoosh, R., & Farahnaky, A. (2020). Ohmic heating as a promising technique for extraction of herbal essential oils: Understanding mechanisms, recent findings, and associated challenges. In *Advances in Food and Nutrition Research* Vol. 91, pp. 227-273.
23. Gavahian, M., Tiwari, B. K., Chu, Y. H., Ting, Y., & Farahnaky, A. (2019). Food texture as affected by ohmic heating: Mechanisms involved, recent findings, benefits, and limitations. *Trends in Food Science & Technology, 86*, 328-339.
24. Gong, C., Zhao, Y., Zhang, H., Yue, J., Miao, Y., & Jiao, S. (2019). Investigation of radio frequency heating as a dry-blanching method for carrot cubes. *Journal of Food Engineering, 245*, 53-56.
25. Gunasekaran, S., & Yang, H. W. (2007). Effect of experimental parameters on temperature distribution during continuous and pulsed microwave heating. *Journal of Food Engineering, 78*(4), 1452–1456. https://doi.org/10.1016/j.jfoodeng.2006.01.017
26. Guo, C., Mujumdar, A. S., & Zhang, M. (2019). New development in radio frequency heating for fresh food processing: A review. *Food Engineering Reviews, 11*(1), 29-43.
27. Guo, Q., Sun, D. W., Cheng, J. H., & Han, Z. (2017). Microwave processing techniques and their recent applications in the food industry. *Trends in Food Science & Technology, 67*, 236-247.
28. Horuz, E., & Maskan, M. (2015). Hot air and microwave drying of pomegranate (Punica granatum L.) arils. *Journal of Food Science and Technology, 52*(1), 285-293.
29. Huang, E., & Mittal, G. S. (1995). Meatball cooking - modeling and simulation. *Journal of Food Engineering, 24*(1), 87–100. https://doi.org/10.1016/0260-8774(94)P1610-A
30. Jiang, H., Zhang, M., Mujumdar, A. S., & Lim, R. X. (2013). Analysis of temperature distribution and SEM images of microwave freeze drying banana chips. *Food and Bioprocess Technology, 6*(5), 1144-1152.
31. Jiang, H., Zhang, M., Mujumdar, A. S., & Lim, R. X. (2016). Drying uniformity analysis of pulse-spouted microwave–freeze drying of banana cubes. *Drying Technology, 34*(5), 539-546.
31. Jiao, Y., Tang, J., Wang, Y., & Koral, T. L. (2018). Radio-frequency applications for food processing and safety. *Annual Review of Food Science and Technology, 9*, 105-127.
32. Jouquand, C., Tessier, F. J., Bernard, J., Marier, D., Woodward, K., Jacolot, P., ... & Laguerre, J. C. (2015). Optimization of microwave cooking of beef burgundy in terms of nutritional and organoleptic properties. *LWT-Food Science and Technology, 60*(1), 271-276.

33. Kappe, C. O. (2013). How to measure reaction temperature in microwave-heated transformations. *Chemical Society Reviews, 42*(12), 4977-4990.
34. Kaur, N., & Singh, A. K. (2016). Ohmic heating: concept and applications—a review. *Critical Reviews in Food Science and Nutrition, 56*(14), 2338-2351.
35. Kim, J. E., Oh, Y. J., Won, M. Y., Lee, K. S., & Min, S. C. (2017). Microbial decontamination of onion powder using microwave-powered cold plasma treatments. *Food Microbiology, 62*, 112-123.
36. Knoerzer, K., Juliano, P., & Smithers, G. (2016). Innovative Food Processing Technologies: Extraction, Separation, Component Modification and Process Intensification. In *Innovative Food Processing Technologies: Extraction, Separation, Component Modification and Process Intensification*. Elsevier Inc.
37. Koutchma, T., Popović, V., Ros-Polski, V., & Popielarz, A. (2016). Effects of Ultraviolet Light and High-Pressure Processing on Quality and Health-Related Constituents of Fresh Juice Products. *Comprehensive Reviews in Food Science and Food Safety, 15*(5), 844–867. https://doi.org/10.1111/1541-4337.12214.
38. Kowalski, S. J., Pawłowski, A., Szadzińska, J., Łechtańska, J., & Stasiak, M. (2016). High power airborne ultrasound assist in combined drying of raspberries. *Innovative Food Science & Emerging Technologies, 34*, 225-233.
39. Krishnamurthy, K., Khurana, H. K., Soojin, J., Irudayaraj, J., & Demirci, A. (2008). Infrared Heating in Food Processing: An Overview. *Comprehensive Reviews in Food Science and Food Safety, 7*(1), 2–13. https://doi.org/10.1111/j.1541-4337.2007.00024.x.
40. Kumar, C., Saha, S., Sauret, E., Karim, A., & Gu, Y. (2016). Mathematical modelling of heat and mass transfer during Intermittent Microwave-Convective Drying (IMCD) of food materials. In *Proceedings of the 10th Australasian Heat and Mass Transfer Conference: Selected, Peer Reviewed Papers:* (pp. 171-176). School of Chemistry, Physics and Mechanical Engineering, Queensland University of Technology.
41. Li, R., Huang, L., Zhang, M., Mujumdar, A. S., & Wang, Y. C. (2014). Freeze drying of apple slices with and without application of microwaves. *Drying Technology, 32*(15), 1769-1776.
42. Liu, Z., Qiao, L., Yang, F., Gu, H., & Yang, L. (2017). Brönsted acidic ionic liquid based ultrasound-microwave synergistic extraction of pectin from pomelo peels. *International Journal of Biological Macromolecules, 94*, 309-318.
43. Lind, I. (1991). The measurement and prediction of thermal properties of food during freezing and thawing - A review with particular reference to meat and dough. In *Journal of Food Engineering* (Vol. 13, Issue 4, pp. 285–319). Elsevier. https://doi.org/10.1016/0260-8774(91)90048-W.
44. Lopez-Iturri, P., de Miguel-Bilbao, S., Aguirre, E., Azpilicueta, L., Falcone, F., & Ramos, V. (2015). Estimation of radiofrequency power leakage from microwave ovens for dosimetric assessment at nonionizing radiation exposure levels. *BioMed Research International, 2015*.

45. Lung, R. B., Masanet, E., & Mckane, A. (2006). The Role of Emerging Technologies in Improving Energy Efficiency: Examples from the Food Processing Industry. In *2006 Industrial Energy Technology ConferenceProceedings, New Orleans, LA, 05/10-11/2008*. COLLABORATION-ResourceDynamicsCorporation/ Virginia. https://digital.library.unt.edu/ark:/67531/metadc898508/.
46. Makroo, H. A., Rastogi, N. K., & Srivastava, B. (2020). Ohmic heating assisted inactivation of enzymes and microorganisms in foods: A review. *Trends in Food Science & Technology, 97*, 451-465.
47. Marszałek, K., Mitek, M., & Skąpska, S. (2015). Effect of continuous flow microwave and conventional heating on the bioactive compounds, colour, enzymes activity, microbial and sensory quality of strawberry purée. *Food and Bioprocess Technology, 8*(9), 1864-1876.
48. Ma, Y., Liu, S., Wang, Y., Adhikari, S., Dempster, T. A., & Wang, Y. (2019). Direct biodiesel production from wet microalgae assisted by radio frequency heating. *Fuel, 256*, 115994.
49. Menéndez, J. A., Arenillas, A., Fidalgo, B., Fernández, Y., Zubizarreta, L., Calvo, E. G., & Bermúdez, J. M. (2010). Microwave heating processes involving carbon materials. *Fuel Processing Technology, 91*(1), 1-8.
50. Mohammad Reza Zareifard. (2014, January). *Electrical conductivity data for foods.* |. Ohmic Heating in Food Processing. https://www.researchgate.net/publication/280532621_Electrical_conductivity_data_for_foods
51. Moreno-Vilet, L., Hernández-Hernández, H. M., & Villanueva-Rodríguez, S. J. (2018). Current status of emerging food processing technologies in Latin America: Novel thermal processing. *Innovative Food Science and Emerging Technologies, 50*, 196–206. https://doi.org/10.1016/j.ifset.2018.06.013.
52. Musto, M., Faraone, D., Cellini, F., & Musto, E. (2014). Changes of DNA quality and meat physicochemical properties in bovine supraspinatus muscle during microwave heating. *Journal of the Science of Food and Agriculture, 94*(4), 785-791.
53. Nowak, D., & Lewicki, P. P. (2004). Infrared drying of apple slices. *Innovative Food Science and Emerging Technologies, 5*(3), 353–360. https://doi.org/10.1016/j.ifset.2004.03.003
54. Ohlsson, T., & Bengtsson, N. (Eds.). (2003). *Minimal processing technologies in the food industry.* CRC.
55. Ozkahraman, B. C., Sumnu, G., & Sahin, S. (2016). Effect of different flours on quality of legume cakes to be baked in microwave-infrared combination oven and conventional oven. *Journal of Food Science and Technology, 53*(3), 1567-1575.
56. Ozkoc, S. O., & Seyhun, N. (2015). Effect of gum type and flaxseed concentration on quality of gluten-free breads made from frozen dough baked in infrared-microwave combination oven. *Food and Bioprocess Technology, 8*(12), 2500-2506.

57. Öztürk, S., Şakıyan, Ö., & Özlem Alifakı, Y. (2017). Dielectric properties and microwave and infrared-microwave combination drying characteristics of banana and kiwifruit. *Journal of Food Process Engineering*, *40*(3), e12502.
58. Pereira, R. N., & Vicente, A. A. (2010). Environmental impact of novel thermal and non- thermal technologies in food processing. *Food Research International*, *43*(7), 1936–1943. https://doi.org/10.1016/j.foodres.2009.09.013
59. Piyasena, P., Dussault, C., Koutchma, T., Ramaswamy, H. S., & Awuah, G. B. (2003). Radio frequency heating of foods: principles, applications and related properties—a review. *Critical Reviews in Food Science and Nutrition*, *43*(6), 587-606.
60. Pradeep, P., Abdullah, S. A., Choi, W., Jun, S., Oh, S., & Ko, S. (2013). Potentials of microwave heating technology for select food processing applications-a brief overview and update. *Journal of Food Processing and Technology*, *4*(11).
61. Priyadarshini, A., Rayaguru, K., & Nayak, P. K. (2020). Influence of Ohmic Heating on Fruits and Vegetables: A Review. *Journal of Critical Reviews*, *7*(19), 1952-1959.
62. Rahman, M. S. (2007). Handbook of Food Preservation. In *Food Science and Technology*. https://doi.org/10.1017/CBO9781107415324.004
63. Richardson, P. (2001). Thermal technologies in food processing. In *Food Science and Technology*. https://doi.org/10.1017/CBO9781107415324.004.
64. S.-S. Kim, D.-H. K. (2017). Synergistic effect of carvacrol and ohmic heating for inactivation of *E. coli* O157_H7, *S. Typhimurium, L. monocytogenes*, and MS-2 bacteriophage in salsa _ Elsevier Enhanced Reader.pdf. *Food Control*, 300–305.
65. Sensoy, I., & Sastry, S. K. (2007). Ohmic blanching of mushrooms. *Journal of Food Process Engineering*, *27*(1), 1–15. https://doi.org/10.1111/j.1745-4530.2004.tb00619.x.
66. Si, X., Chen, Q., Bi, J., Yi, J., Zhou, L., & Wu, X. (2016). Infrared radiation and microwave vacuum combined drying kinetics and quality of raspberry. *Journal of Food Process Engineering*, *39*(4), 377-390.
67. Song, Y., Wu, L., Li, N., Hu, M., & Wang, Z. (2015). Utilization of a novel microwave-assisted homogeneous ionic liquid microextraction method for the determination of Sudan dyes in red wines. *Talanta*, *135*, 163-169.
68. Soysal, Y., Arslan, M., & Keskin, M. (2009). Intermittent microwave-convective air drying of oregano. *Food Science and Technology International*, *15*(4), 397-406.
69. Ştefănoiu, G. A., Tănase, E. E., Miteluţ, A. C., & Popa, M. E. (2016). Unconventional treatments of food: Microwave vs. Radiofrequency. *Agriculture and Agricultural Science Procedia*, *10*, 503-510.
70. Stephen, N. M., Shakila, R. J., Jeyasekaran, G., & Sukumar, D. (2010). Effect of different types of heat processing on chemical changes in tuna. *Journal of Food Science and Technology*, *47*(2), 174-181.
71. Sun, D. W. (2005). *Emerging technologies for food processing*. Elsevier.

72. Tang, J. (2015). Unlocking Potentials of Microwaves for Food Safety and Quality. In *Journal of Food Science* (Vol. 80, Issue 8, pp. E1776–E1793). https://doi.org/10.1111/1750-3841.12959
73. Tao, Y., & Sun, D. W. (2015). Enhancement of food processes by ultrasound: a review. *Critical Reviews in Food Science and Nutrition*, *55*(4), 570-594.
74. Thirumdas, R., Sarangapani, C., & Annapure, U. S. (2015). Cold plasma: a novel non-thermal technology for food processing. *Food Biophysics*, *10*(1), 1-11.
75. Thomas Ohlsson and Nils Bengtsson. (2002). Minimal Processing Technologies in the Food Industry. In *Minimal Processing Technologies in the Food Industry*. https://doi.org/10.1201/9781439823132
76. Uysal, N., Sumnu, G., & Sahin, S. (2009). Optimization of microwave–infrared roasting of hazelnut. *Journal of Food Engineering*, *90*(2), 255-261.
77. Vadivambal, R., & Jayas, D. S. (2010). Non-uniform temperature distribution during microwave heating of food materials-A review. *Food and Bioprocess Technology*, *3*(2), 161–171. https://doi.org/10.1007/s11947-008-0136-0
78. Valero, A., Cejudo, M., & García-Gimeno, R. M. (2014). Inactivation kinetics for Salmonella Enteritidis in potato omelet using microwave heating treatments. *Food Control*, *43*, 175-182.
79. Valerie Orsat & Vijaya G.S. Raghavan. (2005). Radio frequency processing. *Emerging Technologies for Food Processing*, 445-468. https://doi.org/10.1016/B978-012676757-5/50019-0
80. Vinatoru, M., Mason, T. J., & Calinescu, I. (2017). Ultrasonically assisted extraction (UAE) and microwave assisted extraction (MAE) of functional compounds from plant materials. *TrAC Trends in Analytical Chemistry*, *97*, 159-178.
81. Wang, S., Luechapattanaporn, K., & Tang, J. (2008). Experimental methods for evaluating heating uniformity in radio frequency systems. *Biosystems Engineering*, *100*(1), 58–65. https://doi.org/10.1016/j.biosystemseng.2008.01.011
82. Wongsa-Ngasri, P. (2004). Ohmic heating of biomaterials: Peeling and effects of rotating electric field. In *ProQuest Dissertations and Theses*. https://search.proquest.com/docview/305140014?accountid=27575
83. Won, M. Y., Lee, S. J., & Min, S. C. (2017). Mandarin preservation by microwave- powered cold plasma treatment. *Innovative Food Science & Emerging Technologies*, *39*, 25-32.
84. Zielinska, M., & Michalska, A. (2016). Microwave-assisted drying of blueberry (Vaccinium corymbosum L.) fruits: Drying kinetics, polyphenols, anthocyanins, antioxidant capacity, colour and texture. *Food Chemistry*, *212*, 671-680.

2

Microbial Inactivation with Heat Treatments

Sushree Titikshya*, Monalisa Sahoo, Vivek Kumar and S.N. Naik

Centre for Rural Development and Technology, IIT Delhi, Delhi, India

Abstract

Heat utilization in food processing is a century-old tradition utilized for the preservation, extraction, cleaning of different food products. Food quality and safety are the two major concerns in the industrial sector besides targeting different microorganisms found in the food. Techniques for food preservation are always improving at an increasing pace to avoid degradation of the food product, and novel methods to improvise microbial inactivation are also trending. With the rising customer demands for fresh products, industries are giving special attention to the heat resistance of spores from sensitive cold growing spores and finding ways for minimal processing with an amalgamation of facilities for cold storage and improved shelf life. Intensive studies were investigated for developing new alternative conventional techniques which look after better preservation of food, maintaining the color, texture, nutritional quality, and flavor of the food product. In the accumulation of environmental and food safety concerns, the techniques which have evolved as an innovative thermal preservative method in agricultural and food applications are ohmic heating, radiofrequency heating, microwave heating, infrared heating, and instant control pressure drop technology.

***Keywords*:** Food quality, minimal processing, ohmic heating, food safety

2.1 Introduction

Food quality and safety have a long history and are continuing as the major concerns for preservation and targeting microorganisms. The chief cause for any type of deterioration of food products is microorganisms,

**Corresponding author*: titikshya.sushree@gmail.com

Nitin Kumar, Anil Panghal and M. K. Garg (eds.) Thermal Food Engineering Operations, (45–74)

and therefore inhibition of these are targeted using different methods by the food industries [1]. Varied techniques of food preservation have been utilized for centuries to avert poisoning and degradation of the food. With the passing phase, the new models and techniques are evolved promising to avert the degradation by inactivating microorganisms. But then again slowing down or terminating the growth of microorganisms will not ensure the proper safety of the food until and unless the environment is taken care of, like maintaining a cold chain throughout. Processing the food by application of heat has extended antiquity for the inactivation of microorganisms and is still considered as the utmost significant method of preservation [2]. Louis Pasteur introduced pasteurization in the late 19th century to eradicate spoilage-causing organisms and with the passing of time, new science-based heat processing emerged.

In the old heating process, the stability of the microbes was often quite high but the knowledge of the kinetics for different microbes was relevant. Traditional canning methods aim at the destruction of all spores (sterilization) or of all spores that can grow in the container below 40°C (commercial sterilization or appreciation) [3]. With the rising customer request for fresh products, industries are giving special attention to the heat resistance of spores from sensitive cold growing spores and finding ways for minimal processing with an amalgamation of facilities for cold storage and improved shelf life [4]. Quite a few sterilization methods have already been utilized by most of the food industry, the objective of which is to inactivate the microorganism from which the most common form of destruction is targeting the DNA of the microbial cell and destroying it in both liquid and solid food products through denaturation [5] did a study and concluded that the denaturation process requires high heat for 15 s at 71.06 °C accompanied by drying with air inlet temperature extending from 135 to 205 °C for 5–6 s.

Several works have been carried out during the past decades aiming to inhibit the growth of the foodborne pathogens in varied products. However, some of the traditional thermal treatments, al though they carefully inactivate all the microbes and extend the shelf life of the food, are still not favored as they cause degrading effects to the nutritional properties of the food product such as deterioration of the antioxidant compounds, flavor, proteins vitamins and volatile oils [6]. Therefore, intensive studies were investigated for developing new alternative conventional techniques which offer better preservation of food, maintaining the color, texture, nutritional quality, and flavor of the food product. In the accumulation of environmental and food safety concerns techniques which evolved as an innovative thermal preservative method in agricultural and food applications are ohmic heating, radiofrequency heating, microwave heating, infrared heating, and instant control pressure drop technology [7–9]. These physical factors can cause the inactivation of microorganisms at ambient or sublethal temperatures.

The major objective is to gather the knowledge of all the techniques, mechanisms of inactivation, and factors affecting the physical and orogenetic parameters of the particular food products. Furthermore, predicting kinetics and microbial nature during the application of varieties of different techniques through mathematical modeling is also considered the most vital tool in the food matrix. Therefore, the chief goal of this chapter is to provide a record that will provide data of the food degrading organisms, its quantitative behavior portrayal for different food matrix with varied environmental setup and besides with the pre-assumed mechanism of different alternative thermal techniques involved.

The final goal, however, is to answer the question of whether the microorganisms are inactivated after nonthermal processing. Although mathematical descriptions of heat inactivation date already from the early 20s, the development of new software during the last decades enables a better description of inactivation kinetics under different conditions. Combined with better control of temperature in each container a milder heat process can be designed without compromising microbial safety and stability [10]. The principles of modeling death kinetics will be addressed within the framework of quantitative microbiological risk assessment. The effect of environmental conditions and their extrapolation to real food situations will also be discussed.

2.2 Innovate Thermal Techniques for Food Reservation

Pasteurization and sterilization are the leading and finest processing techniques in the industrial sector. These techniques stop the rapid multiplication of spoilage and microbial organism present in the food and eliminate it by the application of heat in an acidic medium [11]. Temperature also plays a vital role in the thermal processing method. The higher the temperature of the food product while entering the process, the more speedily it helps in killing the microbes present in the food [12].

Though high-temperature traditional processing confirms the microbial safety of the food products by reducing it into 3 log cycle as per the 21 Code of Federal Regulation Part 114 [13], the limitations which arise during the application in terms of organoleptic and nutritional properties compels to limit the usage of these traditional techniques. These limitations also extend to non-uniform heating, degradation of the final quality of product, and low heat transfer. But currently, potential heating systems have come up as an alternative to replace the exciting conventional system, some of which are microwave heating, radiofrequency heating, ohmic heating, infrared heating, and instant control pressure drop technology (DIC) [14]. These alternative techniques of heat processing provide the advantages of volumetric and uniform heating while maintaining the quality of the final product and reducing the surplus treatment time by heating directly the food sample [15].

Few bacteria display an advanced heat resistance, subsequently being exposed to temperatures which only stress them [16, 17]. The growth medium of the organism, the growth temperature, and the phase of growth are significant issues concerning their aptitude to withstand heat. It is a very common phenomenon that the bacteria surviving different stress conditions have quite a high tolerance level to the different environmental situations and heat [18]. Inactivation does not mean that all the microbial cells are destroyed but it depends upon the number of cells present in the food sample. The maximum number of cells indicates the increased consumption of time for inactivation of the microorganisms in the food sample. The design of the thermal inactivation process for a given food depends on (i) the heat resistance of the most resistant microorganism (in the product); (ii) the food products dimensions; and (iii) thus the rate of heat penetration within the food matrix [19].

Therefore, this section brings, in particular, all the possible novel heat processing technologies for decontamination of spoilage and pathogenic microorganism present in the food. The table mentioned below provides a summary of the studies done on the microbial techniques and outcome which were achieved for the different food items and their change in properties after the treatment.

2.3 Inactivation Mechanism of Targeted Microorganism

The effective application of innovative technology for food preservation depends on the development in the field of inactivation mechanisms. A requisite acquaintance with the physiological behavior of microorganisms in the direction of decontamination factors is indispensable for the advancement of secure food products [20]. It is vital to consider and understand the crucial environmental factor for identifying the resistance in the food product. This will support easy construction of the mathematical model and interpret the kinetics which is done based on the parameters with an effective prediction of the decontamination in a wider range [21]. Therefore, designing a process becomes easier by gathering information on different preservation agents. The efficiency of any method utilized depends upon product type, the process used, and targeted microorganisms [22]. Inactivation mechanisms depend on the technique utilized for the above action and are also influenced by the structure and number of the microbial cell. Environmental stress kills or injures cells of the microorganism but in some cases, it might cause sublethal injury or also might cause total lethal cell death [23].

Destruction of the cell wall is the main factor which leads to the inactivation of microorganism and it also depends upon the cell morphology and shape of the cell. Destruction of the cell wall can be achieved by either chemical process by modifying the wall leading to leakage of the cellular content or by altering the constituents chemically. As already known, heat is the major factor that causes destruction of the membrane as well as protein denaturation but what type of heat, it is tough to state [24].

Application of significant high pressure with heat can lead to disorientation of the cellular structure of the microorganism and also lead to protein denaturation, whereas this is hardly seen in the bacterial spore because of their morphology. As visually seen nevertheless, that hydrostatic pressure treatment can induce for growth of bacterial spores and a combination of swift decompression pressure and elevated temperature leads to the destruction of the spores which re-germinated [25].

Tolerance of microbial stress is affected both by intrinsic and extrinsic factors as resistance during stationary growth and during exponential growth varies because of varied stress sigma factors [26, 27]. With the intention of endurance, micro-organisms incline to produce biofilms, the utmost prevailing microbial defensive structure. Bacteria produce carbohydrate matrices and might consequently condense the efficiency of approaches for inactivation [28].

2.3.1 Action Approach and Inactivation Targets

Damage to cell structure or any type of physiological change can indicate the death of the cell. So, during the process of cell structure disruption, cell envelopes break, change the anatomy of DNA, alters the ribosome, and disintegrates the protein [29]. Whereas during the physiological change fluctuation of permeability in addition to the loss of enzyme functionality leads to death of the cell. So, during the inactivation process the above changes can occur solely or combinedly, and therefore, identifying the particular event becomes problematic for the researchers while coming across events in this area. Therefore, keeping in mind, the following situation, this can be possible to have multiple inactivation actions that combine to cause the death of the cell. It is likewise probable that the important aim solitary affected when a secondary structure is beforehand injured. For example, due to heat, there is nutrient loss, ions loss, the disintegration of DNA, denaturation of crucial proteins and enzymes [30]. Therefore, exactly capturing the real event of death of the cells becomes difficult as the structure of the cell is also affected by higher temperature. The only possible way to find out the exact reason is by examining the relation between inactivation degree and the modification dress of the targeted microorganism concerning various environmental conditions.

2.4 Environmental Stress Adaption

The cell contains some extracellular protein, which warns or senses the probability of danger when there is any environmental stress for instance heat [31]. These processes comprise fluctuations of events of protein and gene expression with the persistence of averting and/or lessening injury to cells. As already discussed, resistance is higher in the stationary phase of growth rather than the exponential one because of the sigma factor [32]. If a bacterium survives, it can exist in a VBNC state. This denotes a state in which the cells cannot be detected by standard culture on enriched agar media, although remaining viable and capable of resuscitation under favorable conditions. VBNC was mainly observed among Gram-negative bacteria and has been proposed as a strategy for survival in natural environments [33]. More than 60 bacterial species are described as being able to enter into a VBNC state, among them Gram-positive (e.g., L. *monocytogenes, Enterococcus, Micrococcus luteus*) and Gram-negative (e.g., *E. coli, Vibrio cholerae, Vibrio vulnificus, Legionella pneumophila, Campylobacter jejuni, Salmonella enterica, Pseudomonas aeruginosa, Helicobacter pylorii*) bacteria can be found [34].

2.4.1 Sublethal Injury

Sometimes when microorganisms survive any environmental stress, they might again revive and regrow themselves with available appropriate conditions [35]. This shortcoming can cause the wrong estimation of the exact lethality after treatment as it might not detect properly only the cells which are injured. There is a chance therefore to get repair during the treatment and reviving phase. So, this can be avoided by adding some further agents for presentation so the reviving process of the cell can be containing and better inactivation can be achieved [36]. Hence, uncovering and classifying the sublethal injury by innovative preservation techniques is indispensable for optimizing a varied combination of methods for elevated effects of microbes for inactivation.

2.5 Resistance of Stress

As already discussed, microbes tend to grow resistant for some time when coming in contact with varied environmental stress which causes serious issues when it comes to food safety.

Researchers have been studying for many years now how to deal with varied adaptation techniques to do the inactivation process better. Modifying the sigma factor with different RNA polymerase is probably the most significant controlling method in bacterial cells [37]. This factor governs the transcription of the genes in Gram-negative bacteria which are resistant to oxidative, heat, and osmotic stress. Therefore, inducing these sigma factors would help to activate during the cell undergoing different growth phases, may it be in stationary or exponential state [38]. But for Gram-positive bacteria a substitute sigma factor with alike physiological roles are studied in [39, 40]. This infers that a similar process for multiple stress resistance is seen for cells of Gram-negative and Gram-positive. The utmost problem that should be taken care of is to prevent microorganisms from adapting to the stress because that helps them to create a barrier and create greater protection for different other succeeding stress. This is the reason for the emergence of different novel techniques for preservation; a direct heat and traditional thermal method is causing sublethal injury as well as augmenting the sensitivity of the cells to stress when applied mechanically. This is because of some temperature-induced variations in the cell envelopes of microorganisms [41].

2.5.1 Oxidative Stress

The main cause of the oxidative stress in bacteria is due to the following reasons, i.e., imbalance of macromolecules changes, cellular and intracellular antioxidant and oxidant concentration which is related to lipids, proteins, and DNA repair enzymes [42]. Enzymes are considered to be the shield of microbes and catalyses Hydrogen peroxide which is the prevailing bactericide (also for spores) and oxidant that is capable of generating chemicals that can oxidize hydroxyl radicals ($OH^{\cdot}$) [43]. Peroxidases convert to alcohol and water by reducing hydrogen or organic peroxides.

2.5.2 Osmotic Stress

The requirement of water in the food system varies and is calculated through the water activity (aw). The addition of solutes in the microbial cells changes the water activity which results in lowering the water content, causing osmotic stress. And due to this, microbes increase their cytoplasmic cells through various processes, for instance, humidity, air circulation, limitation of nutrition, and tempura which contribute to the stress [42].

2.5.3 Pressure

Elevated pressure when applied to the microbes results in altering the genes, metabolisms, and morphology of the cell. It is known that microorganisms adapt easily to the environment using various methods. They utilize diverse protection processes by activating the expression of the genes, staying in the dormant state, and producing the resistant mutant. There are different ways microorganisms adapt to the adverse condition, such as they create spores which hardly changes its morphology during application of pressure. As mentioned above, the resistance power to pressure in the stationary phase higher as compared to the exponential phase [44]. As the structure of the cell does not change in the stationary phase due to the protection of the membrane, the tolerance stress level gets elevated [45]. *Listeria innocua* endured in replicated milk with supplementary magnesium, calcium, citrate, and phosphate [27]. Magnesium has the stabilizing agent for ribosomes and calcium aids in maintaining the outer wall of the membrane of the cell. Sucrose guards bacteria by stratifying the functionality of membrane proteins [46].

2.6 Various Techniques for Thermal Inactivation

2.6.1 Infrared Heating

2.6.1.1 Principle and Mechanism

Infrared falls in the electromagnetic spectrum fluctuating between 0.78-1000 μm which is between ultraviolet and microwave radiation. Heat is generated due to the motion of the molecules which is both rational and vibrational in nature. Unlike conventional heating where the heating is done by convection at the first from the surface and from inside of the product, it is done by conduction, here infrared offers radiation from outside the surface and conduction from inside of the food product [47]. Numerous studies have been done to verify the anti-microbial effect of IR when applied to various food products such as honey, cheese, milk, fruits, and other liquid and semiliquid products [48, 49]. Infrared works well even for powder products mainly spices and with time varied of the word have also been performed to analyze the effectiveness of the IR treatment for effectiveness in reducing microbial activation. Several factors significantly affect the microbial log reduction but the utmost responsible factor is the temperature and wavelength of infrared radiation. Some of the other factors are as following power, water activity, moisture content, bandwidth, and depth of the food sample [50].

Infrared offers a similar effect of thermal inactivation mechanism as seen in the microwave heating and ultraviolet light which causes DNA damage along with heating through induction and of course. By thermal heating, the inactivation of microorganisms becomes easier; it can destroy or damage different parts of the cell structure which mention in the order of damage magnitude that is protein >RNA >cell wall >DNA. To confirm the effects and study the mechanism of heating for the inactivation of microbes, various methods like fluorescent and spectroscopic probes were utilized for paprika powder performed by [51]. The result inferred from the treatment that radiation of the infrared waves created all-around injury of the cell which includes inactivation of RNA polymerase of the microbial cells, which simultaneously prevents the transferase reaction by combining it with subunits of the ribosome. Much effective and advanced decontamination is visible in this treatment as the waves of the infrared have extended energy levels, and distribution of this energy is very efficient as compared to any other traditional method of heating which mostly uses the fluorescent probe for molecular-level analysis. Analysis using the digital method showed the best results as compared to the conventional one in terms of analyzing the relationship of the targeted organism and the environment they are present in real time.

2.6.1.2 *Application for Inactivation in Food Sector*

Different studies were conducted which explain the different effect of the parameters of infrared radiation. *Bacillus cereus* on paprika powder was studied by applying radiation at 11 kW/m^2 and 5 kW/m^2 at a temperature of 95 °C [51]. From the studies it showed that maximum injury of microbes was seen at a_w 0.5 and at a_w 0.8 the overall log reduction was seen was 0.7 and 1.6 log 10 CFU/g at 5 and 11 kW/m^2. From this, it was inferred that *Bacillus cereus* is susceptible to heating through infrared, plus also preserving the

Table 2.1 Application of thermal techniques in food industry.

Technique applied	Food product	Targeted microorganism	Treatment parameters	References
Infrared Heating	Rice Powder	Bacteria & moulds	Wavelengths: 3.2, 4.5 and 5.8 Time interval: 10, 20 & 30s	[91]
	Almonds	*Enterococcus faecium*	Temperature: 70 °C Time: 1 h	[92]
	Garlics (shredded)	*Aspergillus niger*	Wavelength: 3.3 μm	[93]
	Oregano	*Bacillus cereus*	Temperature: 90 °C Time: 10 min	[52]
Microwave Heating	Bay leaves	Counts of bacteria	Power density: 32.14–142.85 W/g Time: 150s	[94]
	Peanuts	*Aspergillus flavus*	Power levels: 360, 480 & 600 W	[95]
	Infant Formula Mix	*Cronobacter sakazakii*	Power levels: 800 & 900W	[65]
	Ashitaba leaf powder	Colonies count	Power: 300 W Time: 1 h	[96]

(*Continued*)

Table 2.1 Application of thermal techniques in food industry. (*Continued*)

Technique applied	Food product	Targeted microorganism	Treatment parameters	References
Radiofrequency Heating	Broccoli	Bacteria	Heating: 6kW	[74]
	Barley Grass	*Aspergillus* and *E. coli*	Heating: 6kW Frequency: 27.12 MHz Gap (electrode): 120-240mm	[96]
	Walnuts (shredded)	*Staphylococcus aureus*	Time: 40 min Frequency: 27 MHz Gap (electrode): 19 cm	[97]
	Flour (wheat)	*E. faecium* and *S. enteritidis*	Frequency: 27 MHz Time: 39 min	[71]
Instant Controlled Pressure Drop Technology	Apple Pomace	*Aspergillus niger*	Pressure: 0.2–0.6 MPa	[77, 100]
	Dried Carrots	Flora of the bacteria	Pressure: 3 kPa Temperature: 70 °C Time: 3 h	[98]
	Seaweed	*Salmonella* spp.	Pressure: 0.44 MPa Time: 40s	[76]
	Flour (banana)	Colonies of bacteria	Pressure: 3–5 kPa	[99]

(*Continued*)

Table 2.1 Application of thermal techniques in food industry. (*Continued*)

Technique applied	Food product	Targeted microorganism	Treatment parameters	References
Ohmic Heating	Apple juice	*Salmonella Typhimurium, E. coli* & *L. monocytogenes*	30 V/cm Time: 60 s °Brix: 36	[87]
	Tomato soup	*Geobacillus Stearothermophilus*	Frequencies: 60 Hz & 10 kHz Time intervals: 0, 15, 60, 120 s Temperature: 121°C, 125°C & 130°C	[83]
	Meat balls	*L. innocua*	HTST: 50 Hz 8.33 V/cm Temperature: 95°C Time: 7 min	[88]
	Gochujang	Bacillus cells (vegetative)	Frequency: 60 Hz 30 V/cm Temperature: 100°C Time: 2.5 min	[79]

effects of the main product. A similar effect was studied for oregano powder where disinfection of the microbes was done for investigation [52].

Eradicating the population of *E. coli* and *Salmonella typhimurium* was studied by [53] where 500W heating power was applied to the product, i.e., red pepper. After the treatment, it was seen the presence of 0.23 and 0.32 log CFU/g was seen for both the microbes and also maintain its intercellular component and color of the product. Though the volatile content present in the above products, the volatile oil seems to get affected with some infrared heat setting, there some wavelength was not considered for the preservation process of these types of the volatile content of the product.

Various types of food sustain different wavelengths in infrared radiation. So, it is essential to evaluate the pattern of absorption to make the inactivation process easier and ensure proper deactivation without hampering the quality of food. For sensitive foods, it should be taken care of to heat the microbes selectively so that they would not hamper or adversely affect the components of the food. The wavelength selected for protein denaturation for decontamination of *F. proliferatum* and *A. niger* results in 40% increases in the inactivation process. Overall, it can be decided that though the product temperatures before and after infrared heating identical and injury of spores and cell wall augmented with infrared heating [54]. The applications of thermal processing in food sector has been discussed in Table 2.1.

2.6.2 Microwave Heating

2.6.2.1 Principle and Mechanism

With the advancement of emerging technologies, microwave energy has paved its way and has become an indispensable part of every household system. The use of microwave has expanded from heating and defrosting to thawing, blanching, sterilization, drying, etc., in food industries [55, 56]. Microwave is electromagnetic waves with a frequency which ranges from 300 Mhz to 300 GHz. Frequency of microwave used for domestic purposes is 2.45 GHz, whereas the frequency for industrial one is 915 MHz [57]. In the microwave system, the two oscillating perpendicular fields, i.e., electric and magnetic, act directly on the heating material, converting the part of absorbed energy to thermal energy. The interaction of the microwave radiation with chemically bound water present in the food material generates high pressure and temperature due to the absorption of the characteristic photonic energy of electromagnetic waves. This process causes moisture evaporation resulting in pressure exertion on the plant material to cellular and subcellular level leading to swell up and rupture eventually [58].

Microwave has a varying magnetic field which generates heats on interaction with, and absorption by, certain dielectric materials, and with the positioning of the direction of the electric field, the native thermal motion of the polar molecule changes [59].

The heating principle of the microwave is on the interaction between the sample and electric field, resulting in ionic and dipole interaction, a movement finally converted into heat. According to an FAO/WHO Expert Committee on Food Safety report, "the radiation of any food product up to a total average dose of 1000W has proven to present no radiological hazard". Microwave radiation has been used as an efficient alternative in microbial load reduction of a few food powders. However, despite the advantages of microwave treatment the formation of "cold spots" within food powder where harmful bacteria initiates its growth under favorable conditions limited its industrial application. Numerous researches over the years on microwave discovered that the effectiveness of microwave treatments on microorganism decontamination depends on important attributes, is Microwave=*f* (microwave power, microwave temperature, sample thickness, and treatment time) [60].

Ions present in the solution, when suspended to the electric field, orient themselves, experiencing acceleration and an upsurged kinetic energy. When ions collide with each other it gets converted into heat. This frequent collision increases the density or the concentration of the solution, which is also known as the ionic polarization effect [61]. Where in gases the collision becomes difficult due to the spacing between the molecules. In food material, cations are generated by the presence of salts of sodium, potassium, or calcium whilst chlorine produces anions.

When the polar molecule strives to situate itself into the fluctuating electric field caused by the microwave, the dipole rotation is created, where the oscillation of the dipolar species leads to the collision with the surrounding producing heat [62]. With the increase in temperature, the dipole movement decreases whereas ionic conduction increases hence, food samples with both the compounds when heated by the microwave, first governed by the dipole rotation and then with the increase in temperature governed by ionic conduction. The comparative involvement of these methods of heating hinges on the concentration and flexibility of sample ions, plus the sample's relaxation time [63].

2.6.2.2 Application for Inactivation in Food Sector

Log reduction of *E. coli, Shigella,* and *Salmonella* present in black pepper were studied by [64] which programmed for 2 min at 50 °C with a power level of 800W. These three-microorganism got reduced to a log cycle of 6 which would hold up to 3 mins with a sample thickness of 3cm. Microwave treatment was performed on infant formula powder targeting *Cronobacter sakazakii* studied by [65]. At 800 and 900W the log reduction 5 log cycle. [66] currently studied the inactivation rate of *Aspergillus niger* present in garlic which resulted in reduction varying between 1.12 and 1.16 at a microwable temperature of 50 °C. This experiment was conducted using a microwave drum dryer using Weibull, Page, and Bigelow models. Therefore, this model proved to be perfectly fitted to predict the reduction rate of inactivation of *Aspergillus niger*. From all the above case studies it was inferred that treating the food products with microwave proves to be promising for a wide range of products on an industrial scale as it maintains the quality and preserves the nutritional aspects of the products. Though there are studies with different vegetables, more research has to be conducted to verify the exact nature retained at the end of the experiment for various other food powder products and which could be done on a wider scale level.

2.6.3 Radiofrequency Heating

2.6.3.1 Principle and Mechanism

In the electromagnetic spectrum, radiofrequency waves fall in a frequency range of 30–300 MHz. This method works depends upon the dielectric properties of the edible sample, where with help of an indirect mode of heating they treat the target product initially with electrical energy which is then converted to radiation mildly releasing the heat [67]. Radiation from radiofrequency are non-ionizing so therefore the food molecules are not disoriented, which is the main bonus point in the process of microbial inactivation. The advantages of this process are the time consumption which is reduced remarkably, making it a feasible application in food industries for different unit operation.

Some of the chief factors on which efficacy is highly dependent for the treatment of microbial decontaminations are moisture content of the sample, temperature a frequency, targeted microorganism, and sample depth. The current study suggested that the frequency of the microwave makes the condition favorable for microbial decontamination even if it exists in food with lower moisture content. This method favors all types of food material as it works as electrothermal pasteurization which is chemical-free [68].

The principle involved in inactivating microbes by heating through radiofrequency is that the cells of the microbes get thermally injured even at a reduced heating rate as the heat develops within the cells of the microbes causing eventual destruction of the microbes [69]. RF energy combining with other different thermal methods shows synergistic effects, notably making the RF pasteurization efficient particularly for agricultural materials having less moisture content [70]. Due to the radiation mode of heating DNA of the microbial cells absorbs directly fluctuating the structure of the cell and its functionality. Radiofrequency is used for uniform heating of food products even for power products as the treatment usually depends upon the cell structure of the focused organisms.

2.6.3.2 Application for Inactivation in Food Sector

Authentication of microbial content using radiofrequency heating was done for wheat by [71]. The log reduction microorganisms *Enterococcus faecium* and *Salmonella enteritidis* were investigated. It was calculated that it took 18 min for *Salmonella enteritidis* and 25mins for *Enterococcus faecium* to reach 5 log reduction at 85°C. Similarly, [72] studied the effect of radiofrequency heating the same pepper to decontaminate *Salmonella typhimurium* by adjusting the aw to 0.57-0.71. The outcome was a reduction of the log cycle between 2-3. Another study was carried by [73] utilizing radiofrequency for the 80s at 90 °C for dried black pepper and the outcome was a remarkable reduction 7 log cycle of *S. Typhimurium* and *E. coli*. When this log reduction is compared with infrared and microwave, it was the best and highest, but got a little demerit as the moisture content was reduced drastically. Similarly, [74] used broccoli for their experimental research to treat it at 6 kW which showed a reduction by 4.2 and maintaining the quality attributes and preserve the color from degradation. Therefore, mesophilic bacteria work best in radiofrequency heating by attaining the desired log reduction and also maintaining and preserving the quality, sensory and nutritional aspects of the food product for decontamination of the microbes.

2.6.4 Instant Controlled Pressure Drop Technology (DIC)

2.6.4.1 Principle and Mechanism

This technique is the thermochemical process which is built on the notion of auto-vaporization and thermodynamics of instantaneity. This concept is introduced by exposing the sample to a short period of saturated steam and then an immediate drop of pressure is applied until it gets into the vacuum. Prominent mechanical stress is applied after the pressure drop which subsequently results in cooling the water promptly and then auto vaporizing it [75]. The DIC technology testifies as a strong and promising system for inactivation and decontamination in food products and works best with the food products with increased surface area contributing to the reduction of log cycle to an estimated level. The efficiency of this technology depends upon the critical factors like expansion property of food, level of vacuum developed, size of the particles, mechanical and thermal stress, microbes variety volatile content, and pressure drop.

This technique alters the cellular structure and component of the microorganisms and this change is also irreversible therefore and dominating process in the food sector for decontamination. The major factor of this mechanism has two main features which are relaxation of pressure and controlled thermomechanical process of the targeted cell of the microbes. The process of the techniques is as follows: exposing the sample to the vacuum stage, homogenizing it, flowed by an instant drop of pressure, and initiation mechanical stress by releasing the atmospheric pressure causing the cell to explode [76]. Not only mechanical stress but also thermal stress is applied which is effective for the inactivation of the microbes. The procedure of auto-vaporization has likewise been stated to generate a mechanical constraint that performs on the focused cell of the microbes especially the walls of the spores. This technique is known for conserving the nutritional content of the specific food products as well as preserving the organoleptic qualities.

2.6.4.2 *Application for Inactivation in Food Sector*

Mycotoxin decontamination from apple pomace was studied by [77], where the treatment was performed to inactivate *Aspergillus niger* using DIC at 0.2–0.6 MPa. The outcome of the treatment showed remarkable results by increasing the flavonoids to an extreme of 800% after the process was complete and as well as improving the drying kinetics, log reduction, and quality characteristics of the product significantly. Similarly, *Salmonella* and *Staphylococcus aureus* inactivation was evaluated by [76] in which they checked the response of instant pressure drop in skim milk and seaweed. The total time taken for treatment was 40 s and taking into the counting of the targeted organism was done through the LASAT method, and with this, the DIC method proved to be best, giving desirable outcomes in inactivating the above-mentioned microorganisms which were 100% for seaweed and more than 87% for skim milk. Another special combination case experimented with a heated air temperature of 50 °C plus the DIC by [77]. This combination was performed to confirm that the count of bacterial injury was high in the case of DIC application as compared to when no DIC treatment was done. Additionally, the scientists confirm that this process helped in preserving all the nutritional components like vitamins and minerals and hence giving the best possible results in terms of quality. Subsequently gathering and investigation the above-mentioned work we can infer that DIC plays a vital role and effectively contributes to the decontamination process of microorganism because of its mild heating system and in addition, without hampering the qualitative part of the food product.

2.6.5 Ohmic Heating

2.6.5.1 Principle and Mechanism

The traditional method of heat processes like pasteurization and sterilization follows conduction and conduction methods to injure the microorganisms and reducing the load. As mentioned earlier, these methods also consume much time and energy. To mend these disadvantages and meet the demand of the food and increase the quality of the food ohmic heating is also one of the techniques used for better inactivation methods of microbes [78]. Other names of ohmic heating, electro heating, or joules heating, have gained much interest in the food industry. Any conductive food when heat is passed from within, heat is straightaway produced and simultaneously temperature is increased. This technique is utilized for power products, fruits, and vegetables. The pattern in which heat is generated during ohmic heating causes food products to heat up rapidly in a uniform manner [79]. Overheating is avoided of any food product in this situation as there is not much temperature change or gradient. Therefore this technique consumes less heat and also increases the shelf life of the sample as compared to the traditional method.

The major process in this method is the thermal effect which is used for the destruction of the cellular structure of the microbes. With this thermal effect, an additional effect added to the mechanism is the chemical effect which helps in free oxygen and hydrogen formation [80]. It also produces radicles of hydroperoxyl and metal ions. Injury of microbes is usually due to the buffer solution used which makes the solution or the product toxic. This toxic formation is the main principle agent to cause the deactivation of microorganisms, and later after the desired state is achieved, toxicity also reduces gradually [81].

When compared to the traditional water bath and ohmic shows a better death effect which is created by electric current. The heating is performed under similar temperature and time parameters. During this process, the permeability of the cell is higher whereas transudation of intracellular materials is also elevated which eventually leads to cell injury of the microorganisms [82]. Cell permeability depends upon several factors such as electric current, increased frequency, and higher potential of threshold, processing time. These factors play a vital role in progressing the inactivation process.

The augmented permeability might cause irreversible damage to the cell over seepage of cellular compounds, together with proteins, enzymes, amino acids, and nucleic acids. But some studies have also stated that the electroporation method has a specialized effect on the cellular and intracellular effect [83]. Eventually, this can be maintained by fluctuating the parameters so that leakage doesn't happen beyond the threshold level which automatically leads to death. A connection is formed between thermal and chemical effects because the heat is generated due to the generation of electric current passing through the sample of food [84]. In ohmic heating, with temperature, the heating rate also increases. So it is necessary to carefully experiment and design to match the conditions for thermal and chemical effect. Consequently, there will be an extended method to explain the thermal properties owing to the spirit of ohmic heating where heat is formed due to electric current.

2.6.5.2 Application for Inactivation in Food Sector

The effect of inactivation rate and food quality while using the ohmic heating process has been widely studied in several reports. *E. coli* inactivation was done for goat milk using ohmic heating at 50 Hz where the electric field was adjusted and the whole process was compared with the traditional method [85]. The log reduction was significantly reduced when it was done using ohmic heating as compared to the conventional water bath process. For orange juice, the decontamination was done using ohmic heating for moulds, yeast, and bacteria at 50 Hz with varying temperatures. This was also compared with the traditional one which was done using 90°C for 50 s. The results showed that the reduction or the inactivation processing ohmic heating was 98% with only 15% reduction of vitamin C plus the product did not get degraded any of its qualities and was fresh when compared with the conventional pasteurized one [86]. Shelf life during storage of both the sample treated differently at 48°C extended in case of ohmic heated one up to 100 days which was almost twice as seen in traditionally treated pasteurized process product plus preserving the flavors and nutritional attributes.

[87] showed comparable outcomes in which apple juice was treated through ohmic heating for *L. monocytogenes* and *Salmonella Typhimurium*. It was observed the log reduction was 5 CFU/mL at 20Hz which was treated for 30 s without disturbing the superiority of the juice. These conclusions specify that ohmic heating is acceptable for the decontamination of microbes in juices.

When meatballs were subjected to ohmic heating to inactivate the mesophilic moulds, bacteria, and yeast with the following parameters (75°C, 50 Hz, 0 s holding time, 15.26 V/cm). The inoculation was done for *Listeria innocua* and the outcome was compared with the conventional process. The traditional process consumed much time, i.e., 150 min as compared to sample treated through ohmic heating which varied from 7-15 min [88]. During and after the storage the sample treated with ohmic heated showed a longer shelf life of 21 days with maintained product quality. From all the above studies it was testified that ohmic heating is considered prominent to yield a higher-end product in terms of quality and safety for both solid and semisolid food with minimum heating time as compared to traditional heating or water bath.

Similarly, ohmic heating works significantly in inactivation spores from food products. Ohmic heating was used to treat *B. licheniformis* spores present in cloudberry jam at 50 Hz [85]. The sample was also heated with the water bath and the electric field was adjusted according to that. It was observed that the reduction of log cycles was observed perfectly with the sample heated with ohmic heating rather than a simple water bath at high temperature. *B. cereus* spores found in doenjang were shrunken when ohmic heating with parameter 26.7 V/cm at 25 kHz was applied for the 60 s [81], though the temperature was maintained up to 105 °C, the quality and color were yet suitable for the customers. The main advantage of ohmic heating is that the time required is reduced to a significant level adding to the efficacy of the inactivation process; additionally, it provides quality preservation assurance, and hence thereby enhanced shelf life. Therefore, gathering all the above context with every study is done to improve the process to attain higher consistency.

2.7 Forthcoming Movements of Thermal Practices in Food Industry

To validate the innovative techniques and products developed from that is the main focus of the food industries to scale it up for the next generation and match the pace. With the continuing year, there are various techniques which emerged and were marketed in varied countries in the world, such as microwave, high-pressure process but commercialization of DIC was very narrow [89]. Many applications are emerging especially for powder food products for debacterization and decontamination which are gaining a huge interest in the food industrial sector. The technology stated in this chapter proved to be great and promising for the inactivation process of the microbes with each of them having some speciality with some advantages and disadvantages. Consequently, it is important to identify and build a specialized amalgamation of these techniques and the specific food products to improve the product characteristics and increase the lethality of the cell of the microbes. The major demerit of these techniques is elevated initial cost for installation and minimum data available for data processing parameters. To minimize this drawback, treatment state is optimized and efficacy and high-end products are processed by adjusting the process parameters. This helps in preserving the nutritional balance and product quality plus ensures the safety of the product while been exposed to high temperature. And also, the result of the studies clarify that these techniques help to enhanced the decontamination process and eradicate the microbial load, ensuring the safety of the product [90]. Henceforward, a detailed calculation of charges related to the setting of the treatment plant to operating and maintain the whole line of the process gives a very authentic and practical approach of the innovative techniques leading to better value products.

2.8 Conclusion

Progressive innovative thermal technologies have paced their way to stand and proved their potential as a powerful method to reduce the microbial load in food products while maintaining the balance between nutrition content, safety, and quality attributes of the product, also improving the shelf life of the food with minimum still limiting the high investment in the processing line. Besides, these technologies confirm the decontamination technique at the higher end. All the technologies mentioned have their potential stand in the food industries; they perform robustly, replacing the traditional heating process by applying different hurdle technology. Some other techniques utilized in the food industrial sectors are cold plasma, pulse light, and ultrasound technology, which are also extensively studied for the inactivation of several microorganisms like *Salmonella typhimurium*, *Bacillus cereus* and, *E. coli* plus the yeast. Additionally, a promising outcome was seen: integrating the hurdle effects of these innovative methods with different parameters proves to be beneficial for a variety of food products and also for powder products. However, studies are still going on and some notable changes need to be done on the food sector to scale up the process more quantitatively. The spore should not repair itself during the storage of the food product, plus the main highlight which is expected is the system to be environmentally friendly which will automatically help in the success of the industrial plant.

References

1. J. P. P. M. Smelt and S. Brul, "Thermal Inactivation of Microorganisms," *Crit. Rev. Food Sci. Nutr.*, vol. 54, no. 10, pp. 1371–1385, 2014, doi: 10.1080/10408398.2011.637645.
2. E. Ağçam, A. Akyildiz, and B. Dündar, "Thermal Pasteurization and Microbial Inactivation of Fruit Juices," in *Fruit Juices: Extraction, Composition, Quality and Analysis*, 2018.
3. J. Van Impe *et al.*, "State of the art of nonthermal and thermal processing for inactivation of micro-organisms," *J. Appl. Microbiol.*, vol. 125, no. 1, pp. 16–35, 2018, doi: 10.1111/jam.13751.
4. C. Jiménez-Sánchez, J. Lozano-Sánchez, A. Segura-Carretero, and A. Fernández-Gutiérrez, "Alternatives to conventional thermal treatments in fruit-juice processing. Part 1: Techniques and applications," *Crit. Rev. Food Sci. Nutr.*, vol. 57, no. 3, pp. 501–523, 2017, doi: 10.1080/10408398.2013.867828.
5. X. Li and M. Farid, "A review on recent development in non-conventional food sterilization technologies," *Journal of Food Engineering*, vol. 182. 2016, doi: 10.1016/j.jfoodeng.2016.02.026.
6. P. Mañas and R. Pagán, "Microbial inactivation by new technologies of food preservation," *J. Appl. Microbiol.*, vol. 98, no. 6, pp. 1387–1399, 2005, doi: 10.1111/j.1365-2672.2005.02561.x.
7. S. Roohinejad, M. Koubaa, A. S. Sant'Ana, and R. Greiner, "Mechanisms of microbial inactivation by emerging technologies," in *Innovative technologies for food preservation: Inactivation of spoilage and pathogenic microorganisms*, 2018.
8. C. N. Horita, R. C. Baptista, M. Y. R. Caturla, J. M. Lorenzo, F. J. Barba, and A. S. Sant'Ana, "Combining reformulation, active packaging and non-thermal post-packaging decontamination technologies to increase the microbiological quality and safety of cooked ready-to-eat meat products," *Trends in Food Science and Technology*, vol. 72. 2018, doi: 10.1016/j.tifs.2017.12.003.
9. J. B. Portela *et al.*, "Predictive model for inactivation of salmonella in infant formula during microwave heating processing," *Food Control*, vol. 104, 2019, doi: 10.1016/j.foodcont.2019.05.006.

10. D. Bermúdez-aguirre, T. Mobbs, and G. V Barbosa-cánovas, "Ultrasound Technologies for Food and Bioprocessing," pp. 65–105, 2011, doi: 10.1007/978-1-4419-7472-3.
11. B. H. Lado and A. E. Yousef, "Alternative food-preservation technologies: Efficacy and mechanisms," *Microbes and Infection*, vol. 4, no. 4. 2002, doi: 10.1016/S1286-4579(02)01557-5.
12. S. Gaillard, I. Leguerinel, and P. Mafart, "Model for combined effects of temperature, pH and water activity on thermal inactivation of Bacillus cereus spores," *J. Food Sci.*, vol. 63, no. 5, 1998, doi: 10.1111/j.1365-2621.1998.tb17920.x.
13. E. L. Dufort, M. R. Etzel, and B. H. Ingham, "Thermal processing parameters to ensure a 5-log Reduction of *Escherichia coli* O157:H7, *Salmonella enterica*, and *Listeria monocytogenes* in Acidified Tomato-based Foods," *Food Prot. Trends*, vol. 37, no. 6, pp. 409–418, 2017.
14. F. J. Barba, M. Koubaa, L. do Prado-Silva, V. Orlien, and A. de S. Sant'Ana, "Mild processing applied to the inactivation of the main foodborne bacterial pathogens: A review," *Trends in Food Science and Technology*, vol. 66. 2017, doi: 10.1016/j.tifs.2017.05.011.
15. R. N. Pereira and A. A. Vicente, "Environmental impact of novel thermal and non-thermal technologies in food processing," *Food Res. Int.*, vol. 43, no. 7, 2010, doi: 10.1016/j.foodres.2009.09.013.
16. J. P. Huertas *et al.*, "High heating rates affect greatly the inactivation rate of *Escherichia coli*," *Front. Microbiol.*, vol. 7, no. AUG, 2016, doi: 10.3389/fmicb.2016.01256.
17. W. L. Nicholson, N. Munakata, G. Horneck, H. J. Melosh, and P. Setlow, "Resistance of Bacillus Endospores to Extreme Terrestrial and Extraterrestrial Environments," *Microbiol. Mol. Biol. Rev.*, vol. 64, no. 3, 2000, doi: 10.1128/mmbr.64.3.548-572.2000.
18. L. da Cruz Cabral, V. Fernández Pinto, and A. Patriarca, "Application of plant derived compounds to control fungal spoilage and mycotoxin production in foods," *International Journal of Food Microbiology*, vol. 166, no. 1. 2013, doi: 10.1016/j.ijfoodmicro.2013.05.026.
19. M. C. Pina-Pérez, A. Rivas, A. Martínez, and D. Rodrigo, "Effect of thermal treatment, microwave, and pulsed electric field processing on the antimicrobial potential of açaí (*Euterpe oleracea*), stevia (*Stevia rebaudiana* Bertoni), and ginseng (Panax quinquefolius L.) extracts," *Food Control*, vol. 90, 2018, doi: 10.1016/j.foodcont.2018.02.022.
20. A. Rodriguez-Palacios and J. T. LeJeune, "Moist-heat resistance, spore aging, and superdormancy in *Clostridium difficile*," *Appl. Environ. Microbiol.*, vol. 77, no. 9, 2011, doi: 10.1128/AEM.01589-10.
21. Evelyn and F. V. M. Silva, "Resistance of *Byssochlamys nivea and Neosartorya fischeri* mould spores of different age to high pressure thermal processing and thermosonication," *J. Food Eng.*, vol. 201, 2017, doi: 10.1016/j.jfoodeng.2017.01.007.

22. D. Millan-Sango, A. McElhatton, and V. P. Valdramidis, "Determination of the efficacy of ultrasound in combination with essential oil of oregano for the decontamination of *Escherichia coli* on inoculated lettuce leaves," *Food Res. Int.*, vol. 67, 2015, doi: 10.1016/j.foodres.2014.11.001.
23. A. Métris, S. M. George, B. M. Mackey, and J. Baranyi, "Modeling the variability of single-cell lag times for *Listeria innocua* populations after sublethal and lethal heat treatments," *Appl. Environ. Microbiol.*, vol. 74, no. 22, 2008, doi: 10.1128/AEM.01237-08.
24. W. Zhao, R. Yang, X. Shen, S. Zhang, and X. Chen, "Lethal and sublethal injury and kinetics of *Escherichia coli, Listeria monocytogenes* and *Staphylococcus aureus* in milk by pulsed electric fields," *Food Control*, vol. 32, no. 1, 2013, doi: 10.1016/j.foodcont.2012.11.029.
25. S. K. Wimalaratne and M. M. Farid, "Pressure assisted thermal sterilization," *Food Bioprod. Process.*, vol. 86, no. 4, 2008, doi: 10.1016/j.fbp.2007.08.001.
26. P. Loypimai, A. Moongngarm, P. Chottanom, and T. Moontree, "Ohmic heating-assisted extraction of anthocyanins from black rice bran to prepare a natural food colourant," *Innov. Food Sci. Emerg. Technol.*, vol. 27, 2015, doi: 10.1016/j.ifset.2014.12.009.
27. G. Lehrke, L. Hernaez, S. L. Mugliaroli, M. von Staszewski, and R. J. Jagus, "Sensitization of *Listeria innocua* to inorganic and organic acids by natural antimicrobials," *LWT - Food Sci. Technol.*, vol. 44, no. 4, 2011, doi: 10.1016/j.lwt.2010.09.016.
28. Z. Xu *et al.*, "Inactivation effects of non-thermal atmospheric-pressure helium plasma jet on *staphylococcus aureus* biofilms," *Plasma Process. Polym.*, vol. 12, no. 8, 2015, doi: 10.1002/ppap.201500006.
29. J. Zhu *et al.*, "Combined effect of ultrasound, heat, and pressure on *Escherichia coli* O157:H7, polyphenol oxidase activity, and anthocyanins in blueberry (Vaccinium corymbosum) juice," *Ultrason. Sonochem.*, vol. 37, pp. 251–259, 2017, doi: 10.1016/j.ultsonch.2017.01.017.
30. D. Ziuzina, S. Patil, P. J. Cullen, K. M. Keener, and P. Bourke, "Atmospheric cold plasma inactivation of *Escherichia coli, Salmonella enterica serovar Typhimurium* and *Listeria monocytogenes* inoculated on fresh produce," *Food Microbiol.*, vol. 42, pp. 109–116, 2014, doi: 10.1016/j.fm.2014.02.007.
31. V. D. Farkade, S. Harrison, and A. B. Pandit, "Heat induced translocation of proteins and enzymes within the cell: An effective way to optimize the microbial cell disruption process," *Biochem. Eng. J.*, vol. 23, no. 3, 2005, doi: 10.1016/j.bej.2005.01.001.
32. A. J. Brodowska, A. Nowak, and K. Śmigielski, "Ozone in the food industry: Principles of ozone treatment, mechanisms of action, and applications: An overview," *Crit. Rev. Food Sci. Nutr.*, vol. 58, no. 13, 2018, doi: 10.1080/10408398.2017.1308313.
33. M. Salma, S. Rousseaux, A. Sequeira-Le Grand, B. Divol, and H. Alexandre, "Characterization of the Viable but Nonculturable (VBNC) State in

Saccharomyces cerevisiae.," *PLoS One*, vol. 8, no. 10, 2013, doi: 10.1371/jour nal. pone.0077600.
34. I. Albertos *et al.*, "Effects of dielectric barrier discharge (DBD) generated plasma on microbial reduction and quality parameters of fresh mackerel (*Scomber scombrus*) fillets," *Innov. Food Sci. Emerg. Technol.*, vol. 44, 2017, doi: 10.1016/j.ifset.2017.07.006.
35. H. Daryaei, A. E. Yousef, and V. M. Balasubramaniam, "Microbiological aspects of high-pressure processing of food: inactivation of microbial vegetative cells and spores," in *Food Engineering Series*, 2016.
36. J. Raso, I. Alvarez, S. Condón, and F. J. Sala Trepat, "Predicting inactivation of *Salmonella senftenberg* by pulsed electric fields," *Innov. Food Sci. Emerg. Technol.*, vol. 1, no. 1, 2000, doi: 10.1016/S1466-8564(99)00005-3.
37. A. Ait-Ouazzou, P. Mañas, S. Condón, R. Pagán, and D. García-Gonzalo, "Role of general stress-response alternative sigma factors σ S (RpoS) and σ B (SigB) in bacterial heat resistance as a function of treatment medium pH," *Int. J. Food Microbiol.*, vol. 153, no. 3, pp. 358–364, 2012, doi: 10.1016/j.ijfoodmicro.2011.11.027.
38. M. D. Esteban, A. Aznar, P. S. Fernández, and A. Palop, "Combined effect of nisin, carvacrol and a previous thermal treatment on the growth of *Salmonella enteritidis* and *Salmonella senftenberg*," *Food Sci. Technol. Int.*, vol. 19, no. 4, 2013, doi: 10.1177/1082013212455185.
39. C. Hill, P. D. Cotter, R. D. Sleator, and C. G. M. Gahan, "Bacterial stress response in *Listeria monocytogenes*: Jumping the hurdles imposed by minimal processing," in *International Dairy Journal*, 2002, vol. 12, no. 2–3, doi: 10.1016/S0958-6946(01)00125-X.
40. T. Abee and J. A. Wouters, "Microbial stress response in minimal processing," *Int. J. Food Microbiol.*, vol. 50, no. 1–2, 1999, doi: 10.1016/S0168-1605(99)00078-1.
41. G. Cebrián, P. Mañas, and S. Condón, "Comparative resistance of bacterial foodborne pathogens to non-thermal technologies for food preservation," *Frontiers in Microbiology*, vol. 7, no. MAY. 2016, doi: 10.3389/fmicb.2016.00734.
42. A. Chen *et al.*, "Plasma membrane behavior, oxidative damage, and defense mechanism in *Phanerochaete chrysosporium* under cadmium stress," *Process Biochem.*, vol. 49, no. 4, 2014, doi: 10.1016/j.procbio.2014.01.014.
43. J. Dai, A. Gupte, L. Gates, and R. J. Mumper, "A comprehensive study of anthocyanin-containing extracts from selected blackberry cultivars: Extraction methods, stability, anticancer properties and mechanisms," *Food Chem. Toxicol.*, vol. 47, no. 4, 2009, doi: 10.1016/j.fct.2009.01.016.
44. S. Gao, G. D. Lewis, M. Ashokkumar, and Y. Hemar, "Inactivation of microorganisms by low-frequency high-power ultrasound: 1. Effect of growth phase and capsule properties of the bacteria," *Ultrason. Sonochem.*, vol. 21, no. 1, 2014, doi: 10.1016/j.ultsonch.2013.06.006.

45. L. Han, S. Patil, D. Boehm, V. Milosavljević, P. J. Cullen, and P. Bourke, "Mechanisms of inactivation by high-voltage atmospheric cold plasma differ for *Escherichia coli* and *Staphylococcus aureus*," *Appl. Environ. Microbiol.*, vol. 82, no. 2, 2016, doi: 10.1128/AEM.02660-15.
46. A. A. Gabriel, "Inactivation behaviors of foodborne microorganisms in multi-frequency power ultrasound-treated orange juice," *Food Control*, vol. 46, 2014, doi: 10.1016/j.foodcont.2014.05.012.
47. V. Trivittayasil, F. Tanaka, and T. Uchino, "Investigation of deactivation of mold conidia by infrared heating in a model-based approach," *J. Food Eng.*, vol. 104, no. 4, 2011, doi: 10.1016/j.jfoodeng.2011.01.018.
48. E. Eser and H. Ibrahim Ekiz, "Effect of far infrared pre-processing on microbiological, physical and chemical properties of peanuts," *Carpathian J. Food Sci. Technol.*, vol. 10, no. 1, 2018.
49. S. Wilson, "Development of Infrared Heating Technology for Corn Drying and Decontamination to Maintain Quality and Prevent Mycotoxins," *Theses Diss.*, 2016, [Online]. Available: https://scholarworks.uark.edu/etd/1542.
50. R. Abdul-Kadir, T. J. Bargman, and J. H. Rupnow, "Effect of Infrared Heat Processing on Rehydration Rate and Cooking of *Phaseolus vulgaris* (Var. Pinto)," *J. Food Sci.*, vol. 55, no. 5, 1990, doi: 10.1111/j.1365-2621.1990.tb03964.x.
51. N. Staack, L. Ahrné, E. Borch, and D. Knorr, "Effect of infrared heating on quality and microbial decontamination in paprika powder," *J. Food Eng.*, vol. 86, no. 1, 2008, doi: 10.1016/j.jfoodeng.2007.09.004.
52. L. Eliasson, P. Libander, M. Lövenklev, S. Isaksson, and L. Ahrné, "Infrared Decontamination of Oregano: Effects on *Bacillus cereus* Spores, Water Activity, Color, and Volatile Compounds," *J. Food Sci.*, vol. 79, no. 12, 2014, doi: 10.1111/1750-3841.12694.
53. J. W. Ha and D. H. Kang, "Enhanced inactivation of food-borne pathogens in ready-to-eat sliced ham by near-infrared heating combined with UV-C irradiation and mechanism of the synergistic bactericidal action," *Appl. Environ. Microbiol.*, vol. 81, no. 1, 2015, doi: 10.1128/AEM.01862-14.
54. S. Jun and J. Irudayaraj, "A Dynamic Fungal Inactivation Approach Using Selective Infrared Heating," *Trans. Am. Soc. Agric. Eng.*, vol. 46, no. 5, 2003, doi: 10.13031/2013.15435.
55. F. G. Chizoba Ekezie, D. W. Sun, Z. Han, and J. H. Cheng, "Microwave-assisted food processing technologies for enhancing product quality and process efficiency: A review of recent developments," *Trends in Food Science and Technology*, vol. 67. 2017, doi: 10.1016/j.tifs.2017.05.014.
56. G.-A. Ştefănoiu, E. E. Tănase, A. C. Miteluţ, and M. E. Popa, "Unconventional Treatments of Food: Microwave vs. Radiofrequency," *Agric. Agric. Sci. Procedia*, vol. 10, 2016, doi: 10.1016/j.aaspro.2016.09.024.
57. S. Chandrasekaran, S. Ramanathan, and T. Basak, "Microwave food processing-A review," *Food Research International*, vol. 52, no. 1. 2013, doi: 10.1016/j.foodres.2013.02.033.

58. P. Lopez-Iturri, S. De Miguel-Bilbao, E. Aguirre, L. Azpilicueta, F. Falcone, and V. Ramos, "Estimation of radiofrequency power leakage from microwave ovens for dosimetric assessment at nonionizing radiation exposure levels," *Biomed Res. Int.*, vol. 2015, 2015, doi: 10.1155/2015/603260.
59. Q. Guo, D. W. Sun, J. H. Cheng, and Z. Han, "Microwave processing techniques and their recent applications in the food industry," *Trends in Food Science and Technology*, vol. 67. 2017, doi: 10.1016/j.tifs.2017.07.007.
60. H. Jiang, M. Zhang, A. S. Mujumdar, and R. X. Lim, "Drying uniformity analysis of pulse-spouted microwave–freeze drying of banana cubes," *Dry. Technol.*, vol. 34, no. 5, 2016, doi: 10.1080/07373937.2015.1061000.
61. A. Álvarez, J. Fayos-Fernández, J. Monzó-Cabrera, M. J. Cocero, and R. B. Mato, "Measurement and correlation of the dielectric properties of a grape pomace extraction media. Effect of temperature and composition," *J. Food Eng.*, vol. 197, 2017, doi: 10.1016/j.jfoodeng.2016.11.009.
62. M. Vinatoru, T. J. Mason, and I. Calinescu, "Ultrasonically assisted extraction (UAE) and microwave assisted extraction (MAE) of functional compounds from plant materials," *TrAC - Trends in Analytical Chemistry*, vol. 97. 2017, doi: 10.1016/j.trac.2017.09.002.
63. K. Knoerzer, P. Juliano, and G. Smithers, *Innovative Food Processing Technologies: Extraction, Separation, Component Modification and Process Intensification*. 2016.
64. G. C. Jeevitha, H. B. Sowbhagya, and H. U. Hebbar, "Application of microwaves for microbial load reduction in black pepper (*Piper nigrum* L.)," *J. Sci. Food Agric.*, vol. 96, no. 12, 2016, doi: 10.1002/jsfa.7630.
65. M. C. Pina-Pérez, M. Benlloch-Tinoco, D. Rodrigo, and A. Martinez, "*Cronobacter sakazakii* Inactivation by Microwave Processing," *Food Bioprocess Technol.*, vol. 7, no. 3, pp. 821–828, 2014, doi: 10.1007/s11947-013-1063-2.
66. S. Kar, A. S. Mujumdar, and P. P. Sutar, "*Aspergillus niger* inactivation in microwave rotary drum drying of whole garlic bulbs and effect on quality of dried garlic powder," *Dry. Technol.*, vol. 37, no. 12, 2019, doi: 10.1080/07373937.2018.1517777.
67. P. Piyasena, C. Dussault, T. Koutchma, H. S. Ramaswamy, and G. B. Awuah, "Radio Frequency Heating of Foods: Principles, Applications and Related Properties - A Review," *Critical Reviews in Food Science and Nutrition*, vol. 43, no. 6. 2003, doi: 10.1080/10408690390251129.
68. F. Salazar, S. Garcia, M. Lagunas-Solar, Z. Pan, and J. Cullor, "Effect of a heat-spray and heat-double spray process using radiofrequency technology and ethanol on inoculated nuts," *J. Food Eng.*, vol. 227, 2018, doi: 10.1016/j.jfoodeng.2017.12.017.
69. F. Marra, L. Zhang, and J. G. Lyng, "Radio frequency treatment of foods: Review of recent advances," *Journal of Food Engineering*, vol. 91, no. 4. 2009, doi: 10.1016/j.jfoodeng.2008.10.015.

70. S. Ozturk *et al.*, "Inactivation of *Salmonella Enteritidis* and *Enterococcus faecium* NRRL B-2354 in corn flour by radio frequency heating with subsequent freezing," *LWT*, vol. 111, 2019, doi: 10.1016/j.lwt.2019.04.090.
71. S. Liu *et al.*, "Microbial validation of radio frequency pasteurization of wheat flour by inoculated pack studies," *J. Food Eng.*, vol. 217, 2018, doi: 10.1016/j.jfoodeng.2017.08.013.
72. S. Hu, Y. Zhao, Z. Hayouka, D. Wang, and S. Jiao, "Inactivation kinetics for *Salmonella typhimurium* in red pepper powders treated by radio frequency heating," *Food Control*, vol. 85, 2018, doi: 10.1016/j.foodcont.2017.10.034.
73. S. G. Jeong and D. H. Kang, "Influence of moisture content on inactivation of *Escherichia coli* O157:H7 and *Salmonella enterica* serovar Typhimurium in powdered red and black pepper spices by radio-frequency heating," *Int. J. Food Microbiol.*, vol. 176, 2014, doi: 10.1016/j.ijfoodmicro.2014.01.011.
74. Y. Zhao, W. Zhao, R. Yang, J. Singh Sidhu, and F. Kong, "Radio frequency heating to inactivate microorganisms in broccoli powder," *Food Qual. Saf.*, vol. 1, no. 1, 2017, doi: 10.1093/fqs/fyx005.
75. M. Mazen Hamoud-Agha and K. Allaf, "Instant Controlled Pressure Drop (DIC) Technology in Food Preservation: Fundamental and Industrial Applications," in *Food Preservation and Waste Exploitation*, 2020.
76. T. Allaf, C. Besombes, I. Mih, L. Lefevre, and K. Allaf, "Decontamination of Solid and Powder Foodstuffs using DIC Technology," in *Advances in Computer Science and Engineering*, 2011.
77. S. Mounir, C. Besombes, N. Al-Bitar, and K. Allaf, "Study of instant controlled pressure drop DIC treatment in manufacturing snack and expanded granule powder of Apple and Onion," *Dry. Technol.*, vol. 29, no. 3, 2011, doi: 10.1080/07373937.2010.491585.
78. A. Demirdöven and T. Baysal, "Optimization of ohmic heating applications for pectin methylesterase inactivation in orange juice," *J. Food Sci. Technol.*, vol. 51, no. 9, 2014, doi: 10.1007/s13197-012-0700-5.
79. W. Il Cho, J. Y. Yi, and M. S. Chung, "Pasteurization of fermented red pepper paste by ohmic heating," *Innov. Food Sci. Emerg. Technol.*, vol. 34, 2016, doi: 10.1016/j.ifset.2016.01.015.
80. M. Kumar, Jyoti, and A. Hausain, "Effect of ohmic heating of buffalo milk on microbial quality and tesure of paneer," *Asian J. Dairy. Foods Res.*, vol. 33, no. 1, 2014, doi: 10.5958/j.0976-0563.33.1.003.
81. J. H. Ryang *et al.*, "Inactivation of Bacillus cereus spores in a tsuyu sauce using continuous ohmic heating with five sequential elbow-type electrodes," *J. Appl. Microbiol.*, vol. 120, no. 1, 2016, doi: 10.1111/jam.12982.
82. S. H. Park, V. M. Balasubramaniam, S. K. Sastry, and J. Lee, "Pressure-ohmic-thermal sterilization: A feasible approach for the inactivation of *Bacillus amyloliquefaciens* and *Geobacillus stearothermophilus* spores," *Innov. Food Sci. Emerg. Technol.*, vol. 19, 2013, doi: 10.1016/j.ifset.2013.03.005.
83. R. Somavat, H. M. H. Mohamed, Y. K. Chung, A. E. Yousef, and S. K. Sastry, "Accelerated inactivation of *Geobacillus stearothermophilus* spores

by ohmic heating," *J. Food Eng.*, vol. 108, no. 1, 2012, doi: 10.1016/j.jfoodeng.2011.07.028.

84. X. Tian, Q. Yu, W. Wu, and R. Dai, "Inactivation of microorganisms in foods by ohmic heating: A review," *J. Food Prot.*, vol. 81, no. 7, pp. 1093–1107, 2018, doi: 10.4315/0362-028X.JFP-17-343.
85. R. Pereira, J. Martins, C. Mateus, J. Teixeira, and A. Vicente, "Death kinetics of *Escherichia coli in goat milk* and *Bacillus licheniformis* in cloudberry jam treated by ohmic heating," *Chem. Pap.*, vol. 61, no. 2, 2007, doi: 10.2478/s11696-007-0008-5.
86. S. Leizerson and E. Shimoni, "Effect of ultrahigh-temperature continuous ohmic heating treatment on fresh orange juice," *J. Agric. Food Chem.*, vol. 53, no. 9, 2005, doi: 10.1021/jf0481204.
87. I. K. Park, J. W. Ha, and D. H. Kang, "Investigation of optimum ohmic heating conditions for inactivation of *Escherichia coli* O157:H7, *Salmonella enterica serovar Typhimurium,* and *Listeria monocytogenes* in apple juice," *BMC Microbiol.*, vol. 17, no. 1, 2017, doi: 10.1186/s12866-017-1029-z.
88. M. Zell, J. G. Lyng, D. A. Cronin, and D. J. Morgan, "Ohmic cooking of whole beef muscle - Evaluation of the impact of a novel rapid ohmic cooking method on product quality," *Meat Sci.*, vol. 86, no. 2, 2010, doi: 10.1016/j.meatsci.2010.04.007.
89. M. Morales-de la Peña, L. Salvia-Trujillo, M. A. Rojas-Graü, and O. Martín-Belloso, "Impact of high intensity pulsed electric field on antioxidant properties and quality parameters of a fruit juice-soymilk beverage in chilled storage," *LWT - Food Sci. Technol.*, vol. 43, no. 6, 2010, doi: 10.1016/j.lwt.2010.01.015.
90. P. Nath, S. J. Kale, and B. Bhushan, "Consumer Acceptance and Future Trends of Non-thermal-Processed Foods," in *Non-thermal Processing of Foods*, 2019.
91. A. A. Oduola, R. Bowie, S. A. Wilson, Z. Mohammadi Shad, and G. G. Atungulu, "Impacts of broadband and selected infrared wavelength treatments on inactivation of microbes on rough rice," *J. Food Saf.*, vol. 40, no. 2, 2020, doi: 10.1111/jfs.12764.
92. C. Venkitasamy *et al.*, "Feasibility of using sequential infrared and hot air for almond drying and inactivation of *Enterococcus faecium* NRRL B-2354," *LWT*, vol. 95, 2018, doi: 10.1016/j.lwt.2018.04.095.
93. Y. Feng, B. Wu, X. Yu, A. E. G. A. Yagoub, F. Sarpong, and C. Zhou, "Effect of catalytic infrared dry-blanching on the processing and quality characteristics of garlic slices," *Food Chem.*, vol. 266, 2018, doi: 10.1016/j.foodchem.2018.06.012.
94. A. Kapoor and P. P. Sutar, "Finish drying and surface sterilization of bay leaves by microwaves," 2019, doi: 10.4995/ids2018.2018.7822.
95. H. Patil, N. G. Shah, S. N. Hajare, S. Gautam, and G. Kumar, "Combination of microwave and gamma irradiation for reduction of aflatoxin B1 and microbiological contamination in peanuts (*Arachis hypogaea* L.)," *World Mycotoxin J.*, vol. 12, no. 3, 2019, doi: 10.3920/WMJ2018.2384.

96. X. Cao *et al.*, "Radiofrequency heating for powder pasteurization of barley grass: antioxidant substances, sensory quality, microbial load and energy consumption," *J. Sci. Food Agric.*, vol. 99, no. 9, 2019, doi: 10.1002/jsfa.9683.
97. L. Zhang, J. G. Lyng, R. Xu, S. Zhang, X. Zhou, and S. Wang, "Influence of radio frequency treatment on in-shell walnut quality and *Staphylococcus aureus* ATCC 25923 survival," *Food Control*, vol. 102, 2019, doi: 10.1016/j.foodcont.2019.03.030.
98. J. Peng *et al.*, "Freezing as pretreatment in instant controlled pressure drop (DIC) texturing of dried carrot chips: Impact of freezing temperature," *LWT - Food Sci. Technol.*, vol. 89, 2018, doi: 10.1016/j.lwt.2017.11.009.
99. E. J. Rifna, S. K. Singh, S. Chakraborty, and M. Dwivedi, "Effect of thermal and non-thermal techniques for microbial safety in food powder: Recent advances," *Food Res. Int.*, vol. 126, no. September, p. 108654, 2019, doi: 10.1016/j.foodres.2019.108654.
100. Alam, M. S., Kumar, N., & Singh, B. (2018). Development of sweet lime (*Citrus limetta* Risso) pomace integrated ricebased extruded product: Process optimization. *Journal of Agricultural Engineering*, 55(1), 47-53.

3

Blanching, Pasteurization and Sterilization: Principles and Applications

Monalisa Sahoo[1]*, Sushree Titikshya[1], Pramod Aradwad[2], Vivek Kumar[3] and S. N. Naik[3]

[1]Centre for Rural Development and Technology, IIT Delhi, Delhi, India
[2]Indian Agricultural Research Institute, New Delhi, India
[3]Centre for Rural Development and Technology, IIT Delhi, Delhi, India

Abstract
Novel thermal processing technologies, such as radiofrequency heating, microwave heating and ohmic heating, Infrared heating, and inductive heating have been developed and tried successfully for different food products. These techniques improve the efficiency and effectiveness of the heating process while maintaining product quality and safety. These techniques eliminate the factor affecting organoleptic properties and maintain the nutritional profile of food products. Novel thermal technologies are dependent on heat generation directly inside the material, which increases energy utilization, thereby overall energy efficiency of the heating process. These processes are dependent on inherent properties of food material, which are complex in nature such as electrical resistance, thermal conductivity, water content, porosity, pH, and rheological properties. This chapter, encompassing the use of novel thermal technologies, will help to understand the underlying principle, mechanism, advantages, disadvantages, and limitations associated with each technique applied to different food products. It will help to understand suitable technology to inactivate the enzyme, spores, bacteria, microbes in a specific food product.

Keywords: Infrared blanching, microwave & radio frequency assisted pasteurization, advanced retorting, pressure assisted sterilization

**Corresponding author*: saidurga48@gmail.com

Nitin Kumar, Anil Panghal and M. K. Garg (eds.) Thermal Food Engineering Operations, (75–116)

3.1 Introduction

Blanching is a thermal treatment used for fruits and vegetables processing prior to processing of foods such as drying, frying, and freezing [1]. It is required to inactivate enzymes and destroy microorganisms for long-term storage and preservation of fruits and vegetables [2–4]. In the food industry, pasteurization and sterilization processes are mainly used, for their efficacy and to maintain product quality and safety. Excessive heat treatment may alter or damage the product quality, such as denaturation of protein, loss of volatile aroma compounds and vitamins, and non-enzymatic browning. Cooked and sterilized milk, which gives caramelized flavors, are considered a taste defect by many consumers in the USA [5]. Technology advancement in thermal processing has allowed certain advantages over optimization of processes for maximum efficacy against microbial inactivation and minimum product quality deterioration. For example, loss of vitamins was found to be less in high-temperature short-time (HTST) pasteurization and ultra-high temperature (UHT) sterilization processed milk than batch pasteurization and conventional commercial sterilization, respectively [6]. Products processed by modern thermal technologies, however, still lack fresh flavor and texture. Therefore, advanced processing technologies have been developed to overcome the constraints of conventional processing methods.

3.2 Blanching: Principles & Mechanism

Blanching is performed at a predetermined temperature for a specific time, usually 1 to 10 min or less. Afterward, the blanched product is either cooled immediately or processed further. The duration of blanching is dependent upon the nature of enzymes (peroxidase, lipoxygenase, and polyphenoloxidase) [3]. The activity of these enzymes negatively affects taste, color, flavor, and nutritional quality over long-term storage.

3.2.1 Types of Blanching

There are several methods of blanching available including both the traditional and advanced blanching methods which are described in the following sections.

3.2.1.1 Hot Water Blanching

Hot water blanching is a traditional and most popular blanching process widely used for commercial applications, due to its simplicity in nature, establishment, and operation [2]. Products are dipped into hot water (70-100°C) for a fixed time then drained and cooled. Further processing is carried out for value addition and long-term storage. Sometimes sodium metabisulfite, sulfite, and citric acids are used to retain color by preventing browning [3]. Several applications to different fruits and vegetables have been carried out by using hot blanching methods (Table 3.1).

Table 3.1 Application of different blanching methods for processing of agricultural commodities.

Type of blanching	Product	Purpose	References
Hot water blanching	Pepper	Enzyme inactivation	[88]
	Brussels sprouts	Freezing	[89]
	Almond	Removal of pellicle	[90]
	Potato chips	Quality enhancement	[91]
	Carrot	Reduction of microbial load	[92]
Steam blanching	Kiwifruit	Inactivation enzyme	[93]
	Potato	Quality of product	[94, 95]
	Mango slices	Inactivation of enzymes	[96]
	Garlic slices	Inactivation of enzymes	[97]
	Fresh broccoli	Enhancement of antioxidant and phenol	[15]
	Blueberries	Increase in anthocyanin and total phenolic content	[98, 99]
	Vegetable soybean	Preservation of sugar	[100]
	Cabbage	Deactivation of microorganisms	[101]

(Continued)

Table 3.1 Application of different blanching methods for processing of agricultural commodities. (*Continued*)

Type of blanching	Product	Purpose	References
	Cabbage leaves	Preservation of dietary fiber	[102]
High humidity hot air impingement blanching	Yam slices	To prevent enzymatic browning and maintain color	[103]
	Grapes	To increase drying rate	[18]
	Sea cucumber	Enzyme denaturation	[104]
	Chicken skin	Microbial load reduction	[105]
	Lettuce	Microbial load reduction	[106]
	Apple	Polyphenol oxidase inactivation	[107]
	Red pepper	Denature of enzyme and increase in drying rate	[108]
	Sweet potato bar	To get desirable color and texture	[109]
Microwave blanching	Carrot pieces	Quality of the product	[110]
	Mushroom	Inactivation enzymes	[111]
	Asparagus	To retain color and antioxidant	[112]
	Artichokes	Inactivation of chlorophyllase	[113]

(*Continued*)

Table 3.1 Application of different blanching methods for processing of agricultural commodities. (*Continued*)

Type of blanching	Product	Purpose	References
	Peas	To obtain sensory attribute	[114]
	Marjoram & rosemary	To retain volatile oil	[115]
	Pepper	To inactivate PPO	[58]
Ohmic blanching	Artichoke heads	Inactivation of PPO and POD	[116]
	Carrot and red beet	Textural softening kinetics	[36]
	Acerola pulp	Comparison of quality between ohmic and conventional blanching	[117, 118]
	Strawberries	Kinetics of osmotic dehydration and microstructure	[119]
	Strawberries	Quality and microbial load	[120]
	Milk, Apple juice, Cloudberry jam, and mixed vegetable juices	Inactivation of enzymes (pectin methylesterase, Phosphatase & peroxidase)	[121]
	Apple	Inactivation of PPO and microbial stability	[122]
Infrared blanching	Apple	To improve quality	[49, 123]

(*Continued*)

Table 3.1 Application of different blanching methods for processing of agricultural commodities. (*Continued*)

Type of blanching	Product	Purpose	References
	Carrot slices	To get better quality than conventional blanching	[124]
	Red pepper	Enzyme inactivation and antioxidant retention	[125]
	Mango	To retain Vitamin C, Carotene and Inactivate enzyme	[126]

Limitations of hot water blanching

This method of blanching consumes a high amount of water and energy. Besides, it generates a heavy amount of chemical treated blanched water waste, which is difficult to deal with [7]. Blanched water waste contains a high amount of solids caused by leaching out of sugar, protein, carbohydrates, and soluble minerals which causes environmental problems [8]. Loss of water-soluble nutrients by leaching out or diffusion to the blanched water occurs [2]. Simultaneously thermal degradation of some sensitive compounds takes place. Loss of ascorbic acid, phenolic compounds, and total solids are reported by several researchers in cabbage, broccoli, kale, spinach, cauliflower, and potatoes [2, 9–12]. It has been observed that loss of ascorbic acid is a diffusion phenomenon [13]. However, the use of recycled hot water retained higher ascorbic acid [14].

3.2.1.2 *Steam Blanching*

Steam blanching and high humidity hot air impingement blanching is an emerging blanching method. Super-heated steam is used for blanching instead of hot water due to its high enthalpy content. At the preliminary stage, steam is condensed on the surface of the product during steam blanching. Later, a large quantity of latent heat is transferred to the product as the temperature of the product is too low compared with the steam. The product temperature is continuously increased until the critical temperature of the microorganisms and enzymes where they are inactivated.

Limitations of steam blanching
In comparison to hot water blanching, most of the water-soluble components and minerals are retained in steam blanching which may be due to a very low leaching effect [15]. On the other side, undesirable quality changes and softening of tissues have been observed in steam blanching due to slower heat transfer than the hot water blanching.

3.2.1.3 *High Humidity Hot Air Impingement Blanching (HHAIB)*

This is a new blanching technique where both steam and air impingement technologies are combined. Therefore, HHAIB is a rapid, uniform, energy-efficient, and waste water-free process [16]. High rate of heat transfer and more efficient compared with super-heated steam blanching. Similarly, low losses of nutrients were observed compared to traditional hot water blanching [16]. The thermal boundary layer becomes thinner when high-velocity super-heated steam impinges on the surface of the product. Thus, the rate of heat transfer is increased [17] reported that the heat transfer coefficient is about 1403 W/(m^2 K) with a velocity of 14.4 m/s. This blanching method has many advantages over other blanching methods such as reduced oxidation of food, minimum energy requirement, and free from fire and explosive hazards [18, 19].

3.2.1.4 *Microwave Blanching*

Electromagnetic waves in the range of 1mm to 1m wavelength corresponding to 300MHz to 300GHz are referred to as microwaves [20]. Microwaves have several applications such as radar, navigation, radio astronomy, and communications. But in recent years it has been used in food processing. However, only 915 MHz and 2450 MHz microwaves are permitted for scientific, medical, and industrial applications to prevent any interference among communication signals.

Microwave energy is absorbed by the heating products and converts into heat during microwave heating. This is caused by the dielectric heating effect which is caused by agitation of charged ions and dipole rotation in a high-frequency electric field [21]. Polarized dipolar molecules (water molecules) are aligned in the direction of the electromagnetic field (emf) at 2450 or 915 MHz, which may be the result of interactions between high water content products and oscillating electric fields [20]. Volumetric heating is produced by the internal resistance caused by pull, push, and collision of rotating molecules with adjacent molecules or atoms [22].

More microwave heating is generated at 915 MHz than 2450 MHz due to agitation of charged ion in the emf. Microwave heating is generated at both the surface and within of the water-containing biological material. However, in conventional processing heating it takes place due to differences in a temperature gradient, and energy is transferred by conduction.

Compared to traditional heating methods, microwave heating has numerous advantages, which are listed in Table 3.2. Therefore, it was effectively used in blanching, drying, tempering, baking, thawing, pasteurization, etc. [18, 23].

In microwave blanching, there is significantly reduced nutrients loss by leaching compared to hot water blanching. This is because of the direct interaction between food materials and electromagnetic fields for heat generation [23–25]. For example, [26] reported that the microwave blanched green beans, peas, and carrots contain a higher amount of ascorbic acid as compared to conventional hot water blanching.

Limitations of microwave blanching

Despite having numerous advantages, there are some limitations of microwave blanching which restrict its wide application.

a) Water loss during blanching

   Moisture in fruits and vegetable evaporates during microwave blanching. Destruction of microstructure and cell folding is caused by high-intensity microwave power [25]. During blanching, an increase in heat absorption and reduction of water loss can be achieved by heating in immersed water. But the limitation is that the water-soluble nutrients may be lost through diffusion or leaching out.

b) Limited depth of penetration

   The depth of penetration is the function of the dielectric properties of the material, which decides the temperature distribution within the product [27]. Dielectric properties (ε) are directly related to the dielectric constant (ε') and dielectric loss factor (ε'') of the product. The dielectric constant is a measure of the ability to store electromagnetic energy in food material where dielectric loss determines the dissipation of electromagnetic energy in the food after being heated [28]. Depth of penetration (d_p) into the product can be determined using the following equation [29]:

Table 3.2 Advantages and disadvantages of different blanching methods.

Type of blanching	Advantages	Disadvantages
Hot water blanching	• Easy to operate • Small capital investment	• Loss of water-soluble nutrients • Large amount of wastewater • Can cause environmental pollution
Steam blanching	• Low loss of nutrients compared to hot water blanching • Negligible leaching of compounds	• Non-uniform blanching • Longer blanching time • Affects capacity and economic • Undesirable quality changes and softening of tissue
Superheated steam impingement blanching (SSIB)	• Uniform, rapid and energy-efficient process • Reduces nutrient loss • No wastewater generation • High heat transfer rate	• Lower heat transfer • Longer heating time • Tissue softening • Quality changes
Microwave blanching	• Volumetric heating • Short processing time • High heating rate • Rapid and energy-efficient • Easy to clean and install • Short start uptime	• Loss of water by evaporation • Depth of penetration is limited • Non-uniform heating
Ohmic blanching	• Fast and uniform heating • Low operating cost • High energy conversion efficiencies • Less surface fouling • Mild thermal treatment	• Difficulty in blanching temperature control • Oxygen and hydrogen generation • Erosion and corrosion at the electrode
Infrared blanching	• High energy efficiency • Shorter process time • Large heat transfer coefficient	• Poor heat penetration • Non-uniform heating • Surface color degradation • High water loss

$$d_p = \frac{\lambda}{2\pi\sqrt{2\acute{\varepsilon}}}\left[\sqrt{1+\left(\frac{\varepsilon''}{\varepsilon'}\right)^2}+-1\right] \tag{3.1}$$

Where d_p is the depth of penetration, λ in the microwave wavelength, ε'' is the loss factor and ε' is the dielectric constant. Reduction in moisture content reduces the loss factor which causes a reduction in microwave energy conversion into heat during microwave heating. It has been reported that the depth of penetration for the different products such as 1.6cm for mashed potato [27], 12mm for whey protein at 20°C and 915 MHz [30], 1.5-3.5 cm for bell pepper, sweet potato, and broccoli [29]. Additionally, an increase in frequency decreases the depth of penetration of dielectric heating. [31] reported that the depth of penetration is much higher in the range of radiofrequency (27 and 40 MHz) than the penetration depth at microwave frequency range (915 and 2450 MHz). Therefore, it is suitable to use radio frequency technology for thick and large products, while microwave heating is better for thin or small samples.

c) Non-uniform heating

Electromagnetic energy gets converted into heat in microwave heating, caused by friction of molecules and ions, which follow the dipolar polarization at a higher frequency [20]. Sometimes, there is uneven moisture and ion distribution in different sections of materials. Therefore, heating becomes non-uniform at the end. Non-uniform microwave field causes uneven distribution of energy, which creates cold and hot points at different parts or points of samples. Limited penetration depth in microwave heating causes more inhomogeneity. Collectively, these factors produce high-temperature differences during the processing of large sizes and bulky amounts of materials. [29] observed, hot spot at 95°C and a cold spot at 80°C during non-uniform heating occurred during pasteurization of packaged acidified vegetables in a continuous microwave system (915MHz & 4 kW) for 4min. [32] also reported that non-uniform heating caused the charring of edges in microwave dried mushrooms.

d) Difficulty in controlling the temperature
The efficient conversion of electrical energy into thermal energy in a microwave applicator is directly influenced by the dielectric properties of the products. The structure, density, and chemical compositions of the products are the determining factors for the dielectric properties [21]. Free or bound states of water, ionic contents, and water content also play a major role in estimating the dielectric properties of the food materials. Non-uniform allocation of water and ionic concentration in fruits and vegetables also causes uneven heating. This drawback is found in the design of industrial and domestic microwave heating systems, causing irregular cold and hot spots in the samples. Additionally, energy is reduced rapidly as microwaves penetrate the products. Non-uniform moisture distribution in the products affects the standing wave, which causes the rapid decay samples treated with microwave heating. Therefore, it is difficult to predict the temperature fluctuation and accurately control temperature to prevent the resulting inadequate heating or overheating during microwave blanching. Design of a microwave system systematically for products with consistent composition and well-defined shape and size could mitigate these challenges.

3.2.1.5 Ohmic Blanching

Ohmic heating is also referred to as electrical resistance heating, Joule heating, or electroconductive heating. Food products serve as an electrical resistor and are placed in between electrodes, generating heated by passing an electric current through materials. Product temperature rises rapidly as the heat is generated [33, 34]. The generation of heat mainly depends on the electrical conductivity of the product and induced current [35]. Ohmic heating has many advantages, which are mentioned in Table 3.2. Ohmic heating has widespread potential application in food processing industries, such as evaporation, blanching, fermentation, dehydration, sterilization, pasteurization, and extraction [36, 37]. Ohmic heating performance is strongly affected by the frequency of applied voltage [38] found that the heating rate is inversely proportional to the frequency; therefore low frequency is used almost.

Ohmic blanching requires a shorter time than conventional hot water blanching. This is due to the volumetric heating characteristics. Additionally, better quality product is produced by reducing leaching out of solid and nutrients; and preserving the texture and color of the products

[39, 40]. Moreover, this method can be easily used for blanching fruits and vegetables in large quantities (volumetric) than traditional hot water blanching (which causes more degradation f quality due to poor conduction and convention of heat transfer).

Limitations of ohmic blanching

Problems in temperature control in blanching operation

The performance of ohmic heating is affected by electric conductivity which is a temperature dependant factor [41]. Therefore it is essential to design a real-time temperature controlling and reliable feedback monitoring system to direct the blanching temperature. This can be done by adjusting the supply power as per the change in temperature of the processed products. A triple point probe was designed by [42] to improve the performance of ohmic heating by monitoring the change in temperature during ohmic blanching.

Generating oxygen and hydrogen

[43] Reported that the increase in frequency decreases the heating rate. Ohmic heating at low frequencies (ranging from 50-60Hz) could generate hydrogen and oxygen by electrolyzation of water [44]. Generation of hydrogen at cathode and oxygen at the anode during water electrolysis causes degradation of nutrients [45] reported that oxidation of the anthocyanin is caused by the molecular oxygen generated from the electrolysis of water. High content loss of ascorbic acid and greater changes in acerola pulp color was observed at low frequency (10 Hz) [46]. This may be caused by the catalytic action of oxygen.

Corrosion and erosion of electrodes

In the case of fruits and vegetables, cell membrane and water is a poor conductor of electricity. Therefore, acidic solutions or metallic ions are used to enhance the electrical conductivity [37]. But the added substances (salts or acids) accelerate erosion and corrosion of electrodes [44] reported that intense electrode corrosion occurs at pH 3.5. Moreover, quality attributes especially product flavor are affected by added salts or acids. But it is too difficult to attenuate such problems coupled with salt and acid solutions.

3.2.1.6 Infrared Blanching

The application of Infrared (IR) blanching to inactivate enzymes and remove a part of moisture in vegetables and fruit is advanced thermal blanching technology [47, 48]. This technology has several advantages (mentioned in Table 3.2) as compared to conventional blanching systems. Apart from infrared heating, conductive, convective and microwave heating can be also accommodated in this heating system. Both the purpose of blanching and drying can be achieved in a single step of infrared blanching.

Both continuous and intermittent heating modes can be used in infrared blanching. Infrared radiation intensity remains constant in the case of continuous mode, while infrared radiation is operated using on and off modes in intermittent heating. For quick inactivation of enzymes, continuous infrared heating is more suitable as high constant energy is delivered to the products [49]. But a good-quality product is produced by intermittent heating by controlling the processing temperature [50]. This also saves energy. Therefore, infrared blanching is used for various fruits and vegetables.

Electromagnetic radiations generating infrared heating falls in the range of microwaves (1-1000 mm) and visible light waves (0.38-0.78μm) [51]. Infrared radiation can be propagated through both atmosphere and vacuum and absorbed by the food component molecules. The absorption of heat is caused by rotational-vibrational movements which produce heat [51].

The wavelength of the radiation is the main determining factor for infrared heating. Therefore, in line with ISO 20473 (ISO 20473:2007, ISO), infrared heaters are categorized as:

a) Near-infrared (NIR): 0.78-3 μm (Wavelength)
b) Mid-infrared (MIR): 3-50 μm
c) Far-infrared (FIR): 50-1000 μm [52]

The temperature of the radiation in the body is the determining factor and is inversely related to the wavelength of infrared; the more the temperature, the shorter the wavelength. Far infrared energy absorbs the main components of food such as starch, protein, fat, and water superior to near-infrared energy [53]. In infrared radiation, the depth of penetration mostly depends on the wavelengths of radiation, structure, and composition of food. As a result, far-infrared heating is frequently used for food processing.

Infrared heating exhibits higher efficiency and heat transfer rate; therefore it can save heating time and reduce energy consumption when compared to conventional heating. An infrared drying with intermittent facility saves 20 times more energy than convective drying. It means a 200 W/(m^2 K) heat transfer coefficient in convective drying is equivalent to10 W/m^2

energy input in intermittent drying [54]. Opaque and absorbent objects are heated predominantly by infrared rather than the surrounding air. Therefore, it can reduce energy consumption and keep the ambient temperature normal. Additionally, infrared heating has multilevel applications such as baking, drying, roasting, thawing, blanching, and pasteurization. It is environment friendly, contactless heating, simple to construct, easy to operate, and space-saving.

Limitations of infrared blanching
Even though infrared blanching has many advantages, it comes with certain limitations, which are as follows:

- Overheating and surface deterioration
- Non-uniform heating and poor penetrations
- Oxidation and charring of food products
- Severe water loss and low yield

Poor heat penetration
Infrared radiation can penetrate up to a few millimeters deep in product samples [51]. Therefore, IR blanching is unsuitable for thick slices or cubes. To overcome this problem and widen its applications, IR technology can be combined and used with microwave, conductive and convective modes of heating. Therefore, a continuous combined infrared and hot air system is designed by [55] (Figure 3.1) which can be used for various unit operations such as blanching, baking, roasting, and drying of food materials [56] have

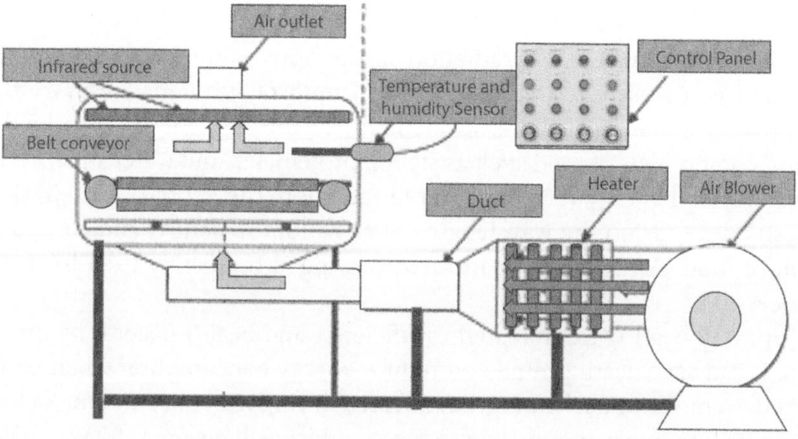

Figure 3.1 Combined Infrared and hot air system [56].

found that the combined IR and hot air technology proves a better alternative to only IR processing by giving synergistic effects. Besides, the combined technology reduced the drying time by 48% and saved 63% energy compared to hot air drying [55].

Non-uniform heating

Wide variation of energy due to poor penetration capacity of infrared radiation causes non-uniform heating of food. Sample surface temperature increases rapidly and heat transfer to inner layers occurs via conduction. However, the temperature of the sample decreases with an increase in sample depth. This phenomena cause overheating of the samples and charring on the sample surface, and insufficient inactivation of PPO and POD due to low heating inside the inner part. Therefore, for blanching of leafy vegetables, infrared blanching is more suitable. Moreover, a continuous moving belt is required to create agitation for exposure of all sides of food for even heating.

Water loss and surface color degradation

The infrared blanching process is not industrialized widely due to some major technical issues in the process. A high quantity of water loss (up to 49%) and severe surface color changes were observed during blanching, especially in the case of sample thickness and for enzyme inactivation (over 90% inactivation of POD) [57]. In vegetable and fruit processing, water loss is a major problem as it affects the product quality by reducing the size (shrinkage), poor texture, and developing undesirable colors.

3.2.2 Application of Blanching

Blanching is a foremost important unit operation in the processing and preservation of vegetables and fruits. It plays a vital role in maintaining product quality and storage of food products. Blanching has many applications as per the need during the processing which are explained below.

3.2.2.1 Inactivation of Enzymes

During transportation, processing, and storage, enzymatic reactions cause degradation of vegetables and fruits [58]. Enzymatic reactions contribute to undesirable changes in color, flavor, odor, and texture including a breakdown of nutrients. Therefore the main function of blanching is to inactivate the quality deteriorating enzymes. Therefore, nutritional quality and texture stability could be maintained during processing and storage [59, 60].

3.2.2.2 *Enhancement of Product Quality and Dehydration*

Drying kinetics and product quality depend upon the pretreatments before drying [61]. A waxy layer naturally present on the fruit surfaces hampers the removal of moisture during drying. During blanching, changes occur in the physical properties of the product, and drying and dehydration rates are enhanced. An increase in cell membrane permeability increases the product quality, which ultimately increases the moisture removal rate [62].

3.2.2.3 *Toxic and Pesticides Residues Removal*

Pesticides are commonly used in agricultural farming to control the weed and diseases to get a better yield. Pesticides are found in raw or semi-processed fruit and vegetables [63]. The presence of pesticide residues in agricultural products has severe toxic effects on human health, causing diseases such as nausea, headache to cancer, etc. Blanching plays a significant role in the removal of pesticide residues in agricultural products. The reduction is caused by the leaching and washing out of toxins to blanched water or degradation of toxic compounds during blanching.

3.2.2.4 *Decreasing Microbial Load*

Food spoilage and poisoning are caused through the contamination of food by microorganisms. Therefore, inhibition or inactivation of microbe is necessary to control microbial growth and ensuring disease risk-free safe foods. Inactivation of microbes can be done by using both thermal and non-thermal technologies. Several advanced thermal technologies are ohmic heating, microwave heating, and radiofrequency heating and non-thermal technologies are ultraviolet light (UV), ozone treatments, high-pressure processing, X-ray or gamma radiations, pulsed electric field, ultrasound, and iodine or chlorine solutions. Conventional methods of microbial inactivation and enzyme inactivation are two different processes with limitations of long processing time and low energy efficiency. Nowadays, thermally decontaminated food products are safer than chemically treated ones. Inactivation of enzyme and microbe can be achieved simultaneously in single thermal blanching. This could reduce process time, control re-contamination or cross-contamination and save energy.

3.2.2.5 *Reducing Non-Enzymatic Browning Reaction*

During different processing of food such as frying, drying, cooking, and storage, non-enzymatic browning reactions occur, especially Maillard reactions. This reaction leads to the degradation of product color. Caramelization and/or Maillard reaction is influenced by the reducing sugar content of the product [64]. Therefore blanching of food could decrease the reducing sugar content, hence could decrease the browning reaction and improve product color.

3.2.2.6 *Peeling*

Peeling is a major unit operation in vegetables and fruit processing. Peeling is done manually, but this method of peeling is time-consuming, laborious, tedious, and inconsistent. Therefore, alternate methods such as mechanical, chemical, and thermal peeling are often used. Although efficiency is high and the system is fully automated, the peeling is high in mechanical peeling. This is due to the uneven geometry of the varying products where it is very difficult to control the depth of peeling. Moreover, the chemical method is costly, not environment friendly, generates chemical wastewater, and has safety and health considerations. On the other hand, steam peeling has low peeling losses and fewer environmental issues.

3.2.2.7 *Entrapped Air Removal*

Blanching can help in expelling intercellular gases entrapped inside the plant tissues. This is essential in the canning because blanching could reduce the stain, expel air and prevent any misshapen cans and faulty seaming. Removal of gas from tissues softens the tissues and makes better texture [65]. Moreover, the removal of air reduces oxidation and corrosion in canning.

3.2.2.8 *Enhancing Bioactive Extraction Efficiency*

Thermal blanching can change the structure of the plant tissues by disrupting the cell membrane, loosening the cellulose, pectin networks, and modifying the cell wall porosity. Thus extractability of the bioactive compounds can be enhanced [66].

3.2.2.9 *Other Applications*

Blanching can also be used to destroy parasites and their eggs, clean plant surface, and remove discolored or damaged seeds. Blanching before frying can reduce oil absorption by reducing the pores and air cells on the starch surface [67].

3.3 Pasteurization: Principles & Mechanism

Pasteurization is a thermal process that uses relatively mild heat to inactivate microbe and enzymes enhancing the storage life and making food safe for consumption [68]. To ensure the desirable inactivation of microbes, the food material is subjected to a fixed temperature for a particular time. Most of the vegetative bacteria are destroyed in this thermal process. Pasteurization is used frequently for the processing of milk and fruit juices. Inactivation of undesirable enzymes in the fruit juices is the main purpose to prevent cloud loss in juices.

Earlier, pasteurization was carried out by conventional thermal cooking methods. But recently, novel thermal (ohmic heating and radio-frequency heating) and non-thermal technologies (ultrasound, pulsed electric field, high hydrostatic pressure, UV treatment, oscillating magnetic field, and ionizing radiation) are used without affecting the nutritional, functional, and sensory values of food products [69]. Although novel thermal and non-thermal technologies have more advantages and save up to 70% energy compared with conventional cooking methods [69], in this chapter, only thermal processes have been discussed with their applications in food.

3.3.1 Thermal Pasteurization

Thermal processing is the conventional and most widely used technology for the pasteurization of milk and fruit juice. Thermal pasteurization's main objective is to reduce the spoilage of organisms by killing the pathogen at a suitable temperature and time [70]. A 5 log reduction of resistance microbes is achieved during pasteurization for public health importance [71]. This process is carried out by the transfer of generated heat to food components via conduction or convection mechanisms [69].

Sterilization, pasteurization, drying, and evaporation are still the most commonly used thermal processing operations in food industries for the inactivation of microbes and enzymes, ensuring the microbiological safety of the product. To heat the product, these methods essentially dependent on the external heat generation by combustion of fuels or by an electricity conversion and transferred by conduction and convection mechanism to the product.

3.3.2 Traditional Thermal Pasteurization

The traditional method of thermal pasteurization is classified into two processes: a) Low-temperature long time (LTLT) pasteurization, and b) High-temperature short time (HTST) pasteurization. During LTLT pasteurization, food is subjected to 63°C for 30 min or more. During HTST pasteurization, food products are subjected to a temperature of 72°C with-holding time 15s or more [71]. However, the color, taste, flavor, and nutritional quality of the food are degraded by both methods [72].

3.3.3 Microwave and Radiofrequency Pasteurization

RF and MW work by the same principle of volumetric heating or dielectric heating, i.e., the interaction between the ionic charges and dipoles within the food and electromagnetic field. Continuous change in the electric field around the food products rearrange or realign the water and ionic molecules constantly, and this molecular movement is severely fast due to the high-frequency electromagnetic field. Both RF and MW range in between 1 to 300 MHz and 300 to 3000 MHz respectively. Dipolar rotation and molecular movement of ionic species due to alternating electromagnetic fields generate heat resulted from molecular friction inside the food. Both RF and MW are considered as non-ionizing radiation as they have low energy (<10 eV) to ionize the atoms [73]. There are limited frequencies bands available to be used as dielectric heating in food processing, because RF and MW range within the radar range, limited by Electromagnetic compatibility (EMC) regulations.

RF and MW offer rapid and non-contact volumetric heating. However, RF energy penetrates more deeply than microwave heating due to the high wavelength in the RF system because of low frequency [74]. The most commonly used frequency values for microwaves are 915 MHz and 2450 MHz. Moreover, it is too difficult to achieve a uniform temperature in the MW system; therefore it is profitable for processing small-sized food [74]. In RF heating, the food material is heated by placing the food (acting as dielectric) in between a pair of capacitor plates where the electricity is generated [73].

The promising of volumetric and rapid heating of MW and RF has drawn significant attention from the food industries. Experimental and industrial applications of microwave pasteurization of milk, juices, and ready-to-eat meals (Table 3.3) for preservation and storage have been conducted. A schematic diagram of a MW-assisted pasteurization system developed by Washington State University has been presented in Figure 3.2.

The application of radiofrequency heating for food pasteurization is still under investigation, because of the rapid, uniform, and producing shelf-stable high-quality food [31]. Before commercial and industrial application of a radiofrequency system for food pasteurization, it is essential to completely address the problem of thermal runaway heating and dielectric arcing [75, 147]. The factors affecting dielectric properties of food are water content, shape, mass, composition, frequency and temperature, etc., which need to be fully understood before industrial set up development [75].

3.3.4 Ohmic Heating Pasteurization

Ohmic heating is based on the principle that heat is generated due to the internal resistance of the food products when an electric current is passed through the food [41]. In ohmic heating, direct contact between the food products and electrodes is required to heat the product. The presence of electrodes, unrestricted frequency applied and waveform makes it different from other electrical heating methods [76, 146].

The main advantages of ohmic heating are mentioned below:

- The required temperature is achieved very quickly for the HTST processes.
- Useful in continuous processing with no heat transfer surfaces.
- Uniform and faster heating rates for liquid foods.
- Very low surface fouling and no overheating.
- Alternative preservation technique with the high-quality product (retains flavor and color and low nutritional degradation).
- No heat transfer and low heat losses after the shut-off of the current.
- Suitable for pre-heating products before canning.
- High energy-saving, low repairing, and maintenance cost, and environmentally friendly.

Industrial applications of ohmic heating include drying, dehydration, evaporation, fermentation, blanching, and pasteurization [77]. However, rapid heating of ohmic heating enables high temperature and quick

Table 3.3 Application of novel thermal pasteurization techniques for the processing of food materials.

Techniques	Food product	Pasteurization conditions	Targeted microbes	Log reduction	Significance	References
Microwave	Apple juice	720–900 W for 60–90 s	*E. coli*	2–4 log	Home microwave oven suitable for inactivation of *E. coli* in apple juice	[134]
Microwave	Apple cider	(900–2000 W), and inlet temperature (3, 21, and 40°C)	*Escherichia coli* 25922	5 log	produce uniform heating throughout the cavity and the use of helical coils would narrow the residence time	[135]
Microwave	Apple juice	(600 W, 720 W) and treatment times (5s, 10s, 15s, 20s, 25s)	*Escherichia coli* O157:H7 and *Salmonella Typhimurium*	7 log	Microwave heating introduces several advantages to produce safe products	[136]
Radiofrequency electric fields	Orange juice	80 kW RFEF 15 and 20 kV/cm at frequencies of 21, 30, and 40 kHz	*Escherichia coli* K12	3.3 log	No loss in ascorbic acid or enzymatic browning	[137]

(*Continued*)

Table 3.3 Application of novel thermal pasteurization techniques for the processing of food materials. (*Continued*)

Techniques	Food product	Pasteurization conditions	Targeted microbes	Log reduction	Significance	References
Radiofrequency (RF) electric fields	Apple juice	0.17 ms to electric field strengths of up to 26 kV/cm peak over a frequency range of 15 to 70 kHz.	*Escherichia coli*	3 log	inactivates bacteria at moderately low temperatures	[138]
Radiofrequency (RF)	In-shell almond	6 kW, 27.12 MH	*Escherichia coli* ATCC 25922	5 log	effective method to control Salmonella *Escherichia coli* ATCC 25922 could be used as a surrogate of pathogenic Salmonella.	[139]
Radiofrequency (RF)	Ground black pepper	6 kW, 27.12 MHz, 120 and 130 s	Salmonella spp. and *Enterococcus faecium* NRRL B-2354	5.98 log CFU/g 3.89 log CFU/g (130 s)	an effective method to Salmonella spp. with insignificant quality deterioration.	[140]

(*Continued*)

Table 3.3 Application of novel thermal pasteurization techniques for the processing of food materials. (*Continued*)

Techniques	Food product	Pasteurization conditions	Targeted microbes	Log reduction	Significance	References
Microwave and ohmic heating	Canta-loupe juice	Microwave (400 and 800 W for 110 s), ohmic (100 and 200 V for 110 s)	*Escherichia coli, Salmonella Typhimurium, S. Enteritidis,* and *Staphylo-coccus aureus* pathogens	complete inac-tivation in a shorter time	higher degradation of carotene and phe-nolic compounds and a lower loss of vitamin C	[141]
Ohmic heating	Fermented red pepper paste	(0.458 W/m·K), by varying frequen-cies (40–20,000 Hz) and applied voltages (20–60 V)	Bacillus strains	99.7% reduction	pasteurization effect was higher due to uniform internal heating produce by low-frequency AC.	[142]
Ohmic heating	Grapefruit and blood orange	5 kW, 50 Hz AC and 0.1 to 3 $kV{\cdot}m^{-1}$	-	-	Protects carotenoids and especially xanthophylls com-pared to conven-tional heating	[143]

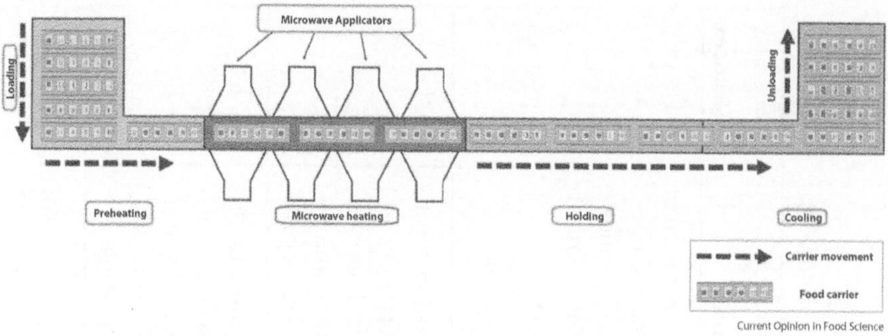

Figure 3.2 Microwave-assisted thermal pasteurization system [145].

pasteurization, without any coagulation or denaturation of constituents [78]. This heating method is suitable for the processing of liquid, viscous foods [79]. Several pieces of research have been carried out on ohmic heating for pasteurization of food products and are mentioned in Table 3.3.

3.3.5 Application of Pasteurization

The application of microwave, ohmic, and radiofrequency has been extensively applied to fluid foods. Pasteurization of milk, grape juice, apple juice, coconut water, apple cider, sweet potato puree, etc., has been studied (illustrated in Table 3.3). It was found that the knowledge of suitable microwave operating conditions and dielectric properties of food materials are required for achieving desirable lethality [80].

3.4 Sterilization: Principles, Mechanism and Types of Sterilization

Sterilization is essential to enhance product safety and extend the shelf life of packaged low acid food products. The main objective of this process is to provide thermal treatment to reduce 12 log cycles of *Clostridium botulinum* at all the cold points in the food products. The total time needed to reach the target microbial load is called 12D (D is the time required to reduce the microbes into one-tenth). These processing parameters are the measure of ensuring the stability of low acid foods. Similarly, the quality parameters of the food products are affected by the extreme thermal processing conditions.

3.4.1 Conventional Sterilization Methods

The most common and conventional technology used for sterilization of different food products in various kinds of containers such as pouches, trays, cups, jars, cans, and bowls is still or static retort processing [81]. The targeted sterilization temperature is 121°C and treated for a specific time. Different heating media such as air, water, or steam are used in the form of immersion or spray. The containers are held in a fixed specific position, and the only variables are temperature and time. This type of treatment has a limitation on food quality [81]. Nutritional quality, texture, flavor, and taste of products deteriorate in severe thermal processing conditions. Therefore, an alternative to static retorting is selected for the processing of containers in agitation mode. This method is suitable for liquid and semi-liquid type food as the convection heating is faster and increases heat transfer rapidly. The degree of convection is affected by headspace, geometry and the way containers are placed, type of retort motion, and the viscosity of the product [81].

Continuous and semi-continuous agitating retorts, hydrostatic cookers and end-over-end processing, etc., are a few commonly existing agitating retorts. They have limited and specific applications as per the type of containers or depend upon some specific symmetry such as jars and cans. Moreover, these types of retorts are criticized, because of limited processing opportunities and restricted thermal processing optimization conditions. With the fast-growing diversity of the products and containers, the trends have changed to process with multi-functionality and flexibility retort, which the above-mentioned retorts lack.

Aseptic processing is another alternative to conventional thermal processing, where the product and containers are sterilized individually and then packed in a sterile environment. This technique is mostly used to sterilize fluids such as fruit juices, milk, soup, stew, and fluid-containing chunks. In comparison to traditional thermal processing methods, aseptic processing offers many advantages, which include shorter processing time, low energy consumption, uniform heating, and the quality of the product improve with less deterioration. Inabilities to process semi-solid foods (for example tuna in chunks), placeable food products such as filets are the major constraints of this process. Moreover, the availability of packaging material is in limited shapes and sizes, and expensive [81].

Some more advanced and novel thermal processing methods are described in the following sections which are recently in use for sterilization of food products.

3.4.2 Advanced Retorting

To overcome the limitations of conventional retorting and other sterilization processes, scientists are continuously investing in new approaches and advancements in design. These advanced designs include better automation and control systems, new agitation modes, reduced carbon footprints and saving of energy, very fast come-up times, versatility for wide application, and various heating media [82].

Types of heating media used in the advanced retort processing include:

- **Water immersion:** Containers submerged completely where overpressure is created by air promoting agitation orinjecting stream and better heat transfer occurs.
- **Water spray:** Containers are directly exposed to an overpressure stream or air. The distribution of nozzles provides uniform heating patterns.
- **Saturated steam:** No overpressure as the retort vessel is evacuated with vent valves.
- **Steam air:** An overpressure process where steam and air come in contact with containers and a fan is used to avoid cold spots by mixing air and steam [81].

As mentioned above, versatility is the main feature of advanced retorting. It can handle containers of different sizes and shape imparting motion. The most common types are:

a) Gentle motion: Reciprocating motion mechanism which can be operated at a stroke of 40 per minute. Flat types of containers such as trays and pouches, horizontally oriented are processed where convection heating is promoted. Containers are placed in a locking system to prevent bumping. Time saved is 5-40% and product quality is improved as compared to static retorting.
b) Shaka: Very similar to gentle motion but strokes are high (in order of 90). Better quality product, faster come up time and less processing time is favored in this agitation process.
c) Rotational: Containers are rotated in a circle in the vertical plane. The rate of heat transfer is increased because of the movement of bubbles inside the container. This method is not used in advanced retorting.
d) Oscillatory or pendulum: In this type of agitation, the metallic tray or rack moves clockwise or anti-clockwise. This

method is suitable for all types of container retorting. The displacement angle can be adjusted as per the processing need. Shorter processing time and better quality product compared to still type retorting.

The primary advantages of advanced retorting compared with conventional retorting are as follows:

- High heat transfer coefficient and uniform heating due to agitation of containers.
- Low processing time and less consumption of energy during processing because of uniform temperature and fast processing.
- Great versatility to accommodate all types of packages (it means shape, size, and types of material) without any complications.
- Retention of nutrition and enhancement of sensory attributes (color, flavor, texture, and appearance).

3.4.3 Microwave-Assisted Thermal Sterilization

Microwave thermal sterilization processing can be explained by different mechanisms such as cell membrane rupture, electroporation, selective heating, and magnetic field coupling [83]. In selective heating mechanism, the temperature of the microbial bodies is higher than the surrounding fluid temperature, which leads to rapid destruction of microorganisms. Formation of pores in cells and leakage of cellular material occurs during the electroporation mechanism. This is caused due to the electric potential across the cell membrane. Moreover, during magnetic field coupling, DNA or proteins coupled with magnetic field and get destroy [84]. Microwave sterilization of microwave-assisted sterilization can be used to reduce the microbes' colony in the food, inactivate pathogens, and control potential microorganisms. Table 3.4 shows the different applications of microwave sterilization in food products. The effectiveness of microwave sterilization is affected by the sterilization time and temperature, and microwave power [85]. Microwave sterilization shows little effect on the nutritional and functional properties of food [84]. Moreover, it is more beneficial as it shows no significant effect on antioxidants and other bioactive compounds, and maintains quality by inactivating enzyme [84].

Washington State University, Pullman, has developed and installed a pilot plant scale microwave-assisted thermal sterilization (MATS) system,

Table 3.4 Sterilization methods and their application in food.

Types of sterilization	Product	Purpose	Significance	References
High-pressure thermal sterilization	Fish	Removal of food processing contaminants (FPC)	71-97% reduction of furan	[127]
	Baby food puree & Fish	FPC removal	41-98% reduction of furan	[128]
Microwave sterilization	Drumettes	Removal of microbes	Elimination of *L. monocytogenes*	[129]
	Salmon and cod	Microbial inactivation	Complete inactivation	[130]
	Potato omelet	*S. enteritidis inactivation*	4.80 log reduction	[85]
	Jalapeño peppers & coriander foliage	Removal of *S. typhimurium*	5.12 log reduction in peppers & 4.45 log reduction in coriander foliage	[131]
	Grape, tomatoes	*Salmonella enteric reduction*	2.05 log reduction in grapes & 1.70 log reduction in tomatoes	[132]
	beef slices	*E. coli* O157:H7 removal	Removal of contamination	[133]

as shown in Figure 3.3. The complete system consists of four sections arranged in process series as preheating, microwave heating, holding, and cooling. Every section has a separate water circulating system consisting of a plate heat exchanger (to control water flow), pressurized tank, and a non-metallic mesh conveyor belt (to convey the food pouches or trays from preheating section to the cooling section).

Vacuum packaged food is submerged in a pressurized vessel (containing water with salt solutions) placed in a microwave cavity (915 MHz) during

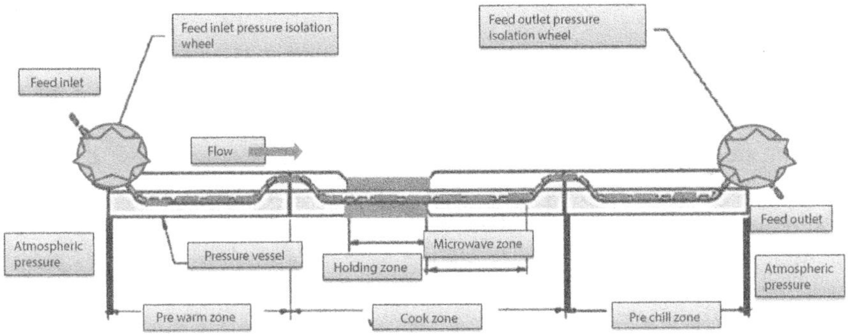

Figure 3.3 Diagram showing components of microwave-assisted thermal sterilization (MATS) (Source: [144]).

sterilization. To prevent or reduce the overheating of the edge, water is distributed inside the cavities. This type of system can be operated at both the continuous and batch mode, and reduce the processing time from one-fourth to one-tenth of processing time as compared to conventional methods. It has also other advantages such as low processing time, high throughput, and better sensory attributes.

3.4.4 Pressure-Assisted Thermal Sterilization

Pressure-assisted thermal sterilization (PATS) or high-pressure thermal sterilization (HPTS) is a promising novel thermal technology for the sterilization of low acidic food which can be shelf-stable at ambient conditions. This intelligent technology is a hybrid technology that is a combination of high-pressure processing (HPP) and thermal energy. The use of this sterilization process in the industry could be a turning point as it achieves superior quality and safe food compared to conventional thermal processes due to shorter process time and the use of lower maximum temperature. PATS uses 60-90°C temperature in the initial chamber, produced due to internal compression heating at 600 MPa or higher pressure, and 90-130°C is in process temperature [86, 87]. This process is considered a high-temperature short-time process, where lethality is achieved by both compression and heat. The major advantages are: the temperature of the sample increases very fast due to compression heating, which helps in making the process shorter and low adversity of thermal effects to samples.

Before processing of products, the low acid products are vacuum packed in a flexible pouch or container with high barrier properties. Then the prepackaged product is heated to a required temperature and processed in a high-pressure vessel. Pressure vessel temperature is controlled to reduce the heat loss to the surrounding environment during processing. High processed unloaded product is cooled to room temperature to put off further thermal degradation of the product. Although PATS processing is more advantageous due to shorter time, lower thermal load, less processing contaminants, better quality product, and suitability for large volumes, the main limitation is that a long preheating time is required before pressurization of products and low processing temperature cannot assure the desired inactivation of *C. botulinum*.

3.5 Conclusions

Thermal processing is a better alternative to chemical treatment, often employed to maintain product quality with an extended shelf life to avoid any chemical residues. Blanching, Pasteurization, and sterilization is an essential operation for food processing and preservation especially fruits and vegetables. These food processing and preservation techniques not only contribute to the inactivation of enzymes, microbes, polyphenol oxidase, and peroxidase but also affect other quality parameters of the product. Traditional thermal treatments may lead to losses of temperature-dependent nutrients and vitamins and organoleptic properties of the product. Consequently, novel thermal techniques such as Steam blanching, microwave blanching, ohmic blanching, high humidity hot air impingement blanching and infrared blanching, and radio frequency (RF) heating have been designed to meet required product safety, while minimizing nutritional losses. It has several advantages like rapid heating, deep thermal penetration, better quality product, and low cost. These techniques have been solely used or in combination with other unit operations. Each processing method has certain advantages and some disadvantages over other methods. This method exhibits good results with specific types of products with specific requirements. Recent improvement in these techniques gives better quality of the product.

References

1. Arroqui, C., Lopez, A., Esnoz, A., & Virseda, P. (2003). Mathematical model of heat transfer and enzyme inactivation in an integrated blancher cooler. *Journal of Food Engineering*, *58*(3), 215-225.

2. Mukherjee, S., & Chattopadhyay, P. K. (2007). Whirling bed blanching of potato cubes and its effects on product quality. *Journal of Food Engineering*, *78*(1), 52-60.
3. Xiao, H. W., Pan, Z., Deng, L. Z., El-Mashad, H. M., Yang, X. H., Mujumdar, A. S., ... & Zhang, Q. (2017). Recent developments and trends in thermal blanching–A comprehensive review. *Information Processing in Agriculture*, *4*(2), 101-127.
4. Cruz, R. M., Vieira, M. C., & Silva, C. L. (2006). Effect of heat and thermosonication treatments on peroxidase inactivation kinetics in watercress (Nasturtium officinale). *Journal of Food Engineering*, *72*(1), 8-15.
5. Blake, M. R., Weimer, B. C., McMahon, D. J., & Savello, P. A. (1995). Sensory and microbial quality of milk processed for extended shelf life by direct steam injection. *Journal of Food Protection*, *58*(9), 1007-1013.
6. Lavigne, C., Zee, J. A., Simard, R. E., & Beliveau, B. (1989). Effect of Processing and Storage Conditions on the Fate of Vitamins B1, B2, and C and on the Shelf-Life of Goat's Milk. *Journal of Food Science*, *54*(1), 30-34.
7. Bingol, G., Wang, B., Zhang, A., Pan, Z., & McHugh, T. H. (2014). Comparison of water and infrared blanching methods for processing performance and final product quality of French fries. *Journal of Food Engineering*, *121*, 135-142.
8. Liu, J., & Yang, W. (2012). Water sustainability for China and beyond. *Science*, *337*(6095), 649-650.
9. Gawlik-Dziki, U. (2008). Effect of hydrothermal treatment on the antioxidant properties of broccoli (*Brassica oleracea* var. botrytis italica) florets. *Food Chemistry*, *109*(2), 393-401.
10. Sikora, E., Cieślik, E., Leszczyńska, T., Filipiak-Florkiewicz, A., & Pisulewski, P. M. (2008). The antioxidant activity of selected cruciferous vegetables subjected to aquathermal processing. *Food Chemistry*, *107*(1), 55-59.
11. Ismail, A., Marjan, Z. M., & Foong, C. W. (2004). Total antioxidant activity and phenolic content in selected vegetables. *Food Chemistry*, *87*(4), 581-586.
12. Haase, N. U., & Weber, L. (2003). Ascorbic acid losses during processing of French fries and potato chips. *Journal of Food Engineering*, *56*(2-3), 207-209.
13. Garrote, R. A. L., Silva, E. N. R., & Bertone, R. I. A. (1988). Effect of freezing on diffusion of ascorbic acid during water heating of potato tissue. *Journal of Food Science*, *53*(2), 473-474.
14. Arroqui, C., Rumsey, T. R., Lopez, A., & Virseda, P. (2002). Losses by diffusion of ascorbic acid during recycled water blanching of potato tissue. *Journal of Food Engineering*, *52*(1), 25-30.
15. Roy, M. K., Juneja, L. R., Isobe, S., & Tsushida, T. (2009). Steam processed broccoli (*Brassica oleracea*) has higher antioxidant activity in chemical and cellular assay systems. *Food Chemistry*, *114*(1), 263-269.

16. Xiao, H. W., Bai, J. W., Sun, D. W., & Gao, Z. J. (2014). The application of superheated steam impingement blanching (SSIB) in agricultural products processing–A review. *Journal of Food Engineering, 132*, 39-47.
17. Du, Z. L., Gao, Z. J., & Zhang, S. X. (2006). Research on convective heat transfer coefficient with air jet impinging. *Transactions of the Chinese Society of Agricultural Engineering, 22*(14), 1-4.
18. Bai, J. W., Sun, D. W., Xiao, H. W., Mujumdar, A. S., & Gao, Z. J. (2013a). Novel high-humidity hot air impingement blanching (HHAIB) pretreatment enhances drying kinetics and color attributes of seedless grapes. *Innovative Food Science & Emerging Technologies, 20*, 230-237.
19. Moreira, R. G. (2001). Impingement drying of foods using hot air and superheated steam. *Journal of Food Engineering, 49*(4), 291-295.
20. Thostenson, E. T., & Chou, T. W. (1999). Microwave processing: fundamentals and applications. *Composites Part A: Applied Science and Manufacturing, 30*(9), 1055-1071.
21. Chandrasekaran, S., Ramanathan, S., & Basak, T. (2013). Microwave food processing—A review. *Food Research International, 52*(1), 243-261.
22. Zhang, M., Tang, J., Mujumdar, A. S., & Wang, S. (2006). Trends in microwave-related drying of fruits and vegetables. *Trends in Food Science & Technology, 17*(10), 524-534.
23. Ramesh, M. N., Wolf, W., Tevini, D., & Bognar, A. (2002). Microwave blanching of vegetables. *Journal of Food Science, 67*(1), 390-398.
24. Brewer, M. S., & Begum, S. (2003). Effect of microwave power level and time on ascorbic acid content, peroxidase activity and color of selected vegetables. *Journal of Food Processing and Preservation, 27*(6), 411-426.
25. Kidmose, U., & Martens, H. J. (1999). Changes in texture, microstructure and nutritional quality of carrot slices during blanching and freezing. *Journal of the Science of Food and Agriculture, 79*(12), 1747-1753.
26. Güneş, B., & Bayindirli, A. (1993). Peroxidase and lipoxygenase inactivation during blanching of green beans, green peas and carrots. *LWT-Food Science and Technology, 26*(5), 406-410.
27. Gunasekaran, S., & Yang, H. W. (2007). Effect of experimental parameters on temperature distribution during continuous and pulsed microwave heating. *Journal of Food Engineering, 78*(4), 1452-1456.
28. Muley, P. D., & Boldor, D. (2013). Investigation of microwave dielectric properties of biodiesel components. *Bioresource Technology, 127*, 165-174.
29. Koskiniemi, C. B., Truong, V. D., Simunovic, J., & McFeeters, R. F. (2011). Improvement of heating uniformity in packaged acidified vegetables pasteurized with a 915 MHz continuous microwave system. *Journal of Food Engineering, 105*(1), 149-160.
30. Wang, Y., Wig, T. D., Tang, J., & Hallberg, L. M. (2003a). Dielectric properties of foods relevant to RF and microwave pasteurization and sterilization. *Journal of Food Engineering, 57*(3), 257-268.

31. Wang, Y., Wig, T. D., Tang, J., & Hallberg, L. M. (2003b). Sterilization of foodstuffs using radio frequency heating. *Journal of Food Science, 68*(2), 539-544.
32. Walde, S. G., Velu, V., Jyothirmayi, T., & Math, R. G. (2006). Effects of pretreatments and drying methods on dehydration of mushroom. *Journal of Food Engineering, 74*(1), 108-115.
33. Assiry, A., Sastry, S. K., & Samaranayake, C. (2003). Degradation kinetics of ascorbic acid during ohmic heating with stainless steel electrodes. *Journal of Applied Electrochemistry, 33*(2), 187-196.
34. Sastry, S. K., & Barach, J. T. (2000). Ohmic and inductive heating. *Journal of Food Science, 65*, 42-46.
35. Reznick, D. (1996). Ohmic heating of fluid foods: ohmic heating for thermal processing of foods: government, industry, and academic perspectives. *Food Technology (Chicago), 50*(5), 250-251.
36. Farahnaky, A., Azizi, R., & Gavahian, M. (2012). Accelerated texture softening of some root vegetables by ohmic heating. *Journal of Food Engineering, 113*(2), 275-280.
37. Halden, K., De Alwis, A. A. P., & Fryer, P. J. (1990). Changes in the electrical conductivity of foods during ohmic heating. *International Journal of Food Science & Technology, 25*(1), 9-25.
38. Lima, M., & Sastry, S. K. (1999). The effects of ohmic heating frequency on hot-air drying rate and juice yield. *Journal of Food Engineering, 41*(2), 115-119.
39. Leizerson, S., & Shimoni, E. (2005). Effect of ultrahigh-temperature continuous ohmic heating treatment on fresh orange juice. *Journal of Agricultural and Food Chemistry, 53*(9), 3519-3524.
40. Mizrahi, S. (1996). Leaching of soluble solids during blanching of vegetables by ohmic heating. *Journal of Food Engineering, 29*(2), 153-166.
41. Sakr, M., & Liu, S. (2014). A comprehensive review on applications of ohmic heating (OH). *Renewable and Sustainable Energy Reviews, 39*, 262-269.
42. Zell, M., Lyng, J. G., Morgan, D. J., & Cronin, D. A. (2009). Development of rapid response thermocouple probes for use in a batch ohmic heating system. *Journal of Food Engineering, 93*(3), 344-347.
43. Amatore, C., Berthou, M., & Hebert, S. (1998). Fundamental principles of electrochemical ohmic heating of solutions. *Journal of Electroanalytical Chemistry, 457*(1-2), 191-203.
44. Samaranayake, C. P., & Sastry, S. K. (2005). Electrode and pH effects on electrochemical reactions during ohmic heating. *Journal of Electroanalytical Chemistry, 577*(1), 125-135.
45. Sarkis, J. R., Jaeschke, D. P., Tessaro, I. C., & Marczak, L. D. (2013). Effects of ohmic and conventional heating on anthocyanin degradation during the processing of blueberry pulp. *LWT-Food Science and Technology, 51*(1), 79-85.
46. Mercali, G. D., Schwartz, S., Marczak, L. D. F., Tessaro, I. C., & Sastry, S. (2014). Ascorbic acid degradation and color changes in acerola pulp during

ohmic heating: Effect of electric field frequency. *Journal of Food Engineering, 123*, 1-7.
47. Pan, Z., Shih, C., McHugh, T. H., & Hirschberg, E. (2008). Study of banana dehydration using sequential infrared radiation heating and freeze-drying. *LWT-Food Science and Technology, 41*(10), 1944-1951.
48. Pan, Z., & McHugh, T. H. (2006). Novel infrared dry-blanching (IDB), infrared blanching and infrared drying technologies for food processing. *Patent Application, 10*, 917797.
49. Zhu, Y., Pan, Z., McHugh, T. H., & Barrett, D. M. (2010). Processing and quality characteristics of apple slices processed under simultaneous infrared dry-blanching and dehydration with intermittent heating. *Journal of Food Engineering, 97*(1), 8-16.
50. Chua, K. J., & Chou, S. K. (2003). Low-cost drying methods for developing countries. *Trends in Food Science & Technology, 14*(12), 519-528.
51. Rastogi, N. K. (2012). Recent trends and developments in infrared heating in food processing. *Critical Reviews in Food science and Nutrition, 52*(9), 737-760.
52. Sandu, C. (1986). Infrared radiative drying in food engineering: a process analysis. *Biotechnology Progress, 2*(3), 109-119.
53. Sakai, N., & Hanzawa, T. (1994). Applications and advances in far-infrared heating in Japan. *Trends in Food Science & Technology, 5*(11), 357-362.
54. Ratti, C., & Mujumdar, A. S. (1995). Infrared drying. *Handbook of Industrial Drying* (A.S. Mujumdar, ed.), Marcel Dekker, New York, 567-588.
55. Hebbar, U. H., & Ramesh, M. N. (2002). Combined infrared and convective heating system for food processing. Indian Patent application, 336/DEL/02.
56. Hebbar, H. U., Vishwanathan, K. H., & Ramesh, M. N. (2004). Development of combined infrared and hot air dryer for vegetables. *Journal of Food Engineering, 65*(4), 557-563.
57. Zhu, Y., & Pan, Z. (2009). Processing and quality characteristics of apple slices under simultaneous infrared dry-blanching and dehydration with continuous heating. *Journal of Food Engineering, 90*(4), 441-452.
58. Dorantes-Alvarez, L., Jaramillo-Flores, E., González, K., Martinez, R., & Parada, L. (2011). Blanching peppers using microwaves. *Procedia Food Science, 1*, 178-183.
59. Bahçeci, K. S., Serpen, A., Gökmen, V., & Acar, J. (2005). Study of lipoxygenase and peroxidase as indicator enzymes in green beans: change of enzyme activity, ascorbic acid and chlorophylls during frozen storage. *Journal of Food Engineering, 66*(2), 187-192.
60. Arroqui, C., Rumsey, T. R., Lopez, A., & Virseda, P. (2001). Effect of different soluble solids in the water on the ascorbic acid losses during water blanching of potato tissue. *Journal of Food Engineering, 47*(2), 123-126.

61. Negi, P. S., & Roy, S. K. (2001). The effect of blanching on quality attributes of dehydrated carrots during long-term storage. *European Food Research and Technology*, *212*(4), 445-448.
62. Severini, C., Baiano, A., De Pilli, T., Carbone, B. F., & Derossi, A. (2005). Combined treatments of blanching and dehydration: study on potato cubes. *Journal of Food Engineering*, *68*(3), 289-296.
63. Claeys, W. L., Schmit, J. F., Bragard, C., Maghuin-Rogister, G., Pussemier, L., & Schiffers, B. (2011). Exposure of several Belgian consumer groups to pesticide residues through fresh fruit and vegetable consumption. *Food Control*, *22*(3-4), 508-516.
64. Villamiel, M., Castillo, del. D., & Corzo, N. (2008). Browning reactions. Hui, Y. H., Nip, W. K., Nollet, L. M., Paliyath, G., & Simpson, B. K. (eds.). *Food Biochemistry and Food Processing*. John Wiley & Sons, 71-100.
65. Chafer, M., Gonzalez-Martinez, C., Fernandez, B., Perez, L., & Chiralt, A. (2003). Effect of blanching and vacuum pulse application on osmotic dehydration of pear. *Food Science and Technology International*, *9*(5), 321-328.
66. Xu, G., Ye, X., Chen, J., & Liu, D. (2007). Effect of heat treatment on the phenolic compounds and antioxidant capacity of citrus peel extract. *Journal of Agricultural and Food Chemistry*, *55*(2), 330-335.
67. Krokida, M. K., Oreopoulou, V., Maroulis, Z. B., & Marinos-Kouris, D. (2001). Deep fat frying of potato strips—quality issues. *Drying Technology*, *19*(5), 879-935.
68. Ahmed, J., & Ramaswamy, H. S. (2004). Microwave pasteurization and sterilization of foods. *Food Science and Technology*, New York, Marcel Dekker, *167*, 691.
69. Pereira, R. N., & Vicente, A. A. (2010). Environmental impact of novel thermal and non-thermal technologies in food processing. *Food Research International*, *43*(7), 1936-1943.
70. Ramaswamy, H. S., Chen, C., & Marcotte, M. (2004). Novel processing technologies for food preservation. In *Processing Fruits* (pp. 211-229). CRC Press.
71. Chen, Y., Yu, L. J., & Rupasinghe, H. V. (2013). Effect of thermal and non-thermal pasteurisation on the microbial inactivation and phenolic degradation in fruit juice: A mini-review. *Journal of the Science of Food and Agriculture*, *93*(5), 981-986.
72. Charles-Rodríguez, A. V., Nevárez-Moorillón, G. V., Zhang, Q. H., & Ortega-Rivas, E. (2007). Comparison of thermal processing and pulsed electric fields treatment in pasteurization of apple juice. *Food and Bioproducts Processing*, *85*(2), 93-97.
73. Piyasena, P., Dussault, C., Koutchma, T., Ramaswamy, H. S., & Awuah, G. B. (2003). Radio frequency heating of foods: principles, applications and related properties—a review. *Critical Reviews in Food Science and Nutrition*, *43*(6), 587-606.

74. Marra, F., Lyng, J., Romano, V., & McKenna, B. (2007). Radio-frequency heating of foodstuff: solution and validation of a mathematical model. *Journal of Food Engineering, 79*(3), 998-1006.
75. Zhao, Y., Flugstad, B. E. N., Kolbe, E., Park, J. W., & Wells, J. H. (2000). Using capacitive (radio frequency) dielectric heating in food processing and preservation–a review. *Journal of Food Process Engineering, 23*(1), 25-55.
76. Vicente, A. A., & Castro, I. (2007). Novel thermal processing technologies.
77. Cho, H. Y., Yousef, A. E., & Sastry, S. K. (1996). Growth kinetics of *Lactobacillus acidophilus* under ohmic heating. *Biotechnology and Bioengineering*, 49(3), 334-340.
78. Parrott, David L. 1992. Use of ohmic heating for aseptic processing of food particulates. *Food Technology*, 68-72.
79. Icier, F., &Ilicali, C. (2005). The use of tylose as a food analog in ohmic heating studies. *Journal of Food Engineering*, 69(1), 67-77.
80. Salazar-González, C., San Martín-González, M. F., López-Malo, A., & Sosa-Morales, M. E. (2012). Recent studies related to microwave processing of fluid foods. *Food and Bioprocess Technology, 5*(1), 31-46.
81. Ibarz, A., & Barbosa-Cánovas, G. V. (2003). *Unit operations in food engineering*. CRC Press.
82. Bermudez-Aguirre, D., Lima, F., Reitzel, J., Garcia-Prez, M., & Barbosa-Canovas, G. V. (2013). Evaluation of total heat transfer coefficient (hT) during innovative retort processing: static, gentle motion and rocking mode. In *IFT Annual Meeting, Abstract number: 031-02.*
83. Kozempel, M. F., Annous, B. A., Cook, R. D., Scullen, O. J., & Whiting, R. C. (1998). Inactivation of microorganisms with microwaves at reduced temperaturas. *Journal of Food Protection, 61*(5), 582-585.
84. Guo, Q., Sun, D. W., Cheng, J. H., & Han, Z. (2017). Microwave processing techniques and their recent applications in the food industry. *Trends in Food Science & Technology, 67*, 236-247.
85. Valero, A., Cejudo, M., & García-Gimeno, R. M. (2014). Inactivation kinetics for Salmonella Enteritidis in potato omelet using microwave heating treatments. *Food Control, 43*, 175-182.
86. Matser, A. M., Krebbers, B., van den Berg, R. W., & Bartels, P. V. (2004). Advantages of high pressure sterilisation on quality of food products. *Trends in Food Science & Technology, 15*(2), 79-85.
87. Sizer, C. E., Balasubramaniam, V. M., & Ting, E. (2002). Validating high-pressure processes for low-acid foods. *Food Technology (Chicago)*, 56(2), 36-42.
88. Schweiggert, U., Schieber, A., & Carle, R. (2005). Inactivation of peroxidase, polyphenoloxidase, and lipoxygenase in paprika and chili powder after immediate thermal treatment of the plant material. *Innovative Food Science & Emerging Technologies, 6*(4), 403-411.
89. Lisiewska, Z., Słupski, J., Skoczeń-Słupska, R., & Kmiecik, W. (2009). Content of amino acids and the quality of protein in Brussels sprouts, both

raw and prepared for consumption. *International Journal of Refrigeration, 32*(2), 272-278.
90. Harris, L. J., Uesugi, A. R., Abd, S. J., & McCarthy, K. L. (2012). Survival of Salmonella Enteritidis PT 30 on inoculated almond kernels in hot water treatments. *Food Research International, 45*(2), 1093-1098.
91. Pimpaporn, P., Devahastin, S., & Chiewchan, N. (2007). Effects of combined pretreatments on drying kinetics and quality of potato chips undergoing low-pressure superheated steam drying. *Journal of Food Engineering, 81*(2), 318-329.
92. DiPersio, P. A., Kendall, P. A., Yoon, Y., & Sofos, J. N. (2007). Influence of modified blanching treatments on inactivation of Salmonella during drying and storage of carrot slices. *Food Microbiology, 24*(5), 500-507.
93. Llano, K. M., Haedo, A. S., Gerschenson, L. N., & Rojas, A. M. (2003). Mechanical and biochemical response of kiwifruit tissue to steam blanching. *Food Research International, 36*(8), 767-775.
94. Sotome, I., Takenaka, M., Koseki, S., Ogasawara, Y., Nadachi, Y., Okadome, H., & Isobe, S. (2009). Blanching of potato with superheated steam and hot water spray. *LWT-Food Science and Technology, 42*(6), 1035-1040.
95. Liu, E. Z., & Scanlon, M. G. (2007). Modeling the effect of blanching conditions on the texture of potato strips. *Journal of Food Engineering, 81*(2), 292-297.
96. Ndiaye, C., Xu, S. Y., & Wang, Z. (2009). Steam blanching effect on polyphenoloxidase, peroxidase and colour of mango (*Mangifera indica* L.) slices. *Food Chemistry, 113*(1), 92-95.
97. Fante, L., & Noreña, C. P. Z. (2012). Enzyme inactivation kinetics and colour changes in Garlic (*Allium sativum* L.) blanched under different conditions. *Journal of Food Engineering, 108*(3), 436-443.
98. Brambilla, A., Maffi, D., & Rizzolo, A. (2011). Study of the influence of berry-blanching on syneresis in blueberry purées. *Procedia Food Science, 1*, 1502-1508.
99. Rossi, M., Giussani, E., Morelli, R., Scalzo, R. L., Nani, R. C., & Torreggiani, D. (2003). Effect of fruit blanching on phenolics and radical scavenging activity of highbush blueberry juice. *Food Research International, 36*(9-10), 999-1005.
100. Saldivar, X. E. A., Wang, Y. J., Chen, P., & Mauromoustakos, A. (2010). Effects of blanching and storage conditions on soluble sugar contents in vegetable soybean. *LWT-Food Science and Technology*, 43(9), 1368-1372.
101. Phungamngoen, C., Chiewchan, N., & Devahastin, S. (2013). Effects of various pretreatments and drying methods on Salmonella resistance and physical properties of cabbage. *Journal of Food Engineering*, 115(2), 237-244.
102. Tanongkankit, Y., Chiewchan, N., & Devahastin, S. (2012). Physicochemical property changes of cabbage outer leaves upon preparation into functional dietary fiber powder. *Food and Bioproducts Processing*, 90(3), 541-548.

103. Xiao, H. W., Yao, X. D., Lin, H., Yang, W. X., Meng, J. S., & Gao, Z. J. (2012). Effect of SSB (superheated steam blanching) time and drying temperature on hot air impingement drying kinetics and quality attributes of yam slices. *Journal of Food Process Engineering*, 35(3), 370-390.
104. Gao, Z., & Xiao, H. W. (2007). Air impingement drying method and apparatus for sea cucumber. China Patent No. ZL200710176389, 5.
105. Kondjoyan, A., & Portanguen, S. (2008). Effect of superheated steam on the inactivation of Listeria innocua surface-inoculated onto chicken skin. *Journal of Food Engineering*, 87(2), 162-171.
106. Rico, D., Martín-Diana, A. B., Barry-Ryan, C., Frías, J. M., Henehan, G. T., & Barat, J. M. (2008). Optimisation of steamer jet-injection to extend the shelflife of fresh-cut lettuce. *Postharvest Biology and Technology*, 48(3), 431-442.
107. Bai, J. W., Gao, Z. J., Xiao, H. W., Wang, X. T., & Zhang, Q. (2013b). Polyphenol oxidase inactivation and vitamin C degradation kinetics of Fuji apple quarters by high humidity air impingement blanching. *International Journal of Food Science & Technology*, 48(6), 1135-1141.
108. Wang, J., Fang, X. M., Mujumdar, A. S., Qian, J. Y., Zhang, Q., Yang, X. H., ... & Xiao, H. W. (2017a). Effect of high-humidity hot air impingement blanching (HHAIB) on drying and quality of red pepper (*Capsicum annuum* L.). *Food Chemistry*, *220*, 145-152
109. Xiao, H. W., Lin, H., Yao, X. D., Du, Z. L., Lou, Z., & Gao, Z. J. (2009). Effects of different pretreatments on drying kinetics and quality of sweet potato bars undergoing air impingement drying. *International Journal of Food Engineering*, 5(5).
110. Lemmens, L., Tibäck, E., Svelander, C., Smout, C., Ahrné, L., Langton, M., ... &Hendrickx, M. (2009). Thermal pretreatments of carrot pieces using different heating techniques: Effect on quality related aspects. *Innovative Food Science & Emerging Technologies*, 10(4), 522-529.
111. Devece, C., Rodríguez-López, J. N., Fenoll, L. G., Tudela, J., Catalá, J. M., de los Reyes, E., & García-Cánovas, F. (1999). Enzyme inactivation analysis for industrial blanching applications: comparison of microwave, conventional, and combination heat treatments on mushroom polyphenoloxidase activity. *Journal of Agricultural and Food Chemistry*, 47(11), 4506-4511.
112. Sun, T., Tang, J., & Powers, J. R. (2007). Antioxidant activity and quality of asparagus affected by microwave-circulated water combination and conventional sterilization. *Food Chemistry*, 100(2), 813-819.
113. Ihl, M., Monsalves, M., & Bifani, V. (1998). Chlorophyllase inactivation as a measure of blanching efficacy and colour retention of artichokes (*Cynara scolymus* L.). *LWT-Food Science and Technology*, 31(1), 50-56.
114. Lin, S., & Brewer, M. S. (2005). Effects of blanching method on the quality characteristics of frozen peas. *Journal of Food Quality*, *28*(4), 350-360.

115. Singh, M., Raghavan, B., & Abraham, K. O. (1996). Processing of marjoram (*Majorana hortensis* Moench.) and rosemary (*Rosmarinus officinalis* L.). Effect of blanching methods on quality. *Food/Nahrung, 40*(5), 264-266.
116. Guida, V., Ferrari, G., Pataro, G., Chambery, A., Di Maro, A., & Parente, A. (2013). The effects of ohmic and conventional blanching on the nutritional, bioactive compounds and quality parameters of artichoke heads. *LWT-Food Science and Technology, 53*(2), 569-579.
117. Mercali, G. D., Jaeschke, D. P., Tessaro, I. C., & Marczak, L. D. F. (2013). Degradation kinetics of anthocyanins in acerola pulp: Comparison between ohmic and conventional heat treatment. *Food Chemistry, 136*(2), 853-857.
118. Mercali, G. D., Jaeschke, D. P., Tessaro, I. C., & Marczak, L. D. F. (2012). Study of vitamin C degradation in acerola pulp during ohmic and conventional heat treatment. *LWT-Food Science and Technology, 47*(1), 91-95
119. Moreno, J., Simpson, R., Baeza, A., Morales, J., Muñoz, C., Sastry, S., & Almonacid, S. (2012a). Effect of ohmic heating and vacuum impregnation on the osmodehydration kinetics and microstructure of strawberries (cv. Camarosa). *LWT-Food Science and Technology, 45*(2), 148-154.
120. Moreno, J., Simpson, R., Pizarro, N., Parada, K., Pinilla, N., Reyes, J. E., & Almonacid, S. (2012b). Effect of ohmic heating and vacuum impregnation on the quality and microbial stability of osmotically dehydrated strawberries (cv. Camarosa). *Journal of Food Engineering, 110*(2), 310-316.
121. Jakób, A., Bryjak, J., Wójtowicz, H., Illeová, V., Annus, J., & Polakovič, M. (2010). Inactivation kinetics of food enzymes during ohmic heating. *Food Chemistry, 123*(2), 369-376.
122. Moreno, J., Simpson, R., Pizarro, N., Pavez, C., Dorvil, F., Petzold, G., & Bugueño, G. (2013). Influence of ohmic heating/osmotic dehydration treatments on polyphenoloxidase inactivation, physical properties and microbial stability of apples (cv. Granny Smith). *Innovative Food Science & Emerging Technologies, 20*, 198-207.
123. Lin, Y. L., Li, S. J., Zhu, Y., Bingol, G., Pan, Z., & McHugh, T. H. (2009). Heat and mass transfer modeling of apple slices under simultaneous infrared dry blanching and dehydration process. *Drying Technology, 27*(10), 1051-1059.
124. Vishwanathan, K. H., Giwari, G. K., & Hebbar, H. U. (2013). Infrared assisted dry-blanching and hybrid drying of carrot. *Food and Bioproducts Processing, 91*(2), 89-94.
125. Wang, J., Yang, X. H., Mujumdar, A. S., Wang, D., Zhao, J. H., Fang, X. M., ... & Xiao, H. W. (2017b). Effects of various blanching methods on weight loss, enzymes inactivation, phytochemical contents, antioxidant capacity, ultrastructure and drying kinetics of red bell pepper (*Capsicum annuum* L.). *LWT, 77*, 337-347.

126. Guiamba, I. R., Svanberg, U., & Ahrné, L. (2015). Effect of infrared blanching on enzyme activity and retention of β-carotene and vitamin C in dried mango. *Journal of Food Science, 80*(6), E1235-E1242.
127. Sevenich, R., Bark, F., Crews, C., Anderson, W., Pye, C., Riddellova, K., ...& Knorr, D. (2013). Effect of high pressure thermal sterilization on the formation of food processing contaminants. *Innovative Food Science & Emerging Technologies, 20*, 42-50.
128. Sevenich, R., Bark, F., Kleinstueck, E., Crews, C., Pye, C., Hradecky, J., ...& Knorr, D. (2015). The impact of high pressure thermal sterilization on the microbiological stability and formation of food processing contaminants in selected fish systems and baby food puree at pilot scale. *Food Control, 50*, 539-547.
129. Zeinali, T., Jamshidi, A., Khanzadi, S., & Azizzadeh, M. (2015). The effect of short-time microwave exposures on *Listeria monocytogenes* inoculated onto chicken meat portions. In *Veterinary Research Forum* (Vol. 6, No. 2, p. 173). Faculty of Veterinary Medicine, Urmia University, Urmia, Iran.
130. Bauza-Kaszewska, J., Skowron, K., Paluszak, Z., Dobrzański, Z., & Śrutek, M. (2014). Effect of microwave radiation on microorganisms in fish meals. *Annals of Animal Science, 14*(3), 623-636.
131. De La Vega-Miranda, B., Santiesteban-Lopez, N. A., Lopez-Malo, A., & Sosa-Morales, M. E. (2012). Inactivation of *Salmonella Typhimurium* in fresh vegetables using water-assisted microwave heating. *Food Control, 26*(1), 19-22.
132. Lu, Y., Turley, A., Dong, X., & Wu, C. (2011). Reduction of *Salmonella enterica* on grape tomatoes using microwave heating. *International Journal of Food Microbiology, 145*(1), 349-352.
133. Jamshidi, A., Seifi, H. A., & Kooshan, M. (2010). The effect of short-time microwave exposures on *Escherichia coli* O157: H7 inoculated onto beef slices. *African Journal of Microbiology Research, 4*(22), 2371-2374.
134. Canumir, J. A., Celis, J. E., de Bruijn, J., & Vidal, L. V. (2002). Pasteurisation of apple juice by using microwaves. *LWT-Food Science and Technology*, 35(5), 389-392.
135. Gentry, T. S., & Roberts, J. S. (2005). Design and evaluation of a continuous flow microwave pasteurization system for apple cider. *LWT-Food Science and Technology*, 38(3), 227-238.
136. Mendes-Oliveira, G., Deering, A. J., San Martin-Gonzalez, M. F., & Campanella, O. H. (2020). Microwave pasteurization of apple juice: Modeling the inactivation of *Escherichia coli* O157: H7 and Salmonella Typhimurium at 80–90°C. *Food Microbiology*, 87, 103382.
137. Geveke, D. J., Brunkhorst, C., & Fan, X. (2007). Radio frequency electric fields processing of orange juice. *Innovative Food Science & Emerging Technologies*, 8(4), 549-554.
138. Eveke, D. G., & Brunkhorst, C. (2004). Inactivation of in apple juice by radio frequency electric fields. *Journal of Food Science*, 69(3), FEP134-FEP0138.

139. Li, R., Kou, X., Cheng, T., Zheng, A., & Wang, S. (2017). Verification of radio frequency pasteurization process for in-shell almonds. *Journal of Food Engineering*, 192, 103-110.
140. Wei, X., Lau, S. K., Stratton, J., Irmak, S., & Subbiah, J. (2019). Radiofrequency pasteurization process for inactivation of Salmonella spp. and *Enterococcus faecium* NRRL B-2354 on ground black pepper. *Food Microbiology*, 82, 388-397.
141. Hashemi, S. M. B., Gholamhosseinpour, A., & Niakousari, M. (2019). Application of microwave and ohmic heating for pasteurization of cantaloupe juice: microbial inactivation and chemical properties. *Journal of the Science of Food and Agriculture*, 99(9), 4276-4286.
142. Cho, W. I., Yi, J. Y., & Chung, M. S. (2016). Pasteurization of fermented red pepper paste by ohmic heating. *Innovative Food Science & Emerging Technologies*, 34, 180-186.
143. Achir, N., Dhuique-Mayer, C., Hadjal, T., Madani, K., Pain, J. P., & Dornier, M. (2016). Pasteurization of citrus juices with ohmic heating to preserve the carotenoid profile. *Innovative Food Science & Emerging Technologies*, 33, 397-404.
144. Barbosa-Cánovas, G. V., Medina-Meza, I., Candoğan, K., & Bermúdez-Aguirre, D. (2014). Advanced retorting, microwave assisted thermal sterilization (MATS), and pressure assisted thermal sterilization (PATS) to process meat products. *Meat Science*, *98*(3), 420-434.
145. Tang, J., Hong, Y. K., Inanoglu, S., & Liu, F. (2018). Microwave pasteurization for ready-to-eat meals. *Current Opinion in Food Science*, *23*, 133-141.
146. Ashwini, S. C., Garg, M. K., Kumar, S., Singh, V. K., Kumar, N., & Attkan, A. K. (2020). Ohmic heating effect on enzyme assisted aqueous extraction of rice bran oil. *Agric Res J*, 57 (6), 892-899. DOI No. 10.5958/2395-146X.2020.00130.1
147. Singh, V. K., Garg, M. K., Kalra, A., Bhardwaj, S., Kumar, R., Kumar, S., ... & Kumar, D. (2020). Efficacy of Microwave Heating Parameters on Physical Properties of Extracted Oil from Turmeric (Curcuma longa L.). *Current Journal of Applied Science and Technology*, 126-136.

4

Aseptic Processing

Malathi Nanjegowda[1], Bhaveshkumar Jani[2*] and Bansee Devani[3]

[1]Department of Plant Food and Environmental Science, Dalhousie University, Nova Scotia, Canada
[2]Department of Food Process Engineering, College of Food Technology, Sardarkrushinagar Dantiwada Agricultural University, Sardarkrushinagar, Banaskantha, Gujarat, India
[3]Department of Processing and Food Engineering, College of Agricultural Engineering and Technology, Junagadh Agricultural University (JAU), Junagadh, Gujarat, India

Abstract

Food, being one of the basic needs for the survival of humankind, has been practiced for centuries; various processing techniques such as drying, salting, fermentation, canning, etc., have been adopted. With the ever-increasing population and change in consumer preference and lifestyle, the innovation and necessity of preserving food with still better nutritional quality and sensory attributes has led us to the advanced processing and preservation methodology. Among various thermal processing methods of foods, aseptic processing takes the credit to preserve the food with minimum loss of nutrients along with incredible extended shelf life, undoubtedly without need of refrigeration as do the other methods. Aseptic processing is a high/ultra-temperature short time thermal process aiming to destroy the spoilage-forming microorganism for the extension of shelf life and optimum recovery of nutrients. The method has been adopted for almost all the types of foods including fruits, vegetables, dairy products, etc. The food is thoroughly maintained in aseptic condition assuring no entry to microorganisms starting from processing of food until intermediate storage and packaging as a final step. Controlled utility of heat for controlled time to process the food for attaining sterility and use of chemical such as hydrogen peroxide individual or in combination to heat, contributes to the success of the process and the product. Aseptic processing and packaging has revolutionized

**Corresponding author*: bhaveshjani@sdau.edu.in

Nitin Kumar, Anil Panghal and M. K. Garg (eds.) Thermal Food Engineering Operations, (117–140)

the preservation of food in various available sizes and shape, from retail and bulk packages attracting consumers in supermarket to global trade and also satisfying the needs of army and space personnel.

***Keywords*:** Aseptic processing, aseptic packaging, thermal treatment, UHT processing, preservation of food

4.1 Introduction

Food has been preserved since ancient times using various methods implying different principles of preservation (Table 4.1). The different preservation methods comprise the use of heat, without the use of heat, or a combination of the two. Drying, Dehydration, Evaporation, Thermisation, Pasteurization, Sterilization, Ultra-High Temperature (UHT), Vacreation, etc., are examples of thermal preservation methods, whilst Freezing, Irradiation, High-Pressure Processing (HPP), Pulsed Electric Field (PEF) Processing, Ultrasound Processing, etc., are low-temperature processes that preserve the food without or little heat. The process selection for the preservation of food largely depends on the composition and nature of food, probability and nature of spoilage microorganisms, and packaging and storage. The latter are the advanced food processing cum preservation methods yielding superior quality, and premium products, but costs are higher as compared to the conventional thermal processing and preservation methods. However, aseptic processing and packaging is a rapid-demand technology for all types of food items. In line with the pasteurization and sterilization processes, Ultra-High Temperature (UHT), often known as Aseptic processing has now covered almost all types of food for preservation. It has advantages over batch-type thermal processing as being a continuous sterilization process making it suitable for industrial applications. Aseptically processed and packed foods undoubtedly have a longer shelf life from months to years and have wide scope for being designed as space foods [1–3].

4.2 Aseptic Processing

Aseptic processing is the filling of a commercially sterilized-cooled product into pre-sterilized containers, followed by hermetic sealing with a pre-sterilized closure in a microorganism-free atmosphere [1]. The term "aseptic" is derived from the Greek word "*septicos*" which means the absence of putrefactive microorganisms. Aseptic processing yields

Table 4.1 History of aseptic processing and packaging.

Year	Process/product	Heating medium used	Packaging material used	Company
1931	Aseptic packaging	-	Metal cans	Nielsen, Denmark
1917	Method of sterilizing cans and lids with saturated steam	Saturated steam	Metal cans	US
1933	Development of aseptically filling machine named Heat-Cool-Fill (HCF) system	Saturated steam under pressure. Mixture of steam and air	Metal cans and ends	American Can Company
1940	Pre Sterilizing empty cans with 210 °C and filling with cold sterilized products	Superheated steam	Metal cans	W.M.Martin, US
1950	Commercial aseptic plant	Superheated steam	Metal cans	Dole Process [4]
1940-53	Ultra-high temperature (UHT) sterilized aseptically canned milk	Superheated steam	Metal cans	Alpura AG, Bern and Sulzer AG, Winterthur
1961	Aseptic packaging based on paperboard cartons	Superheated steam	Tetra Pak based on paperboard cartons	Alpura AG, Bern and Tetra Pak, Sweden

commercially sterile products using heat exchangers ensuring longer shelf life by packaging the food into pre-sterilized containers in a sterile/aseptic environment. It uses higher temperature for a shorter time for sterilization of food which is heat-labile, resulting in less possible damage to nutritional and functional components of the food and ensuring better quality products. The UHT or aseptically processed products have comparatively longer shelf life at ambient temperature than that of traditionally processed products.

Aseptic processing as a thermal process uses High-Temperature Short-Time (HTST) or Ultra-High Temperature (UHT) (Figure 4.1). During this process, very high temperatures are used for a very small amount of time to maintain the sensory and nutritional aspects of the food. HTST treatment uses a temperature of 75 °C for a time period of a few seconds to 6 minutes while the UHT process uses very high temperatures ranging from 135-150 °C for one or more seconds. The temperature employed in UHT processing depends on the viscosity of the products being processed. The heated sterilized product is cooled to a suitable temperature depending upon the viscosity of the product. Low viscous foods such as milk and fruit juices are cooled at 20 °C while high viscous products such as deserts are cooled at 40 °C before filling into a pre-sterilized container. Aseptic processing does not only demand commercial sterility of the product but also the product sterilization system (holding tube) and all equipment downstream from the holding tube including the filler; the packaging equipment; and the packaging material are also commercially sterile.

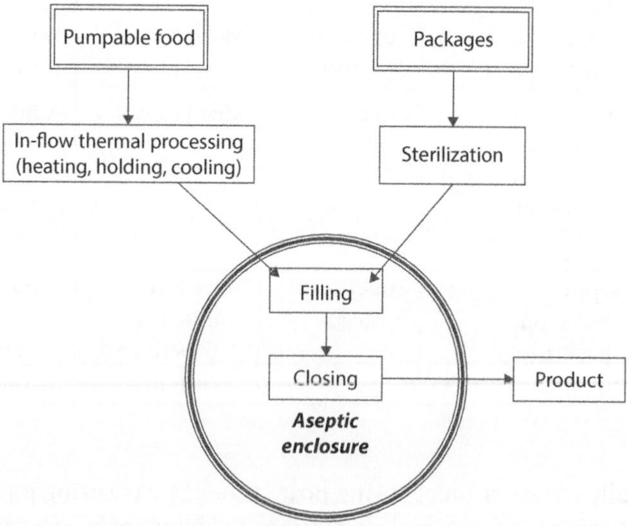

Figure 4.1 Aseptic processing and packaging system.

4.3 Principle of Thermal Sterilization

Sterilization is the application of heat to destroy the microorganisms and thus extend the shelf life of the food. The theory of thermal destruction of microorganisms needs to be understood to design the sterilization process and to use sterilization equipment. The rate of destruction is of the first-order reaction and follows the logarithmic order of death and is described by the Death Rate Curve (Figure 4.2). The rate of destruction of microorganisms as an effect of heat sterilization can be understood by different values such as D value, z value, F value, etc.

Decimal Reduction Time (D-value): The time needed to destroy 90% of the microorganisms (to reduce their numbers by a factor of 10) is referred to as the decimal reduction time or D value (Figure 4.3). It differs for different microorganisms and a higher D-value indicates greater heat resistance of the microorganism. The higher the number of microorganisms present in a product, the longer time it takes to reduce the numbers to a specific level. The destruction of microorganisms is temperature-dependent; cells die more rapidly at higher temperatures. By bringing together D values at different temperatures, a thermal death time (TDT) curve is constructed. The decimal reduction required in bacterial spores for different types of food is shown in Table 4.2.

***z* value:** The slope of the TDT curve is termed the z value and is defined as the number of degrees Celsius required to bring about a ten-fold change in decimal reduction time. The D value and z value are used to characterize

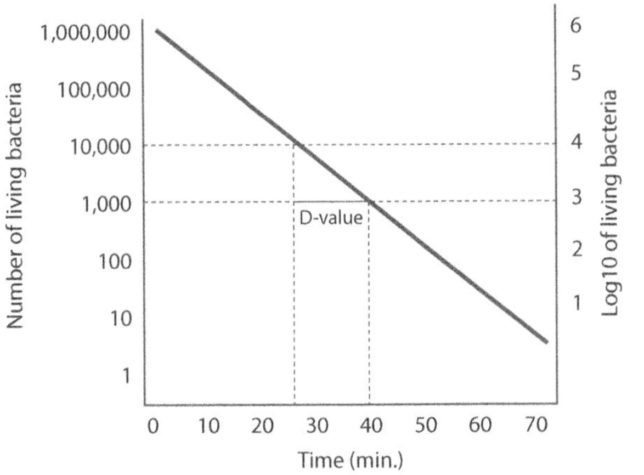

Figure 4.2 Death rate curve for microbial population [3].

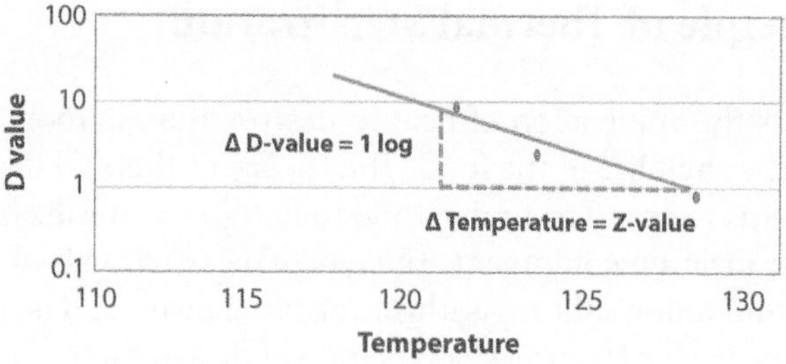

Figure 4.3 z values at different temperatures and D values [3].

Table 4.2 D value for acidic and non-acidic foods.

Types of food	pH	Decimal reduction required
Non-Sterile acidic food	< 4.5	Minimum 4
Sterile, neutral	> 4.5	6
Any probability of growth of *Clostridium botulinum*, a full 12 decimal reduction process is implied.		

the heat resistance of an enzyme, a microorganism, or a chemical component of a food. The common z value for the microorganisms is 10 °C and 33 °C for loss of desirable food nutrients.

F value: It is used as a basis for comparing heat sterilization procedures. F or F value is the time required to destroy a given percentage of microorganisms at a reference temperature and z value. It represents the total time-temperature combination received by food and is quoted with suffixes indicating the retort temperature and the z value of the target microorganism. For example, a process operating at 115 °C based on a microorganism with a z value of 10 °C would be expressed as F_{115}^{10}.

The reference temperature for F_0 value 121.1 °C. $F_{process}$ is the time-temperature effect of a thermal process expressed as time in minutes at reference temperature T_{ref} for microorganisms with a given z value; it is defined by the equation [5]:

$$(F_{Tref}^{z})_{process} = F_{process} = \int_{ti}^{tf} dt/10^{(Tref-T(t))/z}$$

Where, *i* - Initial condition.

F - Final condition.

Ni - Initial concentration of microorganisms at time t_i.

Nf - Final concentration of microorganisms at time t_f.

ti - Time at beginning of the thermal process (minutes).

tf - Time at end of the thermal process, that is, at end of cooling (minutes).

Tref - Reference temperature (°C); common reference temperatures are 121 °C for low-acid foods, 100 °C and 93 °C for high-acid and acidified foods, and 62.8 °C and 71.7 °C for pasteurization of milk.

$T_{(t)}$ Product temperature at the slowest heating point as a function of process time, in °C.

z Temperature difference that causes a tenfold change in the rate of microbial destruction, in °C.

4.3.1 Effect of Thermal Treatment on Enzymes

The thermal process aims to destroy the microbes and/or to inactivate enzymes associated with the food to increase the shelf life of the food. HTST process is enough to destroy many microorganisms but enzymes like peroxidase, bacterial proteases, and psychrotrophic bacteria-induced lipase show high heat resistance and are found not deactivated completely; as a result enzymatic deterioration during storage of processed product increases as the heat t treatment temperature increases. These bacterial enzymes have very high heat resistance (about 4,000 times more than that of most heat resistant spores of *Geobacillus stearothermophilus*), the z value of 20 °C to 60 °C. Such poorly deactivated enzymes can affect the food during storage, i.e., the protease in UHT milk leads to the development of bitter flavors and age gelation.

4.3.2 Effect of Thermal Treatments on Nutrients and Quality

D and z values for nutrients and quality-factor destruction are larger than for microorganisms or heat-labile enzymes ($D_{121.1}$ of up to 150 minutes against 0.1 to 5 minutes; z value of 20 °C to 37.7 °C against 7 °C to 10 °C). D value and z values for the vitamins, pigments, and other food constituents are quite higher than that of the microorganisms and enzymes. This situation forms the basis for high-temperature short-time (HTST) and ultra-high temperature (UHT) processing (132 °C to 150 °C for a few seconds). By 121.1 to 132 °C increase in the temperature, the rate of destruction of microorganisms would be at least tenfold greater, while that of nutrients and quality factors would be only two- to threefold and that is too dependent on exposure time [5].

4.3.3 Effect of Thermal Treatments on the Cooking Index (C_0)

The overall sensory quality deterioration taking place due to thermal processing can be described by the Cooking index. The reference temperature for C_0 index is 100°C. The C_0 value associated with the conventional canning processes is quite higher achieved at lower temperatures long time with the F_0 value of 10 than that of the achieved at high temperature. This implies that with the same level of microbial destruction in LTHT and HTST processes, the HTST process shows comparatively less quality degradation and thus a higher preference for processed food by consumers.

4.3.4 Effect of Heat Treatments on Chemical Reactions in Food

The factor indicating the speed of reactions in food at higher temperatures is considered in this context. Q_{10}-value has defined as the number of times a reaction rate changes with a 10°C change in temperature. If a reaction rate doubles with a 10°C change in temperature, the $Q_{10} = 2$. The value for chemical reactions of food constituents is usually between 2 and 4, the equivalent value for killing bacteria is 10. If the temperature from the usual retorting value of 121°C is raised by 10°C, the time for the same killing rate for bacteria will be 10%, and with a 20°C rise only 1%. If for chemical reactions a Q_{10} of 3 is assumed, they will be speeded up by only a factor of 3 x 3 = 9 when raising the temperature by 20 °C. As a result, a temperature rises from 121 °C to 141 °C for sterilizing foods will theoretically give only about 9% of the damage to quality, compared with retorting at 121 °C.

4.4 Components of Aseptic Processing

Aseptic processing is a continuous processing technique. The primary components in aseptic processing systems include product formulation tank, flow control mechanism, heat exchanger, holding tube sterilizer, continuous heat exchangers, steam-sealed, air-operated valves, PLC-based automation optional aseptic surge tank, and Cleaning in-Place facilities. From the listed components mainly the equipment used in aseptic processing and the sterilization methods of packaging materials are described in detail hereunder.

4.4.1 Equipment Used in Aseptic/UHT Processing

The heating/cooling of products is done with the help of heat exchangers where the heat transfers take place. Indirect heat exchangers such as plates, tubular, and scraped surfaces (Figure 4.4) and direct heating mechanisms such as direct steam injection or infusion plants methods are used in UHT processing to heat the product at different rates and shear conditions. Also, Ohmic heating and Microwave heating [5] are amongst the novel heating technologies being implemented at a small scale in the case of Aseptic processing. There is also a scope of the study for non-thermal treatments like Pulsed Electric Field [6], Irradiation, Cold-plasma, High-Pressure, Ultra-sonication, etc., along with little added heat depending on the material and processing conditions. The selection of heat exchanger and heating method depends on the nature of the products. For viscous products tubular heat exchangers and more suitably scraped surface exchangers are used where a scraper blade continually scrapes the product and sends it away from the heat transfer surface. The following types of heat exchange equipment are used for commercial sterilization of foods and the heat transfer rate and heating and cooling time are greatly affected by the selection of the equipment. Aseptic processing equipment sterilization procedures often use steam or hot water under pressure.

4.4.1.1 *Indirect Heat Exchanger*

In this method of heating, the product and heat exchanging medium is not in contact and is separated by the heating surface.

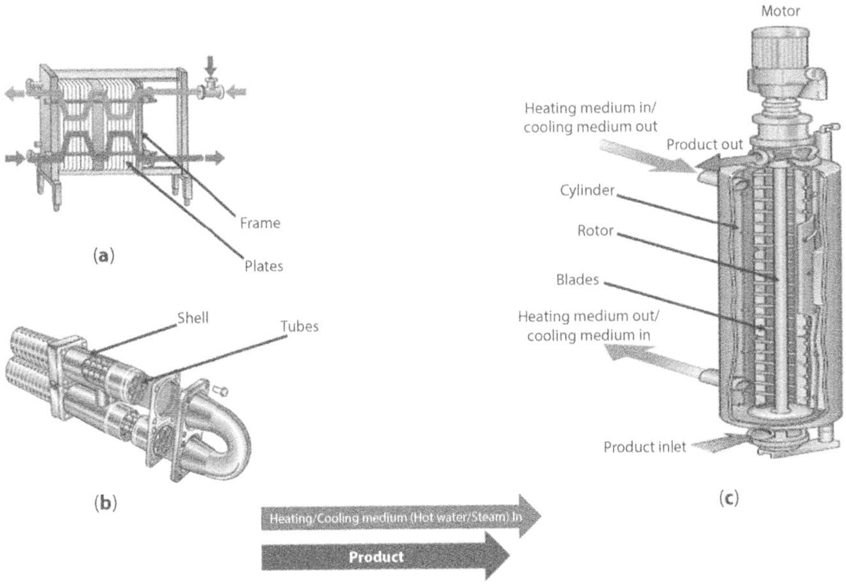

Figure 4.4 (a) Plate heat exchanger, (b) shell and tube heat exchanger, (c) scrapped surface heat exchanger [15].

4.4.1.2 Direct Heat Exchanger

In this method, the heating medium, for example, steam, is injected or infused in the product using the following methods.

Steam injection: The steam is injected into the product to heat it (Figure 4.5a).

Steam infusion: The product is infused with steam by introducing the product in a steam-filled vessel (Figures 4.5b-1 and 4.5b-2).

4.4.1.3 Ohmic Heating (OH)

Ohmic heating or Joule heating or electric resistance heating is an electrical heating technology in which heat is internally generated within the material being processed due to its natural electrical resistance. Heat is produced directly within the fluid as an electric current passes through it, rather than being transferred into the fluid from a hot surface. This has several benefits in comparison to conventional heat exchangers for "difficult" fluids such as slurries, highly viscous materials, liquids containing large or delicate particles, and materials susceptible to damage from hot surfaces. The heating rate is mainly proportional to the electric field strength, electrical conductivity, and temperature of the food being heated. In OH, the electrodes are in contact with food, which acts as a resistance; hence the heating occurs in the form of internal energy transformation from electric to thermal in a very short time. OH is a high-temperature, short-time

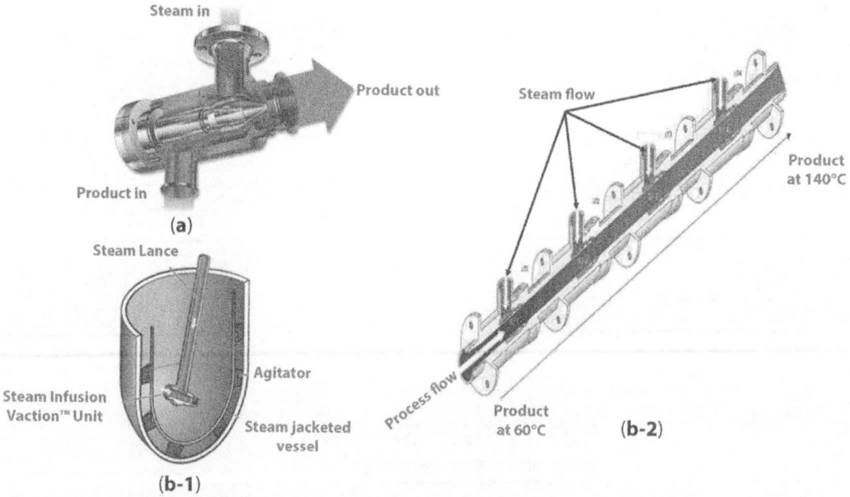

Figure 4.5 (a) Steam Injection system, (b-1) Steam Infusion Vaction™ PumpIn tank (b-2) Steam Infusion Vaction™ Pump inline [15].

method and this fact allows for uniform heating, increasing the final product's quality, especially for thermal-sensitive products like vitamins. The principal advantage claimed for OH is its ability to heat materials rapidly and uniformly, including products containing particulates. The system operates under continuous flow conditions with the product passing over electrodes in one or more heating tubes, followed by product cooling in scraped surface, tube in shell, or plate heat exchangers. The conductivity and electrical resistance of the product influence the heating rate. Because of this, product formulation becomes critical to the process. Food products, which are not good conductors of electrical current, are not better suitable for Ohmic heating.

4.5 Aseptic Packaging

The filling process of UHT/aseptically processed products is quite different from the other thermally processed products. Aseptic packaging is the process of filling the commercially sterile products into sterile containers under sterile conditions to preclude reinfection by hermetically sealing them. While aseptic filling aseptic environments (sterile conditions) must be maintained to exclude any types of undesirable microorganisms and fully airtight (hermetic) sealing is preferred for avoiding any kind of probability of entering microorganisms into packages and their surroundings. The commercial sterile products indicate that the microorganisms capable of reproducing in the food during storage, distribution under normal temperature are not present. Milk and dairy products, desserts, fruits, and vegetable juices soups, sources, etc., can be sterilized and packed aseptically while fresh products such as yogurt which is not pre-sterilized can also be aseptically packed. There are two specific applications of aseptic packaging, i.e., pre-sterilized and sterile products packaging, for example, milk and dairy products, puddings, desserts, fruit and vegetable juices, soups, sauces, and products with particulates; and non-sterile products packaging to avoid infection by microorganisms, i.e., fermented dairy products like yogurt [7, 8].

4.5.1 Types of Packaging Materials Used in Aseptic Processing

The types of packaging materials in various size and shapes have been commercially used for packaging of the aseptically processed food as shown in Table 4.3.

Table 4.3 Methods for the sterilization/decontamination of the packaging materials.

Structure of packaging	Types of materials
Paperboard laminated cartoon FFS carton Prefabricated carton	Polyethylene,bleached paperboard, lamination layer of polyethylene, the adhesive layer of polyethylene (ethylene-co-methacrylic acid), and inner polyethylene layer along with a thin aluminum foil.
Cans	Metal cons
Bottles	Aseptic PET bottles
Sachet and pouch	laminate of LLDPE with a center layer of EVOH copolymer and carbon black
Bag-in-box	EVOH and metalized PET (mPET), paperboard
Cups	HIPS, PP, or coextruded, multilayered polymers,

4.5.2 Methods and Requirements of Decontamination of Packaging Materials

The packaging materials used in aseptically processed foods play an important role in maintaining the shelf life of the foods besides their basic functions of protection, communication, etc. Care must be taken to avoid contamination of the sterile food through the packaging materials and during the filling operation. The following methods are adopted for the decontamination of the packaging materials as shown in Table 4.4.

4.6 Applications of Aseptic Processing and Packaging

Aseptic packaging brings together pre-sterilized products and sterilized packaging materials in sterile conditions. Several food products nowadays have been aseptically processed and packed. The limitless list of products includes milk, condensed milk, edible oils, including butter and margarine based on rapeseed, tomato products, scrambled eggs, tofu, soups and gravies, jelly, chocolate and coffee puddings, fruit juices, fruit salads, etc.

Table 4.4 Methods for sterilization/decontamination of packaging materials.

Method	Medium	Dose	Types of packaging materials	Reference
Irradiation				
Ionizing radiation	Cobalt-60, Cesium-137, Electron beam (80-150 keV)	25 kGy or more	Preforms, bottles, caps	[9]
Pulsed light	White light (200-1000 nm)	For 1 μs to 0.1 s, 1-20 flashes per second	-	[10]
UV-C radiation with hydrogen peroxide	Mercury vapor lamp, UV laser (Excimer) using gas halides such as krypton fluoride	248- 253.7 nm	-	[11]
Plasma	A low-pressure microwave plasma reactor	5 s	PET bottles	[12]

(Continued)

Table 4.4 Methods for sterilization/decontamination of packaging materials. (*Continued*)

Method	Medium	Dose	Types of packaging materials	Reference
Heat				
Saturated steam	Saturated steam under pressure and steam	1. Steam at 165 °C, 600 kPa for 1.4 s 2. Steam at 165 °C, 600 kPa for 1.8 s	Deep drawn plastic containers, Molded PS cups, Foil lids, etc.	[13]
Superheated steam	Saturated steam at normal pressure	220-226 °C for 36-45 s	Tinplate and aluminum cans and lids	[4]
Hot air (suitable only for acidic products)	Hot air	315 °C (surface temperature reaches 145 °C for 180 s)	Paperboard laminated cartons	-
Hot air and steam	Hot air and steam	Hot air is blown through a nozzle	The inner surface of PP cups and lids which is thermally stable up to 160 °C	-

(*Continued*)

Table 4.4 Methods for sterilization/decontamination of packaging materials. (*Continued*)

Method	Medium	Dose	Types of packaging materials	Reference
Extrusion for: ·Pre-sterilization Post sterilization with hydrogen peroxide or peracetic acid (PAA)	Heating in extrusion	180 - 230 °C for up to 30 min	Plastic containers used for acidic products Plastic containers used for	-
Chemical treatments				
Hydrogen peroxide	Hydrogen peroxide	Concentration not greater than 500 ppb at the time of filling and must fall to ≅ 1 ppb within 24 h	Paperboard laminates	-
Hydrogen peroxide by a dipping process	Hydrogen peroxide	30-33% H_2O_2 solution	Hydrophobic plastic from a reel	-

(*Continued*)

Table 4.4 Methods for sterilization/decontamination of packaging materials. (*Continued*)

Method	Medium	Dose	Types of packaging materials	Reference
Hydrogen peroxide by a spraying process	Liquid Hydrogen peroxide A mixture of hot air (130 °C) and vaporized peroxide	Spraying H_2O_2 through nozzle and drying by hot air	Pre-fabricated packages	-
Hydrogen peroxide by rinsing process (when spraying is not suitable)	Hydrogen peroxide or a mixture of Hydrogen peroxide and Peracetic acid (PAA)	Rinsing with Hydrogen peroxide or mixture and drying with hot air	Glass containers, metal cans and blow-molded plastic bottles	-
Hydrogen peroxide combined with UV irradiation and heat	Hydrogen peroxide, UV irradiation, heat	Low peroxide concentration, 0.5-5%	-	-

4.6.1 Milk Processing

Aseptic processes of dairy products, especially milk, are successfully accepted in the market. Many dairy products such as concentrated milk, dairy creams, flavored milk drinks, whey-based drinks, fresh and recombined liquid milk, fermented milk products (yogurt, buttermilk, etc.), ice-cream mix and desserts (custards and puddings) are aseptically processed and packed (Table 4.5). Intensive and brief heating is applied to milk which destroys the microorganisms. The milk is sterilized separately and packaged separately under aseptic conditions so it is important to prevent reinfection during packaging [15].

Commercially, milk is aseptically processed in a complete UHT plant or Aseptic processing plant (2000-3000 l/h) having plant pre-sterilization, production, aseptic intermediate cleaning, and cleaning in place (CIP) facilities. Any kind of risk of supplying an unsterilized product to the aseptic filling machine is strictly eliminated in the UHT plant. The heating mechanism implied in UHT plants is as described in section 4.5.1. The plant is pre-sterilized before starting and entering the product to avoid reinfection of the treated product. Hot water sterilization for 30 minutes is done at the same temperature as the product shall undergo followed by cooling [15, 16].

After pre-sterilization, the milk at about 4 °C is entered from the balance tank and forwarded to the feed pump for preheating using a plate heat exchanger where it reaches around 80 °C. The hot milk is then pressurized to about 4 bar and then the steam is injected into the milk by a ring nozzle injector and the temperature of milk rises to about 140 °C for a few seconds and cooled under vacuum in the expansion chamber. After this, the milk is homogenized and cooled to around 20 °C using a plate-heat exchanger and goes to an aseptic filling machine. Similarly, the milk can be heated by direct heating systems (steam injection and tubular heat exchanger, steam infusion) or indirect heating (plant-based on plate heat exchanger, tubular heat exchanger, or scraped surface heat exchanger) [15].

If the plant used to run for very long, the intermediate cleaning in place for 30 minutes is carried out. It is also done to remove fouling in the production line without losing aseptic conditions. The full CIP cycle takes about 70-90 minutes and is done after production immediately. The CIP cycle for direct or indirect UHT plants comprises sequential processes of pre-rinsing, caustic cleaning, hot-water rinsing, acid cleaning, and final rinsing and done automatically. The UHT processing of milk is shown in Figure 4.6.

Table 4.5 Types of food processed and packed under UHT/Aseptic Condition [14].

Types of product	Temperature-time combination
Milk	135 °C for 1 s
Cream	140 °C for 2 s
Milk-Based products	140 °C for 2 s
Ice cream mix	148.9 °C for 2 s
Chocolate milk	149 °C for 15 s
Pea soup	140 to 150°C for 8.8 s

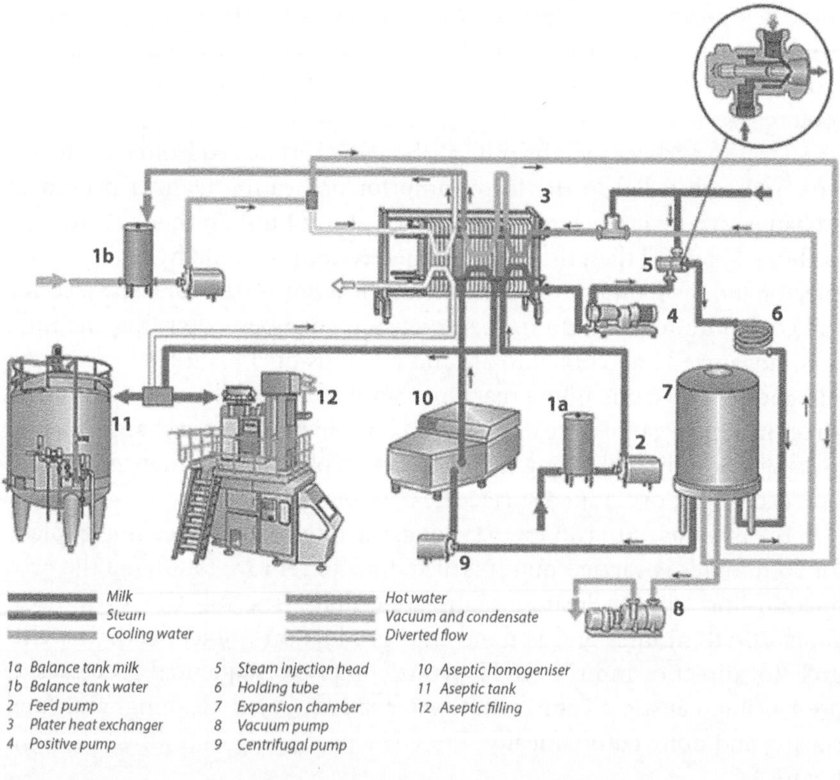

Figure 4.6 UHT process with heating by direct steam injection combined with plate heat exchanger. (Courtesy: Dairy Processing Handbook, Tetra Pak).

4.6.2 Non-Milk Products Processing

Non-milk liquid products, especially fruit juices, and beverages are too aseptically processed and packed. The quality of juices like color, flavor, and texture get least affected while thermal destruction of microorganisms can be achieved such that the products can be stored for a long period at room temperature.

Cold filtration through membranes that have a small enough pore size to exclude microorganisms (<0.45μm) can be used to sterilize clear juices and drinks, such as apple juice, that do not contain particulate matter. Juices that contain pulp or other suspended matter must be prefiltered before sterile filtration. This method of sterilization produces the least chemical alteration of the juice, but it is expensive for large volumes and does not inactivate enzymes. Cold filtration is particularly suitable for sterilizing small amounts of flavor extracts or essences. These sterile flavor essences can be added to aseptically heat-processed juices, after cooling, to enhance the flavor and aroma or to replace any flavors lost during processing.

UHT-treated products have to be packaged under conditions that prevent microbiological contamination, i.e., aseptically packaged. With some high-acid foods ($pH < 4.5$), it may be sufficient to cool the product after UHT treatment to just below 100°C, fill it into a clean container, seal the container and hold it at that temperature for some minutes before cooling it. This procedure will inactivate microorganisms that may have been in the container or entered during the filling operation and which might grow in the product. The filled container may need to be inverted for some or all of the holding period. However, in the case of low-acid foods ($pH > 4.5$), this procedure would not be adequate to ensure the sterility of the product. Consequently, for such products, the aseptic filling must involve sterilizing the empty container or the material from which the container is made, filling it with the UHT-treated product, and sealing it without it being contaminated with microorganisms [17, 18].

Pasteurization of fresh citrus juices requires inactivation of pectinesterase to prevent cloud loss in juices or gelation of concentrates. The thermal destruction of pectinesterase in Hamlin, Pineapple, and Valencia oranges, and Duncan grape fruit was studied and it was reported that a two-log cycle reduction in pectinesterase activity was necessary to achieve commercial stability of citrus juices and concentrates [19] while one minute at 90 °C and 1 min at 85.6 °C were reported necessary to achieve a two-log cycle reduction in pectinesterase activity in oranges and grapefruit, respectively [17, 18].

The results of this technology come in all shapes and sizes, from the consumer packages of milk on the shelves of the supermarket to the huge containers full of orange juice transported around the world by cargo ships. Over the last couple of decades, aseptic bulk storage and distribution has revolutionized the global food trade. For example, more than 90% of the approximately 24 million tons of fresh tomatoes harvested globally each year are aseptically processed and packaged for year-round remanufacture into various food products. A similar trend is growing in the case of mango pulp processing to replace canning and storing the pulp in large metal containers [17, 18].

4.7 Advantages of Aseptic Processing and Packaging

1) Packaging materials, which are unsuitable for in-can sterilization, can be used, such as large containers such as drums, tanks, and tankers, which are filled through the continuous flow process.
2) Therefore, lightweight materials consuming less space offering convenient features and with the low cost such as paper and flexible and semi-rigid plastic materials can be used gainfully.
3) The sterilization process of high-temperature-short time (HTST) for aseptic packaging is thermally efficient and generally gives rise to products of high quality and nutritive value compared to those processed at lower temperatures for a longer time.
4) Extension of shelf-life of products at normal temperatures by packing them aseptically.
5) Sensory point of view, the product remains superior in all aspects.
6) Suitable for homogeneous heat-sensitive and nutritional products.
7) Comparatively minimum nutrient loss.
8) Low labor requirements.
9) Packaging to product ratio is much less, such that 90% of the product and 10% of packaging material is present.

4.8 Challenges of Aseptic Processing and Packaging

1) High expenditure on equipment.
2) Need for trained people.
3) The specific demands made on packaging and filling operations.
4) Keeping all of the components free of contamination.
5) Product, processing, and filling lines, surge tanks, packaging equipment, and materials.
6) Aseptic processing systems have to be sterilized before processing, generally using hot water or saturated steam.
7) Packaging materials are sterilized separately using peroxides, or other chemicals together with sterile air.
8) The process can be complicated and time-consuming and requires good engineering of the lines.

4.9 Conclusion

Aseptic processing technique has been successfully applied to liquid foods (milk and fruit juices) and acid foods. The technology has already been in use for years and still needs development and cost-effectiveness in the processing and packaging. The sterility of the products due to UHT and other heat-inducing techniques has been maintained well but chances of over-cooking smell or taste, exact freshness, etc., are the niche area to be studied. The use of non-thermal liquid processing techniques like Pulsed Electric Field (PEF) processing, High-Pressure Processing, Irradiation, Cold-plasma, and others will revolutionize the aseptic processing industries. Already, HPP has been being studied and utilized at a small-scale for liquid foods decontamination. PEF has also been adopted for decontamination of solid and semi-solid foods along with improved texture and other characteristics as added advantages; the same can be researched and optimized for milk, juices, and other liquid foods with or without introducing thermal treatments. The non-destructive testing of aseptically processed and packaged goods and intelligent aseptic packaging has a scope of exploration by in-depth learning the minute characteristics of pre- and post-processed products and packages. There is an abundant scope of involving bio-degradable layers within the laminate, and recyclable packaging materials along with the utilization of bio-wastes for selective barrier properties. The environment-friendly applications of aseptic processing are the need of the hour.

References

1. *Canned Food: Principles of Thermal Process Control, Acidification and Container Closure Evaluation*, 7th ed. 2007. Washington, DC: GMA Science & Education Foundation.
2. Lewis, M., & Heppell, N. 2000. *Continuous thermal processing of foods.* Gaithersburg: Aspen Publishers. 447 pp.
3. Fellows, P., ed., *Food Processing Technology; Principles & Practice*, Ellis Horwood, England, 221, 1988.
4. White, F. S. 1993. *The Dole Process, in Aseptic Processing and Packaging of Particulate Foods*, ed. E.M.A. Willhoft, Blackie Academic and Professional, London, pp 148–154.
5. David, J. R. D., Graves, R. H. and Szemplenski, T. 2013. *Handbook of Aseptic Processing and Packaging.* CRC Press.
6. Toepfl, S., Heinz, V. and Knorr, D. 2006. Applications of Pulsed Electric Fields Technology for the Food Industry. *Pulsed Electric Fields Technology for the Food Industry - Fundamentals and Applications.* 7: 212-217.
7. Hsu, C-L and Chang, K-S. 2006. Evaluation of the integrity of aseptic packages containing various filling products. *International Journal of Food Science and Technology.* 41: 1061–1066.
8. Moruzzi, G., Garthwright, W.E. and Floros, J.D. 2000. Aseptic packaging machine pre-sterilisation and package sterilisation: Statistical aspects of microbiological validation. *Food Control.* 11: 57–66.
9. Haji-Saeid, M., Sampa, M.H.O., Chmielewski, A.G. 2007. Radiation treatment for sterilization of packaging materials. Radiation Physics and Chemistry 76: 1535–1541.
10. Oms-Oliu, G., Martín-Belloso, O., Soliva-Fortuny, R. 2010. Pulsed light treatments for food preservation. A review. *Food and Bioprocess Technology* 3: 13–23.
11. Warriner, K., Movahedi, S., Waites, W.M. 2004. Laser-based packaging sterilization in aseptic processing. In: *Improving the Thermal Processing of Foods*, Richardson P.S. (Ed.). Boca Raton, FL: CRC Press, pp. 277–304.
12. Deilmann, M., Halfmann, H., Bibinov, N., Wunderlich, J. and Awakowicz, P. 2008. Low-pressure microwave plasma sterilization of polyethylene terephthalate bottles. *Journal of Food Protection.* 71: 2119–2123.
13. Robertson, G. L. 2013. *Food Packaging Principles and Practice.* CRC Press. pp. 381-382.

14. Datta, N. and Deeth, H.C. 2007. *Advances in Thermal and Non-Thermal Food Preservation.* Ames, Iowa. Blackwell Publishing. pp. 63-90.
15. Bylund, G. 1995. *Handbook of Dairy Processing.* Tetra Pak Processing Systems. AB S-221 86 Lund, Sweden.
16. *Handbook of Cleaning in Place: A guide to cleaning technology in the food processing industry,* Tetrapak. pp. 1-40.
17. Graumlich, T. R., Marcy, J. E. and Adams, J. P. 1986. Aseptically Packaged Orange Juice and Concentrate: A Review of the Influence of Processing and Packaging Conditions on Quality. *J. Agric. Food Chem.* 34: 402-405.
18. Hotchkiss, J. H. 1989. Aseptic Processing and Packaging of Apple Juice. *Processed Apple Products.* pp. 189-213.
19. Eagerman, B. A. and Rouse, A. H. 1976. *J. Food Sci.* 41: 1396-1397.

5

Spray Drying: Principles and Applications

Sukirti Joshi[1], Asutosh Mohapatra[1*], Lavika Singh[2] and Jatindra K Sahu[1]

[1]Food Customization Research Laboratory, Centre for Rural Development and Technology, Indian Institute of Technology Delhi, Hauz Khas, New Delhi, India
[2]Bureau of Indian Standards, New Ashok Nagar, New Delhi, India

Abstract
The process technology of spray drying is frequently employed in the food industry for various applications. The advantages offered by this technology, such as sustainable nature and low-cost economics, have increased its utilization as compared to other techniques of drying. In general, spray drying involves the conversion of liquid feed solution into fine droplets on contact with a directed flow of dry and hot air inside a chamber. This chapter summarizes the key features of state of the art, the effect of core and wall materials on drying efficacy, processing conditions affecting the spray drying kinetics and the application of spray drying as a promising method of encapsulation to formulate matrices with improved physicochemical, structural, and technological attributes. This chapter also addresses the encapsulation of different bio-actives by spray drying, highlighting the strategies to enhance the encapsulation efficiency with better retention of the core. The brief review of the spray drying process, the parameters determining the process efficacy and the suitability of the final product would contribute to further improvements in the overall technology.

Keywords: Spray drying, processing conditions, core characteristics, wall materials, encapsulation

**Corresponding author*: asutoshmohapatra3@gmail.com

Nitin Kumar, Anil Panghal and M. K. Garg (eds.) Thermal Food Engineering Operations, (141–178)

5.1 Introduction

The history of spray drying dates back to the mid-19th century with flourishing things still to explore in terms of operational parameters and various perpetual applications in food industries [1]. Industrial application of the spray drying process began during World War II, for the production of milk powders in dairy industries, and has gradually expanded its horizon to other genres of powder production by modifying the processing conditions. According to literature, spray drying is defined as the transformation of feed from a fluid state into a dried particulate form by spraying the feed into a hot drying medium [2]. The feed can be in the form of solution, dispersion, emulsion or suspension. Depending on the physicochemical characterization of feed and the design consideration of dryer, the dried product can be either powders, granules or agglomerates [3]. The principle behind the spray drying mechanism involves fine spraying of liquid feed into a stream of hot air and drying of the atomized droplets by rapid evaporation of moisture because of constant and falling rate diffusion of water molecules from the feed droplets. The mode of heat and mass transfer during the drying operates on the convection process by maintaining the relative humidity of the drying chamber. Intensive evaporative cooling causes a rapid decrease in the hot air temperature and maintains the quality of thermally sensible products.

Therefore, proper knowledge of different processing steps involved in spray drying is crucial to have a clear picture and understanding about the concept, technology and the hardware operations, which are illustrated in this chapter. The mechanism behind the production of fine powder which causes the transformation in the feed solution could be explained by segmenting the drying process into various unit operations, as given below:

The complete process of spray drying of food products can be generally divided into five important categories, such as

- Concentration of the feed solution
- Atomization of the feed
- Droplet-hot air contact
- Drying of droplets
- Particle separation

5.2 Concentration of Feed Solution

The feed solution usually is concentrated either by manually using a glass rod or by magnetic stirrer prior to introduction into the spray dryer.

The total solid percentages of the emulsion need to be fixed based on literature study or trial basis to make the feed possible to enter into the atomizer for spraying without any blockage in the nozzle.

5.3 Atomization of Concentrated Feed

Being the heart of the spray drying process, atomization causes minute division of the liquid feed while passing through the spray dryer. The internal resistance to moisture removal from droplet to the adjacent medium gradually reduces due to breakage of bulk liquid into tiny droplets. The gradual increase in the intensity of atomization causes high droplet fission and a more considerable increase in surface area of droplets. As a result, instability of feed droplets occurs, leading to immediate moisture removal.

The process of atomization very much influences the morphology of particles and their desired physicochemical characteristics in spray drying. According to the reported literature, from a cubic meter of liquid, approximately 2×10^{12} particles are formed having a droplet size of 100 μm and a total surface area of 60,000 m^2 [4]. The rate of drying depends on the ratio of surface-to-volume and directly proportional to each other. Therefore, the shape, structure and size distribution of particles is significantly influenced by the process of atomization and type of atomizer used.

5.3.1 Principle of Atomization

Several researchers have explained the governing principle of liquid disintegration to understand the atomization process. Among these, the 'Liquid jet theory' was proven to be the most accurate technique to explain this phenomenon. The jet falls vertically under gravity, and when its length reaches the critical value, it loses its cylindrical shape by decomposing into a stream of droplets, mostly spherical, due to a gradual decrease in surface tension [5, 6]. It is also called the Plateau-Rayleigh instability, which describes the breaking up of the fluid stream and its conversion into smaller droplets having equal volume but less surface area.

5.3.2 Classification of Atomizers

An atomizer is basically used in all types of spray dryers to convert the feed solution into droplets of a predetermined size. So, the selection of atomizer plays an important role in deciding the efficiency of a spray dryer from the virtue of its techno-economic functionality. The atomizers generally use hot air or nitrogen gas for the process of atomization depending on the type of feed solution. It is further classified into various types based on the factors such as the requirement of atomization energy; desired spray; the size of the final droplets; physical properties of feed and its capacity of intake [7]. Different kinds of atomizers used in the pilot and industrial-scale are rotary, single-fluid high pressure, two-fluid and multiple nozzles. Ultrasonic and electrohydrodynamic nozzles are the new advancements in nozzle variety, now used in industries based on the volume specificity. Characterization of atomizers can be done based on various factors like the volume of liquid for atomization, size of the droplet, desired pattern of size distribution, physiochemical properties of the fluid and the design parameters of the drying chamber.

5.3.2.1 *Rotary Atomizers*

The governing principle of atomization in rotary atomizers is based on the high-velocity centrifugal acceleration of feed solution to the centre of rotating wheel at a velocity of 200 m/s. Rotary atomizers are usually used due to the free-flowing nature of feed through the nozzle without any clogging and production of uniform size droplets ranging from 30-120 μm; but they are difficult to use for the feed with high viscosity. In rotary atomizers, the mean size of droplets has a direct relationship with the feed rate and viscosity while inversely proportional to the diameter and velocity of the rotating wheel. The liquid feed is carried out to the centre of the disc and spread over the disc surface as a layer of thin-film with varying thickness depending on the peripheral speed, viscosity of feed and its flow rate. The atomization of the droplets is carried out at a speed lower than the peripheral speed of the rotating wheel.

5.3.2.2 *Pressure Nozzle/Hydraulic Atomizer*

Governing principle behind the pressure atomizer is based on the conversion of pressure energy into kinetic energy for the disintegration of the feed solution through nozzle orifice with a high-speed film lining. The pressure chamber is designed like a swilling pattern for the droplets to pass through and fall like a hollow cone structure. The feed flow rate and size of the orifice are the deciding factors for the mean particle size of the droplets, which varies between 120 to 250 μm. The diameter of a particle is directly proportional to the viscosity and flow rate of the feed while inversely proportional to the atomization pressure. In general, the pressure varies from 250-10,000 psi depending on the types of feed to be atomized. To increase the amount of flow rate and to have a broad particle size distribution, the pressure nozzles can also be integrated into multiple arrangements of nozzles; however, it produces particles with coarser size. Another advantage of the pressure nozzle is related to the production of dense powders with better flow behaviour due to the presence of particles with less occluded air in the intermolecular and intramolecular void spaces.

5.3.2.3 *Two-Fluid Nozzle Atomizer*

Two-fluid atomizers consist of an orifice and an air cap to allow the interaction of the liquid feed with the atomizing gas/Nitrogen stream. The governing principle involves the disintegration of the liquid feed by high-velocity air, shortening the optimal wavelength of the formed drop, and hence resulting in smaller droplet size [8]. Two-fluid nozzle atomizers are generally used in lower volume applications or if the viscosity of feed solution is very high with abrasive solid content. Particles produced using two-fluid nozzle atomizers are very fine in size distribution with a diameter ranging from 30-150 μm and even the average particle size could go down below one-micron under high pressure with lower solid concentration. The liquid is usually pumped into the atomizer at a pressure of 250-10,000 psi which comes in contact with the compressed hot air when sprayed. As compared to the rotary atomizers, two-fluid nozzle atomizers usually carry lower solid content in the feed solution.

However, the twin-fluid nozzle atomizer has many limitations in terms of its applications. The fixed cost is a bit higher due to the requirement of the compressed air. Particles produced by this process have a lower density due to the presence of occluded air inside the particles. The occurrence of clogging is the major problem in this atomizer whenever a feed of mucilaginous or fibrous nature is needed to be atomized. Higher and tedious maintenance process with time-consuming cleanup procedure makes this atomizer challenging to use in commercial applications [9].

More recently, ultrasonic and electrohydrodynamic atomizers have been used successfully in different dryers for academic and industrial research which carries a higher percentage of dominance to replace the conventional rotary and two-fluid nozzle atomizers in the future advancement of spray drying technology.

5.4 Droplet-Hot Air Contact

Following the previous stage of drying, i.e., atomization of bulk feed into tiny droplets, this stage involves the intimate contact between the hot drying air and the atomized feed droplets resulting in the evaporation of >95% moisture from the droplets in a few seconds. However, maintaining a uniform gas flow throughout the drying chamber to enable proper and rapid evaporation of water is the essential criteria for this stage.

The process of droplet-air contact is usually carried out using three types of process configurations such as:

- Co-current flow
- Counter-current flow
- Mixed flow

Dryers having a co-current configuration are usually recommended for the drying of heat-sensitive products and bioactive compounds. Here, the atomized feed and drying air are passed into the dryer in the same flow direction. The evaporation of moisture takes place at the wet-bulb temperature of hot air to maintain a low-temperature profile. Evaporative cooling occurs due to transfer of moisture from the droplets to its surrounding atmosphere by allowing the dried particles to maintain a temperature below the outlet dry air temperature. The hot air continuously gains the moisture relieved from the droplets until it reaches its saturation point, and its temperature gradually decreases due to moisture transfer in a continuous and parallel fashion.

In contrast, with the counter-current configuration, the atomized feed and hot drying air oppositely enter the dryer from both ends of the drying chamber with the atomizer positioned at the top and the hot air entering at the bottom. Dryers with counter-current configuration are best suited for drying of heat-resistant products as the outlet temperature of the product is always higher than the outlet temperature of exhaust air.

Dryers having mixed-flow configurations include both co-current and counter-current fashion of air-droplet contact. The hot air enters from the top, and the atomizer is situated at the bottom of the drying chamber. This type of arrangement is usually adopted to dry and produce coarse free-flowing powder; however, the outlet temperature of the dried product is always higher than the exhaust air temperature.

5.5 Drying of Droplets

This stage includes the evaporation of moisture from the atomized feed droplets. It is considered as one of the critical stages of the spray drying process concerning the morphological and physicochemical characterization of the dried products. The removal of moisture from the droplets follows the fundamental convective drying kinetics and is characterized in two steps: a) constant rate period, and b) falling rate period.

During the constant rate drying period, rapid evaporation of moisture at a relatively stable rate takes place, keeping the surface sufficiently cool. The temperature of the droplet is maintained the same as the wet-bulb temperature by keeping the surface of the droplet saturated [10]. Peclet number (Pe) plays a controlling parameter here with regards to the quantification of evaporation rate in the constant rate drying period (Eq. 5.1). The rate of evaporation has a direct relationship and is proportional to the surface area of the droplet having a diameter d which is also termed as 'd^2 law' [11].

$$Pe = \frac{K}{D} \tag{5.1}$$

Where,
Pe = Peclet number;
K = Evaporation rate;
D = Diffusion rate.

An important phenomenon of 'crust formation' happens during the constant rate drying period where the dissolved solute gradually forms a thin shell around the surface of the droplet by exceeding its saturation

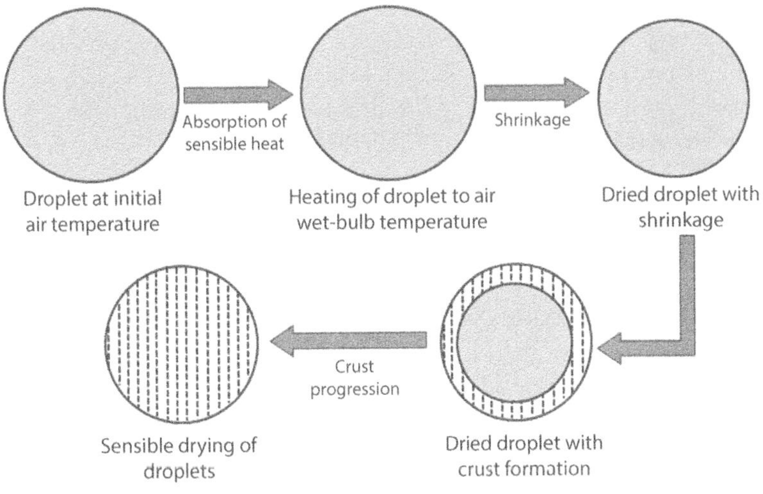

Figure 5.1 Schematic diagram of the droplet drying process.

concentration, which facilitates the drying kinetics to shift into a diffusion-controlled process. Hence, the falling rate period starts. The unavailability of enough surface moisture for maintaining the saturated condition on the droplet surface enables this process. The major controlling factor to this period of drying is the rate of water diffusion from the inner parts to the surface of the dried shell (Figure 5.1). By the starting of this phase, particle tends to be situated at the coolest part of the dryer which hinders the effect of the increase in inlet temperature on outlet temperature of the product and the product temperature is always less than the exhaust dried air temperature (approximately 20°C) [12].

5.6 Particle Separation

Particle separation takes place usually in two stages, namely, primary and secondary separation. Primary separation includes the collection of dried powder at the base section of the conical drying chamber, attached by a separating bottle for the ease of collection of dried heavy particles. Secondary separation system consists of a pneumatic system with cyclone separator or a screw conveyor which pumps out the dried air and finer particles from the mainstream as shown in Figure 5.2. The pneumatic system helps to suck out the exhaust hot air stream with the evaporated moisture from the centre of the cone and discharges it through a side outlet above the conical bottom. Additionally, it is also connected with a scrubber bag system where the finest particles/mist like products get trapped inside the void spaces of the bag. The removal of moisture mostly associated with the low efficiency of powder collection and to increase the product yield, additional particle collection system comprising dry collectors and wet scrubbers are generally used. Also, the use of a bag filter or an electrostatic precipitator could aid in the increase of the product yield as per the final product specifications and type of particles carried away with the hot exhaust air.

5.7 Effect of Process Parameters on Product Quality

The quality of a spray-dried product is a function of the process parameters, design criteria and the material characteristics of the sample to be dried. These factors have a significant affect in determining the morphological and physiological characteristics of the spread. Therefore, understanding the process variables and their techno-physical interactions with

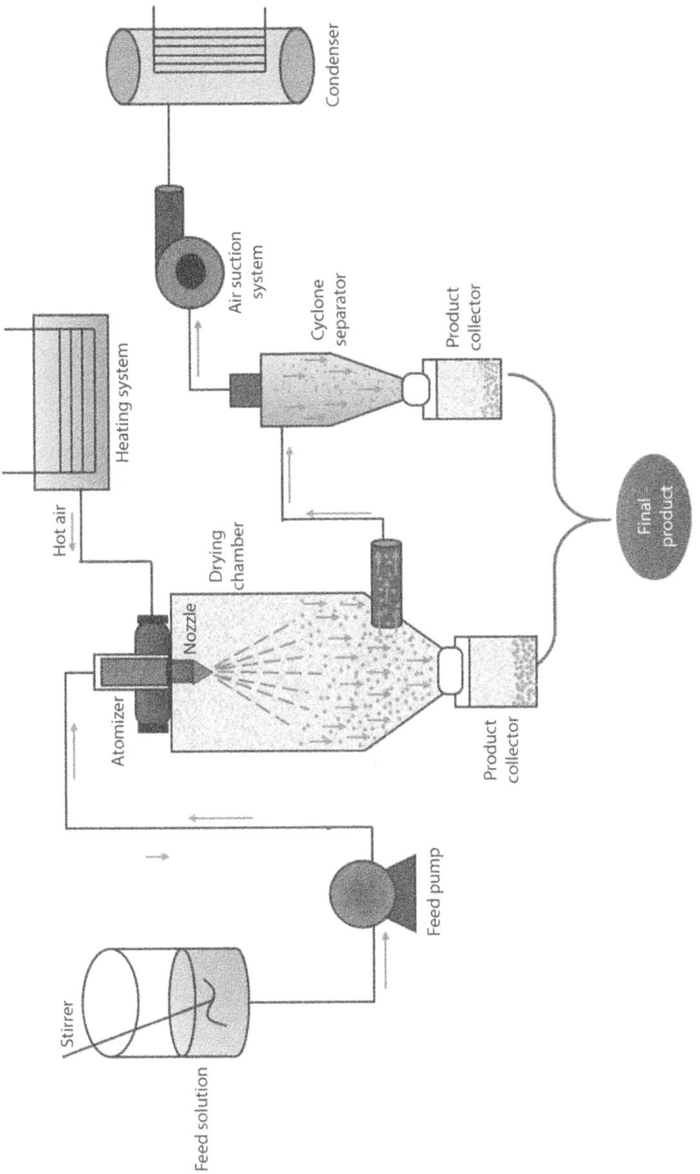

Figure 5.2 Schematic diagram of a spray dryer and drying process.

the product is very much required to optimize the spray drying process and to facilitate intelligent operation. The impact of different spray drying process variables on the product quality is emphasized in the following section.

5.7.1 Process Parameters of Atomization

The efficacy of an atomizer is a direct function of atomization pressure, feed flow rate, density and viscosity of the liquid to be dried. Together, these parameters decide the size of the droplet and have a direct relationship determining the final droplet diameter [2]. According to the literature, the diameter of a droplet decreases with an increase in atomization pressure at a constant feed rate (Eq. 5.2).

However, with an increase in feed flow rate, the droplet size increases at constant pressure, because of insufficient droplet fission and less interaction of the liquid with atomization energy. Also, an increase in diameter speed of the wheel in a rotary atomizer results in reduced droplet size.

Both feed viscosity and density have a direct relationship with the droplet diameter (Eq. 5.3). With an increase in feed viscosity, higher atomization energy needs to be required to overcome the viscous force and break the droplets. If atomization energy is kept constant, then the particle size would become bigger.

The process of atomization is greatly affected by the surface tension of the feed solution. Surface tension has an inverse relationship with the droplet diameter of dried particles. Hence, the need for the emulsification and homogenization of the feed solution prior to atomization is essential for spray drying.

$$\frac{D_2}{D_1} = \left(\frac{P_1}{P_2}\right)^{0.3} \tag{5.2}$$

$$\frac{D_2}{D_1} = \left(\frac{\mu_1}{\mu_2}\right)^{0.2} \tag{5.3}$$

Where,

D_1 and D_2 are the initial and final diameter of droplet;

P_1 and P_2 are initial and final atomization pressure;

μ_1 and μ_2 are the initial and final feed viscosity.

5.7.2 Parameters of Spray-Air Contact and Evaporation

Being an important process step of spray drying, the phenomena of liquid and hot air contact affects the product quality significantly. The essential parameters that need to be addressed during this phase are

- Spray angle
- Inlet and outlet air temperature
- Aspirator flow rate
- Glass transition temperature
- Aspirator rate
- Air humidity
- Residence time of the spread

5.7.2.1 Spray Angle

Spray angle is the angle of spray shape and indicates the index for simplifying spray coverage. When the spinning liquid is divided into fine droplets after getting out of the nozzle orifice, due to the effect of gravitational acceleration and air resistance, it is not able to travel in the tangential direction. The size of the droplet has a direct relationship with the tangential velocity and increases with widening the spray angle. The type of airflow (co-current or counter-current) is an important parameter to decide the width of the spray angle. With an airflow of co-current pattern, the spray angle is usually increased while in counter-current pattern, it is generally decreased.

5.7.2.2 Aspirator Flow Rate

Aspirator flow rate determines the amount of drying of the feed by controlling the supply of heated air and is regulated by a motor aspirator. An increased volume of dry air supply leads to higher removal of moisture and aids in faster drying of atomized feed.

5.7.2.3 Inlet Air Temperature

The inlet temperature of the drying air plays a vital role to bring the atomized droplets to their wet-bulb temperature and also maintains the wet-bulb temperature of the surrounding hot air [13]. It has a direct relationship with the evaporative efficiency and thermal effectiveness of the dryer. Therefore, higher throughput can be achieved by increasing the inlet temperature and vice versa. However, low inlet air temperature is favourable for heat-sensitive bioactive compounds. Hence, optimization of inlet air temperature is very much required in a spray drying process to enhance the drying efficiency of the system. Usually, for drying food commodities, the inlet air temperature is kept between 150-200 °C depending on the commodities. However, the temperature difference in inlet and outlet air, also called the 'temperature gradient' (ΔT), significantly affects the quality of the dried products. A higher temperature gradient can initiate a faster drying process by providing a higher driving force and diffusivity, respectively, resulting in a high moisture removal from the atomized feed.

5.7.2.4 *Outlet Air Temperature*

The outlet temperature of the air is defined as the temperature attained by the heated air with a solid mass just before entering into the cyclone separator. However, unlike inlet air temperature, it cannot be regulated and is resulted by the heat and mass equilibrium of the drying system. Outlet temperature determines the surface morphology and residual moisture content of the product. Therefore, maintaining an optimum temperature gradient (ΔT) is a crucial factor for the spray drying process [14] because an increase in outlet temperature leads to the rapid formation of crust on the particle surface with insufficient drying of the inner core and promotes the agglomerate formation. Hence, the moisture is trapped inside the core during falling rate drying, leading to an increase in the residual moisture content of the final product. Also, the wet-bulb temperature of the particles is further increased if the outlet temperature is increased.

5.7.2.5 *Glass Transition Temperature*

Glass transition temperature (T_g) is described as the temperature at which the amorphous regions experience the transition from a rigid state to a more relaxed state. If the temperature goes below the glass transition temperature, the physical properties of plastics change like those of a glassy or crystalline form. Above this temperature, the molecular matrix shifts its phase from a rigid glassy state to a rubbery state. Hence, T_g plays a vital role in determining the product characteristics as it is associated with the stickiness of the product while spraying inside the chamber [15]. Increase in stickiness leads to agglomeration and caking of the product, hampering product quality and its shelf life. Therefore, different polymers are added with the feed solution before spray drying to modify the glass transition temperature of the final product as per the requirement.

5.7.2.6 Residence Time

Residence time is defined as the time taken by the sprayed particles to move from the nozzle outlet to the exit of the spray dryer and divide into two zones, primary and secondary. The primary zone is the time taken by the droplet particles to come out of the nozzle and strike on the surface wall of the drying chamber, and the secondary zone is comprised of the time required by the sprayed particles to slide down and travel to the exit outlet of the spray dryer. Hence, the residence time is calculated by adding primary and secondary residence time. During spray drying of bioactive compounds, residence time plays a vital role in absolute drying of feed droplets to achieve optimum product specifications with minimal thermal degradation of heat-sensitive products by controlling the outlet temperature of the particles.

5.8 Classification of Spray Dryer

Spray dryers are classified into four different categories based on their functionality and design aspect. The detailed classification is shown in Figure 5.3, and each category is further subcategorized into following types.

5.8.1 Open-Cycle Spray Dryer

It is the standard type spray dryer having suction for inlet drying air and exit outlet for the moist exhaust air.

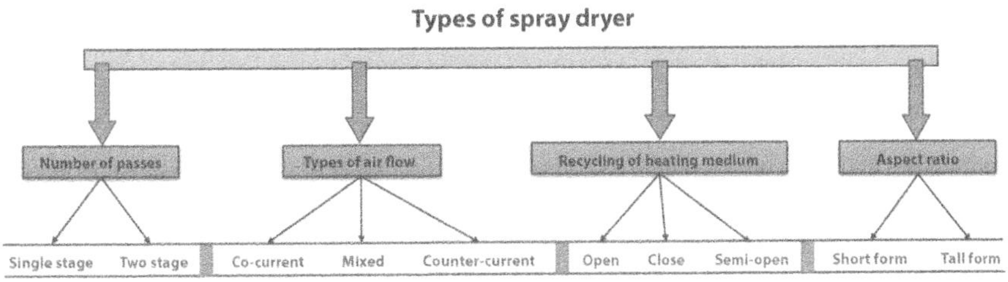

Figure 5.3 Classification of spray dryer.

5.8.2 Closed-Cycle Spray Dryer

In this type, the gaseous medium (mostly Nitrogen gas) is recycled and reused to increase the efficiency of the dryer. It is desirable for the drying of any flammable or toxic feed those are susceptible to oxidation [12].

5.8.3 Semi-Closed Cycle Spray Dryer

In this type, a direct fired-heater is used to supply the required quantity of hot air for the drying process and then the same gas is recycled and could be again used for oxygen-sensitive materials [16].

5.8.4 Single-Stage Spray Dryer

In this type, a single pass is used for the drying of the targeted feed using hot air intake. The average operating inlet and outlet temperature vary from 150-200°C and 90-100°C, respectively; however, the temperature can be changed based on the product requirement. Generally, single-stage dryers are not desirable for heat-sensitive products due to product degradation and agglomerate formation [17].

5.8.5 Two-Stage Spray Dryer

This type of dryer includes the removal of residual moisture of the product in two stages. The first stage comprises lowering of moisture content up to 5-10% and the second stage includes the removal of residual moisture to the predecided level using a fluidized bed dryer. These types of dryers are used for heat-sensitive products and volatile compounds which require high retention of bio-actives with higher evaporative efficiency [18].

5.8.6 Short-Form Spray Dryer

Short-form dryers are widely used in nature, having a height-to-diameter aspect ratio of around 2:1. The residence time of the heated gas is comparatively lower than the tall type dryers and a short form dryer having a bottom outlet, is highly suitable for drying heat-sensitive products [19, 20].

5.8.7 Tall-Form Spray Dryer

A dryer with a height-to-diameter ratio of 5:1 is generally considered as a tall-form spray dryer. The products which are not sensitive to heat in nature are suitable for drying using this type of dryer as the residence time is relatively high here compared to other types [20].

5.9 Morphological Characterization of Spray-Dried Particles

Understanding the effect of process variables on the morphology of spray-dried particles is a crucial and complex phenomenon in the spray drying process. The particle morphology very much influences the quality parameters of a sprayed product such as particle size distribution, moisture content, wettability, dispersibility, flowability, density, etc., during the process. Therefore, optimization of the process parameters is essential to acquire the desired morphology of the dried particles. However, the complex phenomenon of interactions among the process parameters makes the assessment quite tricky. Process variables such as inlet and outlet temperature; aspirator and feed flow rate; total solid content of feed; diameter and type of nozzle used, etc., govern the ultimate morphology of the dried particles. Drying kinetics plays a crucial role in determining the surface morphology of the particles [21]. Figure 5.4 illustrates the effect of different drying patterns on the product morphology. The surface characteristics of spray-dried particles might be directed in two different fashions such as a 'dry shell'

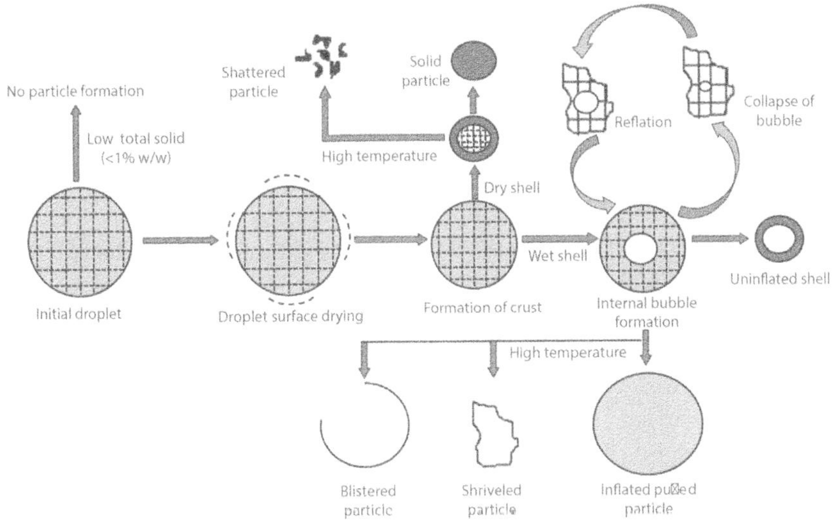

Figure 5.4 Different morphologies due to bubble inflation during spray drying.

with a shrinking hollow core; and a 'wet shell' with a hollow core which may inflate by higher inlet drying temperature. Hence, the morphological characterization is a function of the type of shell formation; henceforth the process of crust formation comes into the light [22].

The understanding of the Peclet number (Pe) is an important criterion to predict the morphology of the dried particles. It is known that Peclet number with a higher value (2-5) indicates pronounced shell formation due to faster evaporation rate and rapid building-up solute concentration on the surface. It produces hollow particles of a larger size. When the Peclet number value is lower, it subjects to a combination of low inlet-outlet temperature which is further characterized by a low evaporation rate, resulting in smaller and solid particles [21]. The rate of shell formation is directly proportional to the final quality of the sprayed product. The faster rate helps in retaining the quality of volatile and heat-sensitive compounds and vice versa [23].

However, the process of atomization also plays a vital role in deciding the particle morphology. Incomplete atomization of the feed leads to the production of non-spherical and elongated particles due to lack of higher impact force to overcome the viscous force of the liquid feed.

5.10 Application of Spray Drying for Foods

Spray drying is adopted as an excellent drying technique for various food commodities on a commercial scale. The foremost applications of spray drying in the food industry include powder production and encapsulation. In the food industry, encapsulation technology has been used for more than 60 years [24]. Encapsulation is a physicochemical or mechanical process of enclosing a targeted bio-active compound in a wall/carrier material to produce particles of nano to micro-range with enhanced protection, controlled release and better immobilization [25, 26].

Encapsulation bears certain technological functions to enhance the material characteristics of a product of interest which are given below.

- To provide stability to reactive materials in external conditions
- Lowering the volatility and thereby losses of material
- Controlling the release of functional material
- Masking of unpleasant odour and taste
- Homogeneous distribution of the material in the formulation
- Enhancing ease of handling materials [27]

Spray drying is the most commonly used encapsulation (micro/nano) method in the food industry since it is economical and flexible. Production of micro or nano-particles is subjected to the choice of requirement, by altering the process conditions which have been already discussed in previous sections. The energy consumption of spray drying is six to ten times lower compared to conventional freeze-drying technique, with the production of improved quality products [28]. The process involves dispersing the core material in the desired carrier material forming an emulsion or dispersion, followed by homogenizing and atomizing the mixture in the drying chamber, which leads to evaporation of the solvent.

Microcapsule consists of a core material/ingredient/substrate/active agent enclosed in a coating/wall/carrier material. There are different types of coating materials available based on the physical and chemical properties of the core material, their functional interactions and the desired properties of the end product. Specific criteria are needed to be fulfilled to select a wall material for the encapsulation process, which are summarized below.

- The most important criteria for the selection of encapsulating material are the type of functionality that could be provided to the final product, potential restrictions for the coating material, concentration of encapsulates, type of release, stability requirements and cost constraints.
- Materials used for the design of the protective shell of an encapsulate must be food-grade irrespective of being natural or synthetic. It should be biodegradable and Generally Recognized as Safe (GRAS), encompassing desired barrier properties.
- The material should provide maximum protection to the active material against harsh environmental conditions, and it should hold the active agent within the capsule-matrix during the processing or storage under various conditions.
- It should be inert without any chemical interaction with the core and should possess good rheological characteristics [29].

5.11 Wall Materials

A wide variety of wall materials are used for encapsulation purpose ranging from high molecular polymers such as starch, inulin, proteins, etc., to low molecular weight polymers such as maltodextrin and cyclodextrin.

The following section addresses the different categories of wall materials used for encapsulation by spray drying for human consumption.

5.11.1 Carbohydrate-Based Wall Materials

Several works have addressed the use of carbohydrates such as starch, inulin, cyclodextrin, etc., as wall materials for encapsulation by spray drying. The configuration of glycosidic linkages in carbohydrates confers specific physical and chemical characteristics that enhance their tolerance to environmental conditions and stability, making them suitable wall materials for encapsulation during spray drying [30, 31]. The most frequently used carbohydrates for encapsulation by spray drying and their characteristics are discussed below.

5.11.1.1 Starch

Starch is an economic biopolymer that contributes to rapid wall formation and good retention characteristics, making it an ideal wall material for encapsulation by spray drying. However, the disadvantage associated with the use of starch as wall material is that it has a poor emulsifying capability. It imparts relatively low viscosity even at high total solids concentration. The starch granules provide a physical barrier to the bioactive core, facilitates their entrapment and aids in their protection by limiting their exposure from the harsh environmental conditions [32, 33].

5.11.1.2 Modified Starch

To overcome the challenges associated with the utilization of starch as wall material, the starch is subjected to esterification with n-octenyl succinyl anhydride. This esterification results in the formation of modified starch which contributes to better emulsifying properties and improved retention capability during spray drying as compared to the native starch. However, the modification might substantially change the rheological properties of starch.

5.11.1.3 Maltodextrins

Maltodextrin is a polysaccharide, produced by starch hydrolysis and composed of molecules of α-D glucose joined together by α-(1,4) glycosidic linkage. Maltodextrin has the advantage of being low cost with low oxygen permeability and therefore aids in protection against oxidation [34].

It provides a physical barrier by reducing the mobility of the bioactive molecule, and promotes rapid wall formation, imparting better stability [32]. Besides, it can resist variations in water activity. The limiting feature of maltodextrin lies in the presence of hydroxyl groups in the biopolymer, hindering the emulsifying capability [35]. However, the utilization of maltodextrin in combination with other wall materials such as whey protein, pectin and gum arabic enhances its emulsifying properties and also prevents the loss of the bioactive compounds [36–38]. Maltodextrin acts as an effective wall material even at harsh processing conditions such as high inlet temperature (>150°C) by creating a barrier around the drop surface, which results in low moisture content and greater yield of the final product [39].

5.11.2 Cyclodextrins

Cyclodextrins are oligosaccharides with a circular structure comprising a hydrophobic interior and hydrophilic exteriors. The cyclodextrins as wall materials offer protection against oxidation and stresses induced due to variation in pH. They also enhance the water solubility of the matrix [40, 41]. Beta-cyclodextrin is the universally recognized wall material for spray drying by FAO/WHO. Several works of literature have reported the use of cyclodextrins on the encapsulation of bioactive compounds of plant origin and essential oils.

5.11.3 Gum Arabic

Gum arabic is a polymer composed of d-glucuronic acid, l-rhamnose, d-galactose, l-arabinose and protein. Gum arabic is regarded as the most frequently used wall material for encapsulation by spray-drying due to its excellent water solubility and glass transition temperature with low viscosity [34, 42]. The protein portion of the gum arabic provides good emulsifying and stabilizing properties facilitating encapsulation of oils. Besides, it offers enhanced stability to the bio-active core at reduced levels of water activity. However, according to the literature, an increase in water activity beyond 0.74 could degrade the core [43]. Furthermore, it is semi-permeable to oxygen and hence only provides partial protection against oxidation.

5.11.4 Inulin

Inulin is a linear polymer of D-fructose units adjoined by β-(2-1) linkages with glucose molecules on each terminal end of the chain. Inulin as wall material has been successfully employed to encapsulate hydrophobic compounds including polyphenols of black currant, anthocyanins, essential oil of rosemary, etc. [44–46]. The susceptibility of inulin to air humidity and temperature results in loss of core caused by water adsorption-promoted agglomeration. However, inulin in combination with other wall materials aids rapid drying due to variation in glass transition temperature and results in the formation of a barrier that entraps the water inside the encapsulated matrix, hence contributing to enhanced shelf stability of the final product [47].

5.11.5 Pectin

Pectin is a heterogeneous polysaccharide consisting of monomers of galacturonic acid joined together by α-(1-4) glycosidic linkage present in the middle lamella of the plant cell wall. This biopolymer is mostly employed for encapsulation to sustain the bioactive core in combination with other wall materials such as maltodextrin and whey protein concentrate [48]. Moreover, the degree of methoxylation dramatically affects the process of encapsulation by spray drying. The pectin with a low degree of methoxylation interacts with other polymers such as protein to create a barrier to protect the core from oxidative stress. The formation of the barrier is induced by the chelation of the hydrophobic portion of the core with the metallic ions. However, the interaction might also lead to the loss of water during the process of encapsulation [49, 50].

5.11.6 Chitin and Chitosan

Chitin is the biopolymer consisting of N-acetyl-D-glucosamine units adjoined by β- 1,4 glycosidic linkage, mainly found in the exoskeleton of crustaceans. The deacetylation of chitin results in the formation of a cationic charged heterogeneous polysaccharide known as chitosan. These high molecular weight substances are employed for encapsulation of hydrophobic and hydrophilic substances [51]. The applications of these biopolymers in spray drying include encapsulation of ascorbic acid and cobalamin with enhanced yield and better antioxidant property; polyphenols of *Camellia sinensis* with improved stability; bioavailability and protection against oxidation; and glutathione to improve the antioxidant effect in wines [52–54].

5.11.7 Protein-Based Wall Materials

Proteins are considered suitable wall materials for encapsulation due to their chemical characteristics. The proteins can be employed as films or polymers singly or in combination with other polymers. They offer retention of the core and provide a physical barrier [55, 56]. A protective high protein film is created due to the migration of proteins to the droplet surface. As the film converts into a glossy skin due to variation in transition temperature during spray drying, a subsequent reduction in the interaction of particles within the spray drying chamber increases the yield of the final product [57].

5.11.7.1 *Whey Protein Isolate*

Whey protein isolate (WPI) is generally recognised as a by-product of the cheese industry. WPI confers low-cost economics, excellent retention and film-forming abilities [58, 59]. WPI has been successfully employed for encapsulation of oils by spray drying using a hydrophobic association between triacylglycerol and the weakly polar sites in protein moiety [60]. However, the challenge associated with them is that they might induce allergic reactions.

5.11.7.2 *Skim Milk Powder*

Skim milk powder comprised of protein and lactose is the most frequently used protein-based wall material for encapsulation by spray drying. The ease in availability and relatively inexpensive nature of skim milk powder has increased the popularity of skim milk powder as a wall material [61]. The protein portion mainly constituting casein imparts amphiphilic nature, and the lactose provides a reduced glass transition temperature, making it suitable for encapsulation of volatile compounds [62].

5.11.7.3 *Soy Protein Isolate (SPI)*

In recent times, growing concerns about the allergens associated with animal proteins have immensely increased the popularity of plant proteins. Hence, vegetable proteins such as wheat proteins, pea proteins, sunflower proteins, rice proteins, oat proteins, and soy proteins are extensively employed as an alternative to animal protein-based wall materials. The easy availability and economic nature of soy protein isolate make it the most utilized wall material of plant origin for encapsulation [63]. Soy protein isolate imparts optimal emulsifying capability and rapid film formation along with osmotic resistance suitable for protection of the core of bioactive compounds [64]. Besides, they also provide a physical barrier for the entrapment of bioactive compounds. Moreover, the SPI associates with hydrophobic interaction resulting in preferential migration, from the polar to weakly polar sites [63, 65].

5.12 Encapsulation of Probiotics

Probiotics are live beneficial microorganisms that are administered to provide health benefits. Probiotics are well known to improve gastrointestinal health and are considered nutraceuticals. Probiotics consist of strains of variable genera of microorganisms such as *Lactobacilli, Bifidobacteria, enterococci* and *yeast.* With the increase in gastrointestinal disorders, the role of these bacteria is coming into the picture by imparting their functionality as a cure of the disease, and it facilitates a large-scale production of shelf-stable probiotics [66]. Lyophilization/freeze-drying is a conventional technique to produce dehydrated starter cultures, but drawbacks like high capital and processing cost along with longer processing time limit its application. Combating all these drawbacks spray drying emerged as a promising technique for producing encapsulated probiotics. Spray drying is a single step flash drying process enabling the production of stable encapsulates. Other advantages of spray-dried encapsulated probiotics include the lesser requirement of storage area for dried probiotics and enhanced protection of enclosed cells with a smooth skin surface.

As per WHO/FAO [67], the concentration of live probiotic bacteria should be more than 10^7 CFU per gram per serve in the final product at the time of consumption to confer adequate health benefits. Therefore, maintaining a healthy viable count of the probiotic microorganism in the food is an essential factor during the encapsulation of probiotics. The viability of probiotic cells over a prolonged storage period and their stability during their passage through the acidic environment of the gastrointestinal tract has to be maintained. However, certain factors affecting the survival rate of bacteria is needed to be considered before it is subjected to spray drying for encapsulation purpose.

5.12.1 Choice of Bacterial Strain

The principle of spray drying involves subjecting the bacteria to different stresses of heat, oxidation, osmotic and desiccation, resulting in a higher inactivation rate [68]. It has been considered that different bacterial strains exhibit varied tolerance towards the stresses involved during spray drying. It raises the need to address the tolerance of bacterial strain towards the stresses conferred in spray drying to enhance the final viability of the probiotic powder produced [69]. The most explored species in spray drying include Bifidobacteria, Lactobacillus and Lactococcus. The prevailing harsh conditions during the process of spray drying results in low viability of the probiotic species in the final products. Propionibacteria typically exhibit higher tolerance, either through their metabolism or a multi-tolerance response, due to their greater capacity for environmental adaptation [70]. Propionibacteria demonstrates greater survivability than strains of Lactobacillus. In comparison, Streptococcus typically exhibits better resistance than Lactobacillus. Thus, the resilience of the bacterial strains to stresses conferred during spray drying follows the order: Propionibacteria > Streptococcus > Lactobacillus. Tolerance, however, still differs widely within the genus. *Lactobacillus plantarum* is known to be a more stress-tolerant species as compared to *Lactobacillus bulgaricus, Lactobacillus casei, Lactobacillus acidophilus, Lactobacillus rhamnosus* and *Lactobacillus paracasei* [68]. A comparison of the resistance between various strains of Bifidobacterium showed that *Bifidobacterium longum* ssp. *longum* compared to *Bifidobacterium longum* ssp. *infantis* has greater viability ascribed to better heat and oxygen resistance. Hence, it could be recognised that the robustness of different bacterial strains to process conditions during spray drying varies greatly and can be considered as species, genus, and strain-dependent.

5.12.2 Response to Cellular Stresses

The tolerance of the probiotics to the environmental stresses during spray drying corresponds to their viability. The innate cellular response system of the bacteria and the growth conditions are the primary determinants to decide the tolerance of the bacteria to these stresses. The growth conditions by which the culture has been produced and the over-synthesis of the stress response proteins significantly affect the spray drying tolerance of the probiotics. The viability of *Lactobacillus casei* BL23 after spray drying enhanced considerably during their growth in concentrated sweet whey [71]. Similarly, the exposure of the bacteria to 0.1% w/v of bile salts for around 30 minutes led to significant improvement in their viability during spray drying [72]. Furthermore, it has been suggested that the accumulation of compatible solutes from the growth media during the growth of the probiotic bacterial strain also boosts their tolerance and thereby their survivability during harsh environmental conditions of spray drying.

Likewise, the overproduction of proteins caused by the exposure of the bacterial strain to the stresses themselves may be an effective method to increase their viability [73]. The *Lactobacillus plantarum* WCFS1 strain on the over-synthesis of the protein ftsH/FtsH exhibited a greater heat tolerance [74]. Given the osmotic robustness of the Opu genes (OpuA, OpuB, OpuC), the viability of *P. freudenreichii* ITG P20 enhanced greatly [71]. Substantial improvements in the oxidative resistance of *B. longum* NCC2705 were reported on the overproduction of the ahpC/AhpC [75]. Also, this improvement in stress tolerance through the overproduction of stress response proteins could also be achieved through genetic engineering. By way of genetic engineering, *Lactobacillus paracasei* strain was able to over synthesize a thermo-tolerant stress protein displaying 10-fold higher heat tolerance and survivability during the spray drying process [76].

5.12.3 Growth Conditions

The growth media of the probiotics is an essential factor with regards to the viability of probiotics during spray drying. Various growth media have been employed for the enumeration of the bacteria as well as wall material for the spray drying process. De Man Rogasa and Sharpe medium, commonly referred to as MRS broth, and dairy media including whey and skim milk, are the most widely reported. The use of MRS broth is limited to industrial scale since it is non-food grade and relatively expensive. Dairy media, however, is inexpensive and food grade and could be effectively employed for mass production. The viability of *Streptococcus thermophilus* MK-10 after spray drying was stated to be 69.5% during their growth in condensed non-fat milk (18% *w/w*) [77]. Similarly, the viability of *Lactobacillus casei* BL23 after spraying at an inlet and outlet temperature of 127 and 47°C, respectively was recorded as 100% during their growth in 30% *w/w* sweet whey [68]. Moreover, to tackle the osmotic stress and maintain viability, the accumulation of compatible solutes such as amino acids, carbohydrates, quaternary amines, etc., from the growth media by the probiotic bacteria has been reported to enhance their survivability during drying [78].

5.12.4 Effect of pH

The effect of pH of the growth media on the survivability of the bacteria in spray drying has not been clearly established. Some researchers have stated that the viability of the *Lactobacillus plantarum* increased by two-fold on maintaining a controlled pH during the growth of the bacteria. On the contrary, the survivability of *Lactobacillus bulgaricus* was stated to be higher when the pH was not controlled during their growth attributing to the overproduction of stress response proteins such as GrosES, Hsp70 and GroEL induced by the reduction in pH (acid stress) [79].

5.12.5 Harvesting Technique

The harvesting of the probiotics with suitable technique at appropriate phases of growth such that maximal yield is achieved is essential to maintain their viability. Generally, the harvesting is carried out through centrifugation during the early stationary phase or late exponential phase. However, the harsh conditions during the early stationary phase induce tolerance to various stresses due to stress response proteins and hence the cells harvested during the early stationary phase exhibit greater viability [80, 81].

5.12.6 Total Solid Content of the Feed Concentrate

The total solids content of the feed concentrate suitable to ensure high viability of the probiotic bacteria after spray drying as stated in the literature lies within the range of 20-30% *w/v* [68]. Any deviation from this range results in a substantial change in the residual viability. Total solids content less than this optimal range might maintain an isotonic condition before the process and result in dried powders with high viability after drying. On the contrary, an increment in the total solids content above this range resulted in low survivability of the probiotics. The short exposure time of spray drying makes it challenging to attain the tolerance to hyperosmotic stress required to increase the survival rate post drying [71]. The high total solids in the feed concentrate are usually employed on a commercial scale to attain high productivity and reduced energy cost [82].

5.13 Encapsulation of Vitamins

Vitamins are generally encapsulated for easy fortification and enrichment of food products. Vitamin encapsulation is done to enhance the stability and effectiveness of vitamins in the desired food matrix. The most widely encapsulated vitamins are Vitamin A, Vitamin C and Vitamin E. Vitamin A is encapsulated to enhance bioavailability [83]. Vitamin E encapsulation is done for preventing degradation due to oxidative reaction.

Similar to all other encapsulate, the compatibility of wall material with the core vitamin is a crucial factor to increase the efficacy of the encapsulation process in spray drying. Therefore, to improve the affinity of vitamin with the wall material, modification of the carrier wall needs to be done. α-tocopherol and ascorbic acid vitamins were encapsulated in native (non-modified) and modified soy protein isolate. Modification of wall material was done by acylation and cationization reactions in aqueous alkaline media to increase the encapsulation efficiency of hydrophobic α-tocopherol from 79.7% to 94.8% [56]. On the other hand, the encapsulation efficiency of hydrophilic ascorbic acid was reduced from 91.8% to 57.3% compared to native soy protein isolates. These results can be owed to the acylation of wall material which enhances the hydrophobic nature.

5.14 Encapsulation of Flavours and Volatile Compounds

Flavour is an organoleptic property that primarily influences acceptance or rejection of the food product. Flavour is the combined perception of taste and aroma. The aromatic component is volatile, while the taste component is usually non-volatile. The presence of volatile organic compound renders the handling of the flavour system difficult. Also, one flavour system may contain various volatile compounds, thus enhancing the complexity of handling it. Flavour encapsulation has emerged as a promising technique to combat these drawbacks of handling flavours. Flavour encapsulation involves entrapping the extracted flavour compound in a wall material to increase its stability against adverse environmental conditions such as temperature, pH, light, humidity, etc., during storage. Prevention of undesirable interactions of flavour compounds with food matrix as well as other flavour compounds is needed to maintain the integrity of the core [66]. Several organic wall materials can be used for flavour encapsulation additional to gum Arabic, maltodextrin, whey protein concentrates or isolates (WP(C/I)) and modified starches, etc., to create a synergistic effect, which has already been emphasized in the previous sections.

5.14.1 Selective Diffusion Theory

Flavour matrix are both polar and non-polar, with essential oils and oleoresins being widely used non-polar compounds. Though flavour compounds are sensitive to high temperatures, spray drying came out to be a widely accepted and effective technique for flavour encapsulation at a commercial scale. The 'Selective Diffusion Theory' can best support this conclusion. Selective Diffusion is a phenomenon that sets in due to the

Table 5.1 Application of spray drying for encapsulation of various bio-active compounds.

Application	Core material	Wall material	Inference	References
Probiotics	*Streptococcus thermophilus*	Condensed non-fat milk (18% *w/w*)	Encapsulation efficiency= 69.5%	[77]
	Lactobacillus casei BL23	Sweet whey (30% *w/w*)	Encapsulation efficiency= 40%	[71]
	Lactobacillus acidophilus LA -5	reconstituted (20% *w/v*) goat's milk	Encapsulation efficiency ~1%	[85]
	Bifidobacterium infantis ATCC 15679	Maltodextrin (15% *w/v*)	Encapsulation efficiency= 91.6%	[86]
	Propionibacterium freuedenreichii ITG P20	Sweet whey (30% *w/w*) with	Encapsulation efficiency ~100%	[68]
Flavour	Pineapple Peel Extract	maltodextrin, inulin, and arabic gum (5% *w/w*)	Enhanced antioxidant and total phenols	[87]
	2-acetyl-1-pyrroline	Gum: maltodextrin in varied ratios	ACPY reduction rate was lower in the acidic form, 30% of ACPY reduction was found in acidic solution after 35 days	[88]

(*Continued*)

Table 5.1 Application of spray drying for encapsulation of various bio-active compounds. (*Continued*)

Application	Core material	Wall material	Inference	References
Color	Curcumin	Inulin (14.5g): Maltodextrin (14.5 g): Tamarind gum (1g)	Encapsulation efficiency= 82.5%	[89]
	Betalain	Maltodextrin: Gum Arabic (1:1)	Better storage stability of the pigment	[90]
	Anthocyanin	Maltodextrin (20% *w/v*)	Stable dried pigment containing powder; however higher inlet, and outlet temperature results in degradation	[91]
Antioxidant	Gallic Acid	Nopal mucilage	Stability of microcapsules during storage	[92]
Enzyme	Beta-galactosidase	Chitosan 1% (*w/v*), sodium alginate 1% (*w/v*) and Arabic gum 1% (*w/v*)	Product yield = 36-59%	[93]
Essential oil	Avocado oil	Whey protein isolate and maltodextrin in varied ratios	Encapsulation efficiency = 45-66%	[94]

(*Continued*)

Table 5.1 Application of spray drying for encapsulation of various bio-active compounds. (*Continued*)

Application	Core material	Wall material	Inference	References
	Rosemary oil	Combinations of Gum Arabic, inulin, maltodextrin, modified starch	Combination of modified starch and maltodextrin exhibited the best retention	[38]
	Flaxseed oil	Maltodextrin in combination with other materials such as whey protein concentrate, um arabic, Modified starch	Encapsulation efficiency = 62.3% to 95.7%	[95]
	Chia seed oil	Whey protein concentrate (WPC) with mesquite gum (MG) (67:33)	Encapsulation efficiency > 70%	[96]
Vitamins and minerals	Vitamin A	Binary and ternary blends of gum arabic, starch and maltodextrin	Encapsulation efficiency = 88-98%	[54]
	Alpha-tocopherol	Sodium alginate	Encapsulation efficiency = 56.7%	[97]
	Potassium iodide and potassiumiodate	Modified starch	Encapsulation efficiency = 89.7-100%	[98]
	Calcium citrate and calcium lactate	Cellulose derivatives and polymethacrylic acid	Modified calcium release, enhanced therapeutic effects with a minimum dose of calcium	[99, 100]

difference in diffusion between water and core compound. The flavours with higher relative volatility than water would have higher losses than those with lower volatility during the initial drying stage [84]. At the onset of drying, the loss of volatile compound occurs due to relative volatility, but crust formation takes place at the surface of droplets with the advancement in drying, leading to the selective diffusion process. At the later stages of drying, selective diffusion is established due to the hardening of the outer surface.

A detailed application of spray drying for encapsulation of different bio-active compounds has been listed in Table 5.1.

5.15 Conclusion and Perspectives

Spray drying is an extensively used process technology in the food industry for the development of several encapsulated and dried powders, vitamins, and bio-actives. This chapter has highlighted the detailed system conditions with the mechanism of the spray drying process; various strategies to improve the efficacy of drying by considering parameters such as core material characteristics and their responses to stresses related to spray drying; wall materials and their optimal properties along with the processing conditions governing the process of drying. It was established that spray drying had been extensively used in the encapsulation of probiotics, flavours, colours, antioxidants, enzymes, vitamins and minerals.

Furthermore, the combination of core-to-wall material and the process conditions of spray drying needs to be optimized to get a better encapsulation efficiency with improved quality products. The strategy of triggering adaptive responses on the exposure of the bio-active core to heat, oxidative, osmotic stresses related to spray drying also improves the efficacy of the process. Therefore, this chapter integrates the different aspects of the process technology of spray drying, which might be more useful for further research in this field.

References

1. Percy, S.R., Improvement in drying and concentrating liquid substances by atomizing. US Patent 125, 406, 1872.
2. Masters, K. (Ed.), *Spray Drying Handbook*, Longman Scientific and Technical, Harlow, 1991.

3. Michael, J.K., Spray drying and spray congealing of pharmaceuticals. In: *Encyclopedia of pharmaceutical technology*. Marcel Dekker INC, New York, 14, 207-221, 1993.
4. Masters, K., *Spray Drying in Practice*, Spray Dry Consult International, Charlottenlund, Denmark, 1-35, 2002.
5. Rayleigh, L., On the instability of jets. *Proc. Lond. Math. Soc.* 1, 4-13, 1878.
6. Plateau, J., *Experimental and theoretical statics of liquids subject to molecular forces only*. Gauthier-Villars, Paris 4, 4-13, 1873.
7. Anandharamakrishnan, C., Reilly, C.D., Stapley, A.G.F., Effects of process variables on the denaturation of whey proteins during spray drying. *Dry. Technol.*, 25, 799-807, 2007.
8. Weber, C., *Zum Zerfall eines Flüssigkeitsstrahles*. Ztschr. F. Angew. Math. und. Mech, 11, 136-154, 1931.
9. Sewell, P.C., *Low pressure atomization nozzle*. EP 0244204 A1, 1987.
10. Dolinsky, A., High-temperature spray drying methods. *Dry. Technol.*, 19, 785-806, 2001.
11. Law, C.K. and Law, H.K., A d^2 law for multicomponent droplet vaporization and combustion. *AIAA J.*, 20, 4, 522-527,1982.
12. Gohel, C.M., Parikh, R.K., Nagori, S.A., Spray Drying: A Review. *Pharm. Rev.*, 7, 5, 2009.
13. Oldfield, D.J., Taylor, M.W., Singh, H., Effect of preheating and other process parameters on whey proteins reactions during skim milk powder manufacture. *Int. Dairy J.*, 15, 501-511, 2005.
14. Maas, G.S., Schaldach, G., Littringer, E.M., The impact of spray drying outlet temperature on the particle morphology of mannitol. *Powder Technol.*, 213, 27-35, 2011.
15. Rahman, M.S. (ed.), *Food Properties Handbook*, 1st edition, 87-177, CRC Press, Boca Raton, 1995.
16. Anundhia, C.J., Raval, J.A., Patel, M.M., Spray drying in the pharmaceutical industry - A Review. *Indo Am. J. Pharm.*, 2, 1, 125-138,2011.
17. Early, R. (ed.) Milk concentrates and milk powders. In: *Technology of Dairy Products*, 2nd edition, Blackie Academic and Professional, UK, 265-266, 1998.
18. Westergaard, V., *Milk powder technology-evaporation and spray drying*. GEA Process Engineering A/S, Soborg, Denmark, 2010.
19. Langrish, T.A.G. and Fletcher, D.F., Spray drying of food ingredients and applications of CFD in spray drying. *Chem Eng. Process.*, 40, 345-354, 2001.
20. Anandharamakrishnan, C., Experimental and computational fluid dynamics studies on spray-freeze-drying and spray-drying of proteins. Ph. D thesis, Loughborough University, UK, 2008.
21. Vehring, R., Pharmaceutical particle engineering via spray drying. *Pharm. Res.*, 25, 5, 999-1022, 2008.
22. Handscomb, C.S., Kraft, M., Bayly, A.E., A new model for the drying of droplets containing suspended solids. *Chem. Eng. Sci.*, 64, 628-637, 2009.

23. Hassan, H.M., Sayed, A.A., Mumford, C.J., Volatiles retention in the drying of skin forming materials. III. Heat sensitive materials. *Dry. Technol.*, 14, 3-4, 581-593, 1996.
24. Desai, K.G.H. and Jin Park, H., Recent developments in microencapsulation of food ingredients. *Dry. Technol.*, 23, 7, 1361-1394, 2005.
25. Chávarri, M., Marañón, I., Villarán, M.C., Encapsulation technology to protect probiotic bacteria. In *Probiotics*. IntechOpen, 2012.
26. Poncelet, D., Microencapsulation: fundamentals, methods and applications. In *Surface chemistry in biomedical and environmental science*, Springer, Dordrecht, 23-44, 2006.
27. Favaro-Trindade, C.S., Patel, B., Silva, M.P., Comunian, T.A., Federici, E., Jones, O.G. and Campanella, O.H., Microencapsulation as a tool to producing an extruded functional food. *LWT*, 128, 109433, 2020.
28. Martín, M.J., Lara-Villoslada, F., Ruiz, M.A., Morales, M.E., Micro encapsulation of bacteria: A review of different technologies and their impact on the probiotic effects. *Innov. Food Sci. Emerg.*, 27, 15-25, 2015.
29. Nedovic, V., Kalusevic, A., Manojlovic, V., Levic, S., Bugarski, B., An overview of encapsulation technologies for food applications. *Procedia Food Sci.*, 1, 1806-1815, 2011.
30. Zhu, C., Krumm, C., Facas, G.G., Neurock, M., Dauenhauer, P.J., Energetics of cellulose and cyclodextrin glycosidic bond cleavage. *React. Chem. Eng.*, 2, 2, 201-214, 2017.
31. Gray, C.J., Schindler, B., Migas, L.G., Pičmanová, M., Allouche, A.R., Green, A.P., Mandal, S., Motawia M.S., Sánchez-Pérez A., Bjarnholt N., Møller B.L., Rijs A.M., Barran P.E., Compagnon I., Eyers C.E., Flitsch S.L., Bottom-up elucidation of glycosidic bond stereochemistry., *Anal. Chem.*, 89, 8, 4540-4549, 2017.
32. Gupta, C., Chawla, P., Arora, S., Tomar, S.K., Singh, A.K., Iron microencapsulation with blend of gum arabic, maltodextrin and modified starch using modified solvent evaporation method–milk fortification. *Food Hydrocoll.*, 43, 622-628, 2015.
33. Jafari, S.M., Assadpoor, E., He, Y., Bhandari, B., Encapsulation efficiency of food flavours and oils during spray drying. *Dry. Technol.*, 26, 7, 816-835, 2008.
34. Tonon, R.V., Baroni, A.F., Brabet, C., Gibert, O., Pallet, D., Hubinger, M.D., Water sorption and glass transition temperature of spray dried açai (Euterpe oleracea Mart.) juice. *J. Food Eng.*, 94, 3-4, 215-221, 2009.
35. Waterhouse, G.I., Sun-Waterhouse, D., Su, G., Zhao, H., Zhao, M., Spray-drying of antioxidant-rich blueberry waste extracts; interplay between waste pretreatments and spray-drying process. *Food Bioproc. Tech.*, 10, 6, 1074-1092, 2017.
36. Soottitantawat, A., Partanen, R., Neoh, T.L., Yoshii, H., Encapsulation of hydrophilic and hydrophobic flavors by spray drying. *Japan J. Food Eng.*, 16, 1, 37-52, 2015.

37. Sansone, F., Mencherini, T., Picerno, P., d'Amore, M., Aquino, R.P., Lauro, M.R., Maltodextrin/pectin microparticles by spray drying as carrier for nutraceutical extracts. *J. Food Eng.*, 105, 3, 468-476, 2011.
38. de Barros Fernandes, R.V., Borges, S.V., Botrel, D.A., Gum arabic/starch/maltodextrin/inulin as wall materials on the microencapsulation of rosemary essential oil. *Carbohydr. Polym.*, 101, 524-532, 2014.
39. Paini, M., Aliakbarian, B., Casazza, A.A., Lagazzo, A., Botter, R., Perego, P., Microencapsulation of phenolic compounds from olive pomace using spray drying: A study of operative parameters., *LWT*, 62, 1, 177-186, 2015.
40. Pasrija, D., Ezhilarasi, P.N., Indrani, D., Anandharamakrishnan, C., Microencapsulation of green tea polyphenols and its effect on incorporated bread quality. *LWT*, 64, 1, 289-296, 2015.
41. Carlotti, M.E., Sapino, S., Ugazio, E., Caron, G., On the complexation of quercetin with methyl-β-cyclodextrin: photostability and antioxidant studies., *J. Incl. Phenom. Macrocycl. Chem.*, 70, 1-2, 81-90, 2011.
42. Comunian, T.A., Monterrey-Quintero, E.S., Thomazini, M., Balieiro, J.C., Piccone, P., Pittia, P., Favaro-Trindade, C.S., Assessment of production efficiency, physicochemical properties, and storage stability of spray-dried chlorophyllide, a natural food colourant, using gum Arabic, maltodextrin, and soy protein isolate-based carrier systems., *Int. J. Food Sci. Technol.*, 46, 6, 1259-1265, 2011.
43. Rascón, M.P., Beristain, C.I., García, H.S., Salgado, M.A., Carotenoid retention and storage stability of spray-dried encapsulated paprika oleoresin using gum Arabic and soy protein isolate as wall materials. *LWT*, 44, 2, 549-557, 2011.
44. Bakowska-Barczak, A.M. and Kolodziejczyk, P.P., Black currant polyphenols: Their storage stability and microencapsulation, *Ind. Crop Prod.*, 34, 2, 1301-1309, 2011.
45. Lacerda, E.C.Q., de Araújo Calado, V.M., Monteiro, M., Finotelli, P.V., Torres, A.G., Perrone, D., Starch, inulin, and maltodextrin as encapsulating agents affect the quality and stability of jussara pulp microparticles., *Carbohydr. Polym.*, 151, 500-510, 2016.
46. Botrel, D.A., de Barros Fernandes, R.V., Borges, S.V., Yoshida, M.I., Influence of wall matrix systems on the properties of spray-dried microparticles containing fish oil., *Food Res. Int.*, 62, 344-352, 2014.
47. Castel, V., Rubiolo, A.C., Carrara, C.R., Brea gum as wall material in the microencapsulation of corn oil by spray drying: Effect of inulin addition, *Food Res. Int.*, 103, 76-83, 2018.
48. Sansone, F., Mencherini, T., Picerno, P., d'Amore, M., Aquino, R.P., Lauro, M.R., Maltodextrin/pectin microparticles by spray drying as carrier for nutraceutical extracts. *J. Food Eng.*, 105, 3, 468-476, 2011.
49. Chan, S.Y., Choo, W.S., Young, D.J., Loh, X.J., Pectin as a rheology modifier: Origin, structure, commercial production, and rheology. *Carbohydr. Polym.*, 161, 118-139, 2017.

50. Tamm, F., Härter, C., Brodkorb, A., Driesch, S., Functional and antioxidant properties of whey protein hydrolysate/pectin complexes in emulsions and spray-dried microcapsules. *LWT*, 73, 524-527, 2016.
51. Aranaz, I., Paños, I., Peniche, C., Heras, Á., Acosta, N., Chitosan spray-dried microparticles for controlled delivery of venlafaxine hydrochloride. *Molecules*, 22, 11, 2017.
52. Estevinho, B.N., Damas, A.M., Martins, P., Rocha, F., Microencapsulation of β-galactosidase with different biopolymers by a spray-drying process. *Food Res. Int.*, 64, 134-140, 2014.
53. Liang, J., Yan, H., Puligundla, P., Gao, X., Zhou, Y., Wan, X., Applications of chitosan nanoparticles to enhance absorption and bioavailability of tea polyphenols: A review. *Food Hydrocoll.*, 69, 286-292, 2017.
54. Ribeiro, A.M., Shahgol, M., Estevinho, B.N., Rocha, F., Microencapsulation of Vitamin A by spray-drying, using binary and ternary blends of gum arabic, starch and maltodextrin. *Food Hydrocoll.*, 106029.
55. Moser, P., Telis, V.R.N., de Andrade Neves, N., García-Romero, E., Gómez-Alonso, S., Hermosín-Gutiérrez, I., Storage stability of phenolic compounds in powdered BRS Violeta grape juice microencapsulated with protein and maltodextrin blends. *Food Chem.*, 214, 308-318, 2017.
56. Nesterenko, A., Alric, I., Silvestre, F., Durrieu, V., Comparative study of encapsulation of vitamins with native and modified soy protein. *Food Hydrocoll.*, 38, 172-179, 2014.
57. Tontul, I. and Topuz, A., Spray-drying of fruit and vegetable juices: Effect of drying conditions on the product yield and physical properties. *Trends Food Sci. Technol.*, 63, 91-102, 2017.
58. Bazaria, B. and Kumar, P., Effect of whey protein concentrate as drying aid and drying parameters on physicochemical and functional properties of spray dried beetroot juice concentrate. *Food Biosci.*, 14, 21-27, 2016.
59. Bhusari, S.N., Muzaffar, K., Kumar, P., Effect of carrier agents on physical and microstructural properties of spray dried tamarind pulp powder. *Powder Technol.*, 266, 354-364, 2014.
60. Encina, C., Vergara, C., Giménez, B., Oyarzún-Ampuero, F., Robert, P., Conventional spray-drying and future trends for the microencapsulation of fish oil. *Trends Food Sci. Technol.*, 56, 46-60, 2016.
61. Aghbashlo, M., Mobli, H., Madadlou, A., Rafiee, S., Influence of wall material and inlet drying air temperature on the microencapsulation of fish oil by spray drying. *Food Bioproc. Tech.*, 6, 6, 1561-1569, 2013.
62. Jarunglumlert, T., Nakagawa, K., Adachi, S., Influence of aggregate structure of casein on the encapsulation efficiency of β-carotene entrapped via hydrophobic interaction. *Food Struct.*, 5, 42-50, 2015.
63. Liu, F., Chen, Z., Tang, C.H., Microencapsulation properties of protein isolates from three selected Phaseolus legumes in comparison with soy protein isolate. *LWT*, 55, 1, 74-82, 2014.

64. Muzaffar, K. and Kumar, P., Parameter optimization for spray drying of tamarind pulp using response surface methodology. *Powder Technol.*, 279, 179-184, 2015.
65. Tapal, A. and Tiku, P.K., Complexation of curcumin with soy protein isolate and its implications on solubility and stability of curcumin. *Food Chem.*, 130, 4, 960-965, 2012.
66. Anandharamakrishnan, C. (ed.), *Handbook of drying for dairy products*. John Wiley & Sons, 2017.
67. FAO/WHO, *Probiotics in food*. Health and nutritional properties and guidelines for evaluation, Rome, Italy: FAO Food and Nutrition Paper No. 85, 2006.
68. Huang, S., Vignolles, M.L., Chen, X.D., Le Loir, Y., Jan, G., Schuck, P., Jeantet, R., Spray drying of probiotics and other food-grade bacteria: A review. *Trends Food Sci. Technol.*, 63, 1-17, 2017.
69. Santivarangkna, C., Kulozik, U., Foerst, P., Alternative drying processes for the industrial preservation of lactic acid starter cultures. *Biotechnol. Prog.*, 23, 2, 302-315, 2007.
70. Huang, S., Rabah, H., Jardin, J., Briard-Bion, V., Parayre, S., Maillard, M.B., Loir Y.L., Chen X.D., Schuck P., Jeantet R., Jan, G., Hyperconcentrated sweet whey, a new culture medium that enhances *Propionibacterium freudenreichii* stress tolerance. *Appl. Environ. Microbiol.*, 82, 15, 4641-4651, 2016.
71. Huang, S., Cauty, C., Dolivet, A., Le Loir, Y., Chen, X.D., Schuck, P., Jan, G., Jeantet, R., Double use of highly concentrated sweet whey to improve the biomass production and viability of spray-dried probiotic bacteria. *J. Funct. Foods*, 23, 453-463, 2016.
72. Desmond, C., Fitzgerald, G.F., Stanton, C., Ross, R.P., Improved stress tolerance of GroESL-overproducing *Lactococcus lactis* and probiotic *Lactobacillus paracasei* NFBC 338. *Appl. Environ. Microbiol.*, 70, 10, 5929-5936, 2004.
73. Zhang, Y., Lin, J., Zhong, Q., Effects of media, heat adaptation, and outlet temperature on the survival of *Lactobacillus salivarius* NRRL B-30514 after spray drying and subsequent storage. *LWT*, 74, 441-447, 2016.
74. Bove, P., Capozzi, V., Garofalo, C., Rieu, A., Spano, G., Fiocco, D., Inactivation of the ftsH gene of *Lactobacillus plantarum* WCFS1: effects on growth, stress tolerance, cell surface properties and biofilm formation. *Microbiol. Res.*, 167, 4, 187-193, 2012.
75. Zuo, F., Yu, R., Khaskheli, G.B., Ma, H., Chen, L., Zeng, Z.,Mao, A., Chen, S., Homologous overexpression of alkyl hydroperoxide reductase subunit C (ahpC) protects Bifidobacterium longum strain NCC2705 from oxidative stress. *Res. Microbiol.*, 165, 7, 581-589, 2014.
76. Corcoran, B.M., Ross, R.P., Fitzgerald, G.F., Dockery, P., Stanton, C., Enhanced survival of GroESL-overproducing *Lactobacillus paracasei* NFBC 338 under stressful conditions induced by drying. *Appl. Environ. Microbiol.*, 72, 7, 5104-5107, 2006.

77. Bielecka, M. and Majkowska, A., Effect of spray drying temperature of yoghurt on the survival of starter cultures, moisture content and sensoric properties of yoghurt powder. *Nahrung*, 44, 4, 257-260, 2000.
78. Kets, E.P.W., Teunissen, P.J.M., Debont, J.A.M., Effect of compatible solutes on survival of lactic acid bacteria subjected to drying. *Appl. Environ. Microbiol.*, 62, 1, 259-261, 1996.
79. Silva, J., Carvalho, A.S., Ferreira, R., Vitorino, R., Amado, F., Domingues, P., Teixeira, P., Gibbs, P.A., Effect of the pH of growth on the survival of *Lactobacillus delbrueckii subsp. bulgaricus* to stress conditions during spray-drying. *J. Appl. Microbiol.*, 98, 3, 775-782, 2005.
80. Hussain, M.A., Knight, M.I., Britz, M.L., Proteomic analysis of lactose-starved *Lactobacillus casei* during stationary growth phase. *J. Appl. Microbiol.*, 106, 3, 764-773, 2009.
81. Mills, S., Stanton, C., Fitzgerald, G.F., Ross, R., Enhancing the stress responses of probiotics for a lifestyle from gut to product and back again. *Microb. Cell Fact.*, 10, 1, S19, 2011.
82. Lian, W.C., Hsiao, H.C., Chou, C.C., Survival of bifidobacteria after spray-drying. *Int. J. Food Microbiol.*, 74, 1-2, 79-86, 2002.
83. Sauvant, P., Cansell, M., Sassi, A.H., Atgie, C., Vitamin A enrichment: Caution with encapsulation strategies used for food applications. *Food Res. Int.*, 46, 469-479, 2012.
84. Bhandari, B., Spray drying - an encapsulation technique for food flavours. In: Mujumdar, A.S. (ed.), *Drying of Products of Biological Origin*. Science Publishers, Enfield, USA, 2005.
85. Ranadheera, C.S., Evans, C.A., Adams, M.C., Baines, S.K., Microencapsulation of *Lactobacillus acidophilus* LA-5, *Bifidobacterium animalis subsp. lactis* BB-12 and *Propionibacterium jensenii* 702 by spray drying in goat's milk. *Small Ruminant Res.*, 123, 1, 155-159, 2015.
86. Burgos-Díaz, C., Opazo-Navarrete, M., Soto-Añual, M., Leal-Calderón, F., Bustamante, M., Food-grade Pickering emulsion as a novel astaxanthin encapsulation system for making powder-based products: Evaluation of astaxanthin stability during processing, storage, and its bioaccessibility. *Food Res. Int.*, 109244, 2020.
87. Lourenço, S.C., Fraqueza, M.J., Fernandes, M.H., Moldão-Martins, M., Alves, V.D., Application of edible alginate films with pineapple peel active compounds on beef meat preservation. *Antioxidants*, 9, 8, 667, 2020.
88. Apintanapong, M. and Noomhorm, A., The use of spray drying to microencapsulate 2-acetyl-1-pyrroline, a major flavour component of aromatic rice. *Int. J. Food Sci. Technol.*, 38, 2, 95-102, 2003.
89. Guo, J., Li, P., Kong, L., Xu, B., Microencapsulation of curcumin by spray drying and freeze drying. *LWT*, 132, 109892, 2020.
90. Janiszewska, E., Microencapsulated beetroot juice as a potential source of betalain. *Powder Technol.*, 264, 190-196, 2014.

91. Ersus, S., Yurdagel, U., Microencapsulation of anthocyanin pigments of black carrot (Daucus carota L.) by spray drier. *J. Food Eng.*, 80, 3, 805-812, 2007.
92. Medina-Torres, L., García-Cruz, E.E., Calderas, F., González Laredo R.F., Sánchez-Olivares, G., Gallegos-Infante, J.A., Rocha-Guzmán, N.E., Rodriguez-Ramirez, J., Microencapsulation by spray drying of gallic acid with nopal mucilage (Opuntia ficus indica). *LWT*, 50, 2, 642-650, 2013.
93. Estevinho, B.N., Carlan, I., Blaga, A., Rocha, F., Soluble vitamins (vitamin B12 and vitamin C) microencapsulated with different biopolymers by a spray drying process. *Powder Technol.*, 289, 71-78, 2016.
94. Bae, E.K. and Lee, S.J., Microencapsulation of avocado oil by spray drying using whey protein and maltodextrin. *J. Microencapsul.*, 25, 8, 549-560, 2008.
95. Carneiro, H.C., Tonon, R.V., Grosso, C.R., Hubinger, M.D., Encapsulation efficiency and oxidative stability of flaxseed oil microencapsulated by spray drying using different combinations of wall materials. *J. Food Eng.*, 115, 4, 443-451, 2013.
96. Rodea-González, D.A., Cruz-Olivares, J., Román-Guerrero, A., Rodríguez-Huezo, M.E., Vernon-Carter, E.J., Pérez-Alonso, C., Spray-dried encapsulation of chia essential oil (*Salvia hispanica* L.) in whey protein concentrate-polysaccharide matrices. *J. Food Eng.*, 111, 1, 102-109, 2012.
97. Yoo, S.H., Song, Y.B., Chang, PS., Lee, H.G., Microencapsulation of α-tocopherol using sodium alginate and its controlled release properties. *Int. J. Biol. Macromol.*, 38, 1, 25-30, 2006.
98. Diosady, L.L., Alberti, J.O., Mannar, M.V., Microencapsulation for iodine stability in salt fortified with ferrous fumarate and potassium iodide. *Food Res. Int.*, 35, 7, 635-642, 2002.
99. Oneda, F. and Ré, M.I., The effect of formulation variables on the dissolution and physical properties of spray-dried microspheres containing organic salts. *Powder Technol.*, 130, 1-3, 377-384, 2003.
100. Kaur, P., Kaur, A. and Kumar, N., Optimization of Spray Drying Conditions for Production of Aloe-Vera Powder. *Chemical Science Review and Letters*, 6, 21, 525-532, 2017.

6

Solar Drying: Principles and Applications

Baher M. A. Amer[1,2]

[1]*Department of Agricultural Systems Engineering, College of Agricultural & Food Sciences, King Faisal University, P.O. Box 420, Al-Hofuf, Al-Ahsa, Saudi Arabia*

[2]*Department of Agricultural Engineering, Faculty of Agriculture, Cairo University, Giza, Egypt*

Abstract

Solar energy is one of the most obtainable renewable energies all over the world. Hence, it is high time for researchers to develop suitable drying technology for the production of high-quality agricultural products coping with the socio-economic condition of the countries. Solar dryers are used to improve the quality of the dried product and reduce the drying time compared to sun drying. There are many types and applications of solar dryers used to dry different types of agricultural products such as crops, fruits, vegetables meat, and fish. These applications have been improved through many years by many researchers, resulting in an increase in efficiency.

***Keywords*:** Solar dryer, construction, classification, application

6.1 Introduction

Drying is one of the important preservation techniques for agricultural products. Anciently, natural/sun drying was employed to remove the moisture from foods. Later on, solar drying was mechanized and nowadays is being used efficiently by employing specially constructed solar dryers. The use of solar dryers could avoid direct contact between the product and the ground during sun drying. Besides, it prohibits the permeation of insects in the dried products [1]. It is also used to control the affecting drying air factors like temperature, relative humidity, and velocity which are difficult

Email: bamer@kfu.edu.sa; baher.amer@agr.cu.edu.eg

Nitin Kumar, Anil Panghal and M. K. Garg (eds.) Thermal Food Engineering Operations, (179–196)

to control by natural open sun drying [2]. There are many types and categories for solar dryers that were developed to improve their performance and to overcome the disadvantages of sun drying. Recently, some computer and mobile applications are being used to simulate and enhance the experimental performance of solar dryers.

6.2 Principle of Solar Drying

Solar drying is better than sun drying for its advantages, according to Esper and Mühlbauer [3], and Sablani [4] which are:

1. Significant improvement of the product quality in terms of color, texture, and taste.
2. No contamination by insects, microorganisms, and mycotoxin.
3. Reduction of drying time up to 50%.
4. Reduction of the drying and storage losses to a minimum.
5. Considerable increment in the shelf life of products.

Kordylas (1990) [5] stated that the principle of solar drying is simple. If a small amount of air is held in a relatively confined space, it will get warmer no matter how much heat is received from the sun. The confined air will get hotter if less air is allowed to move through the confined space.

He mentioned that ideal conditions for sun drying or solar drying are as follows:

- The food items must be out of direct sunlight.
- The food must be heated directly but heated by warm air moving across the food at a constant rate.
- Moisture or water vapor must be constantly removed from the drying food.
- A temperature range of 35 to 38°C must be constantly maintained.

Also, it has been mentioned that the higher temperatures will help to shorten the drying time. As with sun-drying, drying in a solar dryer is at the mercy of the weather. Thus, there is no effective way of controlling the temperature. The relative humidity is another key factor in solar drying. The lower the relative humidity, the greater is the capacity of the air to absorb moisture from drying food. However, this capacity varies with temperature and increases as the temperature rises. If air comes into a solar dryer, it can be heated further by 38°C or more, thus, its moisture absorbing capacity is significantly increased. Under these conditions, moisture can be rapidly removed from the food, which is highly beneficial to the food, especially during the first part of the drying process, when flavor and vital nutrients are at their peak [5, 6].

Using solar energy is one of the oldest types of food preservation techniques which is still practiced today for foods by removing water [7]. The advantages of this process are that the energy is renewable, free, and non-polluting. The disadvantages are that it can be very labor-intensive, suffers a lack of control, and is prone to deterioration by insect infestation or biochemical reactions, or microbiological. Besides, several researches have been conducted to improve simple systems that could be used in developing countries and areas where other energy sources are limited [8, 9].

6.3 Construction of Solar Dryer

The simplest method of carrying out solar drying is by placing the product on the ground in the sun. However, this leaves the material open to spoilage reactions as described earlier. To enhance the drying conditions, simple structures can be cheaply built. Many researchers have reported several studies related to the construction of solar dryer which are as discussed below.

Sodha *et al.* (1987) [10] reported that a solar dryer can be constructed with racks and a drying chamber, where the product to be dried is laying and the drying takes place subsequently. It also includes a thermal device that heats the airflow and the product, a device that produces the airflow through the apparatus, and a thermal storage device.

Kordylas [5] and Imre and Mujumdar [6] mentioned that solar dryers are designed both to concentrate the sun's heat and to provide the airflow necessary for efficient drying. The best results are obtained when drying is done mainly on sunny days. Several researchers have constructed solar dryers; therefore they are not always as well built as they could be. Some have fans inserted, which greatly increases their efficiency by reducing the drying time. Others operate by convection currents, which take longer and have less predictable results. The drying food must be turned and rotated to promote uniform and thorough drying. Various plans for the building of solar dryers are available from international agencies. These include models based on a chimney, a modified cold frame, a barrel, and box dryers. These items can collect the sun's rays thus raising the air temperature inside it by 20 to 25°C above the ambient air temperature.

Imre and Mujumdar [6] and Knorr [11] mentioned that dryers essentially consist of a transparent cover system or black-painted surfaces that absorb most of the heat or the radiation incident on them, and a drying chamber that holds the substance to be dried. Inlet and outlet points must be provided for adequate amounts of airflow by natural flow air or by a fan. Easy accessibility to the material being drying for mixing and replacement is also essential for successful design.

Several designs for solar dryers have been well documented which range from simple cabinet dryers, convection dryers, shelf dryers to more complicated semi-artificial dryers that include some form of a heat storage device. Thus, the main parts of a solar dryer can be listed as being:

1. Drying space, where the material to be dried is placed, and where the drying takes place.
2. Collector to convert solar radiation to heat.
3. Auxiliary energy source (optional).
4. Heat transfer equipment transferring heat to the drying air and/or to the material.
5. A means for keeping drying air inflow may be a fan or a chimney.
6. Unit of heat storage (optional).
7. Measuring and control equipment (optional).
8. Ducts, pipes, and other appliances.

A reflective mirror can also be used to increase heat transfer area and the use of selective coatings on absorber plates to increase the collection of solar radiation. These can improve the thermal performance of the dryer to a great extent [12].

6.4 Historical Classification of Solar Energy Drying Systems

The design of solar dryers provides both amplifying the sun's heat and supplying an airflow for efficient drying. These solar drying systems could be classified into several groups by different researches as follows:

Brenndorfer *et al.* (1985) [13] mentioned that more sophisticated methods (solar drying) collect solar energy and heated air, which in turn is used for drying. Solar dryers are classified into:

1. Direct natural-circulation dryers (a combined collector and drying chamber).
2. Direct dryers with a separate collector.
3. Indirect forced-convection dryers (separate collector and drying chamber).

Sodha *et al.* [10] and Garg [14] stated that based on available designs, solar dryers may be classified into several general categories, depending upon the mode of heating or the operational mode of heat derived from the solar radiation and the subsequent use of this heat to remove moisture from the wet product. They classified the solar dryers into two basic categories:

1. Natural dryers, which do not require any mechanical or electrical power to run a fan.
2. Forced convection dryers, require the use of a fan to pump air through the product.

Both dryers can be classified further as direct-mode and indirect-mode dryers.

Knorr (1993) [11] mentioned that solar energy has been used for drying vegetables, fruits, and grains since antiquity. Several different configurations have evolved through the decades to meet specific demands in dehydration. It is difficult to divide the solar dryers and other dryers into distinct categories, but an attempt could be made to classify them according to their size and way of utilizing solar energy as follows:

1. Open-air, direct radiation dryers.
2. Closed chamber, direct radiation dryers.
3. Preheated air dryers.
4. Industrial-scale crop dryers.

Esper and Mühlbauer (1998) [3] also, mentioned that in simple natural convection type dryers, the insect infestations cannot be avoided especially during long extended periods of adverse weather which leads to crop deterioration. However, using a fan which can force air as in the case of solar cabinet dryer, solar tunnel dryer, solar processing center, and greenhouse solar dryer can provide the airflow required to remove the evaporated moisture. The air circulation breaks down completely during the night and in cloudy weather. There are two drying methods by using solar energy, natural sun drying and solar drying which are divided into two types, namely, the simple natural convection type dryers and the solar convection dryer.

Imre and Mujumdar (2006) [6] reported that there are different methods of thermal energy storage used in solar dryers based on the method of energy, which can be either chemical or thermal

Mustayen *et al.* (2014) [15] submitted that solar drying can be categorized into four categories according to the mechanism used to transfer the energy of evaporation to the product:

1. Natural or sun dryers.
2. Direct or cabinet dryers.
3. Indirect solar dryers.
4. Mixed type solar dryers.

Solar energy drying systems are classified primarily according to their heating modes and how the solar heat is utilized. In broad terms, they can be categorized into two major groups [16]:

1. Active solar energy drying systems: Most types of which are often termed hybrid solar dryers.
2. Passive solar energy drying systems: Conventionally termed natural-circulation solar drying systems.

Three distinct sub-classes of either the active or passive solar drying systems can be identified which vary mainly in the design arrangement of system components and the mode of utilization of the solar heat:

- Integral-type solar dryers
- Distributed-type solar dryers
- Mixed-mode solar dryers.

Kumar *et al.* (2016) [17] mentioned the classification of indirect type solar dryer systems as follows:

a. Natural convection dryer without any thermal storage and with thermal storage in the form of latent heat storage and sensible heat storage.
b. Forced convection dryer without any thermal storage and with thermal storage as in the case of recirculation solar dryer

However, Lingayat *et al.* (2020) [18] also reported that it is two main drying groups according to the utilization of solar power which are: open drying and controlled drying with solar dryer. The open drying is divided into sun drying and rack type. The controlled drying with solar dryer is divided into the natural convection dryer and the forced convection dryer, which is subdivided into cabinet dryer, greenhouse dryer, tunnel dryer, mixed dryer, indirect, and hybrid dryer.

Acar *et al.* (2020) [19] classified of dryers based on their type of operation as follows:

a. Batch: divided into dispersion and layer which includes contact, convective and special.
b. Continuous: divided into dispersion and layer which includes conductive, convective, and special.

However, using the indirect dryers enable better and faster drying as the damage of products is reduced during the drying process. To improve the quality of food products in terms of color, texture, and taste, many types of indirect solar dryers have been developed. Some types of indirect solar dryers were built to decrease the drying time and achieve better efficiency for short periods [20, 21].

6.5 Storing Solar Energy for Drying

Misra *et al.* (1982) [22] reported that the advantage of storing solar heat for several weeks for use in grain drying was to enable drying to proceed independently of the fall weather conditions. This allowed management flexibility in harvesting and drying the crops. The major disadvantage was that it required more hardware, in the form of a large heat storage structure and heat recovery equipment, which could lead to the excessive cost.

Variability and time-dependent characteristic of solar radiation make storage necessary for continuous operations of food processing as drying [23]. It has been found by Goyal and Tiwari (1999) [24] that the thermal energy storage effect can be observed during non-sunshine hours which is very important for drying crops.

The operation of a solar-assisted dryer extended through the night hours and it was found that thermal storage during the day can be used as a heat source during the night for continuing the drying of agricultural products and also preventing their rehydration from the surrounding air, [25–27].

Storing energy is one of the best solutions for solving the problem of long drying time and continued drying even at night [28, 29]. There are two ways for thermal energy storage: chemical or thermal. It is preferable to apply a thermal way approach in solar drying. In this way, sensible heat storage can be used by increasing the temperature of a liquid or a solid. That means increasing their internal energy during the high radiation period and discharging it during the low radiation period [30]. The factors affecting the quantity of stored energy are the characteristics of the stored materials and the diversity of the radiation. The best liquid storage material is water because of its cost, its specific heat properties, and the temperature range of its use [31, 32, 58].

6.6 Hybrid/Mixed Solar Drying System

It has been found that the systems which utilize solar energy beside energy from conventional sources for drying purposes are known as hybrid/mixed drying system. In addition to a fan, an air heater, and a drying chamber; the system consists of two main components as the following [10]:

1. Auxiliary unit: since solar radiation is an intermittent source of energy, some additional items of air heating must be provided for the continuous operation of the system during night-time as well as cloudy-rainy daytime hours. These means could be an electric heating system or an oil-gas-fired system.
2. Storage unit: is occasionally provided to store excess solar energy from peak radiation hours and then to deliver it to the ambient air to dry products during low or non-radiation hours. The item generally used for this purpose is the rock-bed, thermal storage type.

It is reported that auxiliary supply and storage can be used to assess the compatibility of solar energy to meet the drying process temperature [33, 34]. Some researches of latent heat storage for a water solar heating system and electricity as an auxiliary source were conducted for tomatoes [35], banana [36], chamomile [37], and for drying mushroom [38] (Figure 6.1). Otherwise, few researches of latent heat storage for a water solar heating system were conducted to be as the only supplementary source for the solar system for pinewood solar drying [39] and for drying osmotically cherry tomatoes [40] (Figure 6.2).

Figure 6.1 Photograph of solar collector and dryer [35].

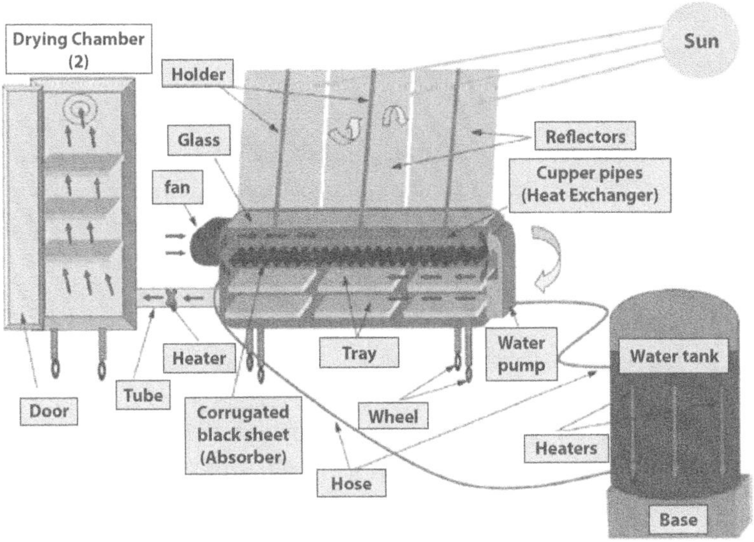

Figure 6.2 Schematic diagram of the components of the integrated solar dryer system [37].

Some hybrid dryers had been constructed to control the drying air conditions throughout the drying time especially at night when it is not possible to use solar energy, such as using steam beside the solar energy, produced by using a biomass stove [41].

Significant improvement can be registered after the heater is added to the solar dryer during periods of low sunshine [42, 43]. Besides, a

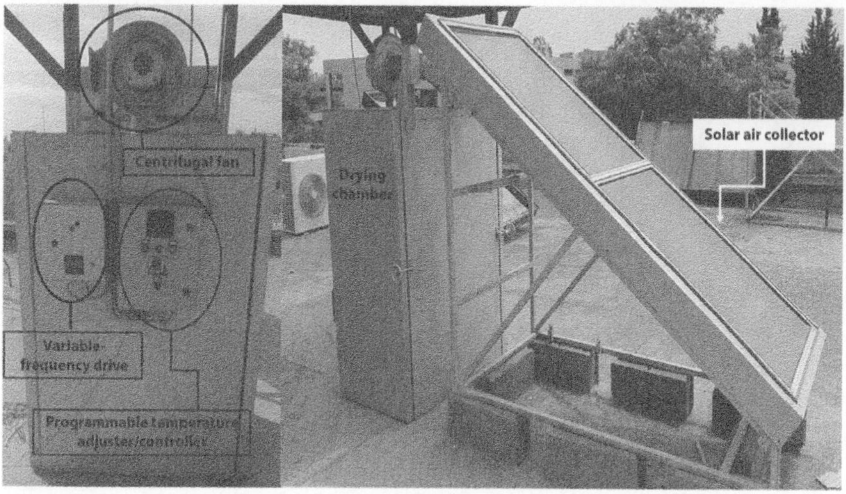

Figure 6.3 The forced convective solar dryer implemented in Marrakech [44].

programmable temperature controller and variable frequency drive can be provided to an indirect-hybrid solar-electrical forced convection dryer for drying apple peels Moussaoui *et al.* (2020) [44]. Besides, some hybrid solar dryers are employing PV modules and solar battery [45] (Figure 6.3).

6.7 Solar Greenhouse Dryer

Considerable research has been done to investigate the performance of solar greenhouse dryers [46–48] (Figure 6.4 and Figure 6.5).

6.8 Solar Drying Economy

Sodha *et al.* (1987) [10] mentioned that conventional drying techniques utilize large-scale, fossil-fuel-fired air dryers. In many cases, the hot combustion gases are passed directly through the grain, which often becomes contaminated by unburned fuel, fumes, and soot. Electrical heating of air for drying is preferred, but it is very expensive and not feasible in rural areas of developing countries. Of the other energy sources, solar energy

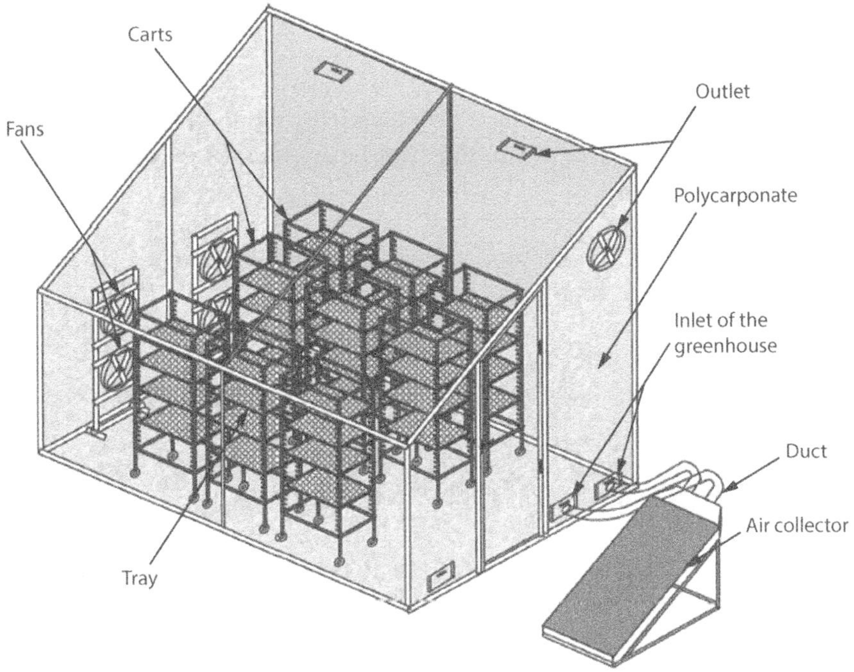

Figure 6.4 Schematic diagram of the new mixed solar dryer [47].

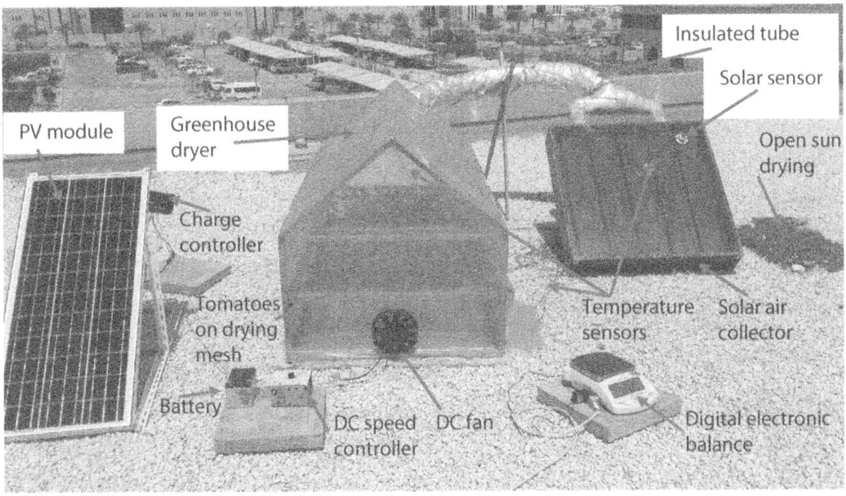

Figure 6.5 Experimental setup of solar PV operated the greenhouse dryer for drying tomatoes [48].

may be the most promising answer to the demands of drying, particularly in developing countries, most of which receive a high degree of solar radiation throughout the year.

[5, 6] mentioned that in a direct contact dehydration plant, between 25 and 40 percent of the total running cost is accounted for by the cost of heating the air to the required temperature. If cheaper energy sources are made available, it can greatly reduce the cost of industrial dehydration. Suitable solar energy collectors are available to heat the air. These help to cut down the use of other fuels like coal, electricity, and oil, which are then conserved. The cost of the operation is thus lowered. Though solar energy collectors are a cheap source of energy, they have some limitations. The most obvious limitation is that they rely on bright sunlight, which is often unavailable during the rainy season. This limitation, however, can be overcome by coupling or by combining the use of solar energy collectors with other conventional sources of energy, which are switched on when there is no sunlight. Solar drying equipment is commercially available and includes solar collectors, racks, trays, ventilated cold frames, and hoods. This equipment is relatively inexpensive when compared with electrical units. The dryers which enclose solar collectors with controlled ventilation are quite good for 100% solar drying. The balance between temperature and airflow is quite delicate and success depends on the operator. For sanitary reasons and efficient drying, the frame should be airtight, except for the vents which help to direct air through the dryer. These must be screened to prevent insects from sampling the drying food.

Adelaja *et al.* [49] and Agrawal *et al.* [50] mentioned that the extra cost and complexity of storage is a negative point to solar systems. The solar system also requires a good solar climate for efficient operation. Where land for solar collectors is abundant and energy demand is high, the solar flux will be also high and the direct beam component will be maximum. Finally, the solar system must meet the criteria of economic competitiveness to be widely accepted. On the other hand, solar energy has an advantageous position compared with scarce fossil fuels.

6.9 New Applications Related to Solar Drying

Some years ago, there were some applications related to solar drying appeared. These applications are computer software or programs used to predict the behavior of the solar drying process or used to calculate the basic equation related to the same process. These applications could be mentioned as the following:

MATLAB program

MATLAB is a very useful tool for developing mathematical models to predict the crop temperature, air temperature, the moisture evaporated and for predicting the thermal performance of the solar dryer. The software developed has been experimentally validated. It has been proven that the analytical and experimental results for jaggery drying are in good agreement [51, 56, 57].

The MATLAB-based software TLDRY has been developed by Wang *et al.* (2004) [52] to examine the range of applicability of equations and several other thin layers drying equations as reported in the literature. The software proved to be a convenient tool to identify the range of applicability and potential uncertainties for these equations.

CFD program

A computational fluid dynamics model was developed by Sanghi *et al.*, 2018 [53] to simulate the corn drying process in a solar cabinet dryer. Incident solar radiation was modeled using a dual-band spectrum to simulate the absorption of shortwave radiation by corn which accounts for the greenhouse effect caused by glazing materials. The model allowed visualization of temperature, humidity, and air velocity profiles in the dryer. The model was validated with experimental results, which showed an overprediction of temperature (8.5%) and humidity (21.4%).

An investigation for the performance of a solar cabinet drying system equipped with a heat pipe evacuated tube solar collector and thermal storage system with application of PCM has been reported by Iranmanesh *et al.* (2020) [54]. CFD modeling of the system and quality evaluation of dried

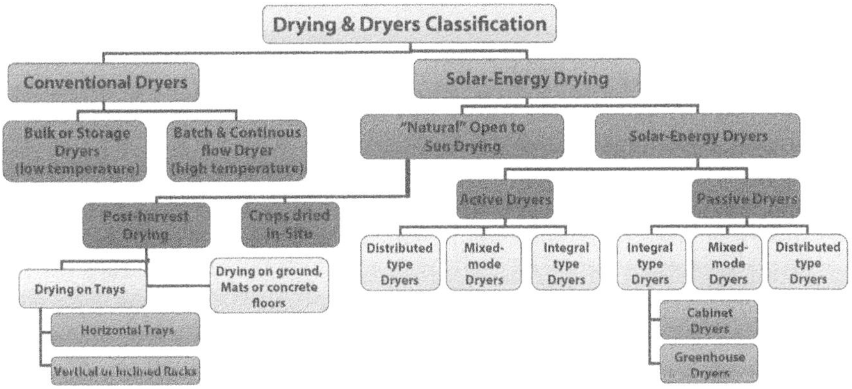

Figure 6.6 Dryers and drying modes and classification.

apple slices was considered. The performance of the dryer was simulated and validated by the experimental data.

Amer and Albaloushi (2020) [45] used the COMSOL program to simulate the behavior of air temperature, velocity and to predict the performance of the developed solar dryer with a water storage heat unit for drying agricultural products such as Cantaloupe.

Mobile software

A new software application by Amer and Chouikhi (2020) [55] was developed on mobile phones depending on some of the mathematical equations used to conduct some of the solar drying characteristics to assist researchers and engineers working in the food engineering field. This innovative program can be installed on cellphones in Android and iOS systems to save effort and time.

References

1. Labed, A., Moummi, N., Aoues, K., Benchabane, A., Solar drying of henna (Lawsonia inermis) using different models of solar flat plate collectors: an experimental investigation in the region of Biskra (Algeria). *J. Clean. Prod.*, 112, 2545-2552. 2016.
2. Amer, B.M.A., Gottschalk, K. Studying the affecting factors on drying rates of plum fruits under varying drying air conditions. In *Proceedings of the 16th Triennial World Congress IFAC, Prague, Czech Republic, 4–8 July 2005*; 52–56.
3. Esper, A., Mühlbauer, W., Solar drying - an effective means of food preservation. *Renewable Energy*, 5, 95–100. 1998.
4. Sablani, S.S., Drying of Fruits and Vegetables: Retention of Nutritional Quality. *Drying Technol.*, 24: 428–432. 2006.
5. Kordylas, J.M., *Processing and Preservation of Tropical and Subtropical Foods.* Macmillan Education Ltd. London. 1990.
6. Imre, L., Mujumdar, A.S., *Solar Drying, Handbook of Industrial Drying.* Boca Raton: CRC Press. Marcel Dekker, Inc. USA. 2006.
7. Lamidi R.O., Jiang L, Pathare P.B., Wang Y.D., Roskilly A.P., Recent advances in sustainable drying of agricultural produce: a review. *Appl. Energy*, 233–234:367–85, 2019. https://doi.org/10.1016/J.APENERGY.2018.10.044.
8. Anderson J.O, Westerlund L., Improved energy efficiency in sawmill drying system. *Appl Energy*; 113:891–901, 2014. https://doi.org/10.1016/j.apenergy.
9. Seyfi S., Design, experimental investigation and analysis of a solar drying system. *Energy Convers Manag*; 68:227–34, 2013. https://doi.org/10.1016/j.enconman.

10. Sodha, M. S., Bansal, N. K., Kumar, A., Bansal, P. K., and Malik, M. A. S. Solar Crop Drying. CRC Press, Boca Raton, Florida, USA. 1987.
11. Knorr, D. *Sustainable Food Systems*. AVI Publishing Company, USA. 1993.
12. Yagnesh B. Chauhan, Pravin P. Rathod, A comprehensive review of the solar dryer, *International J. Ambient Energy*, 41:3, 348-367, 2020. DOI: 10.1080/01430750.2018.1456960.
13. Brenndorfer, B., Kenneddy, L., Oswin-Bateman, C.O. and Trim, D.S., *Solar dryers*. Commonwealth Science Council, Commonwealth Secretariat, Pall Mall, London SW1Y 5HX. 1985.
14. Garg, H.P., *Solar Food Drying. In Advances in Solar Energy Technology. Heating, Agricultural and Photovoltaric Application of Solar Energy*. (3). D. Reidel Publishing Co.: Dordrecht, Holland. 1987.
15. Mustayen, A.G.M.B., Mekhilef, S., Saidur, R., Performance Study of Different Solar Dryers: A Review. *Renewable and Sustainable Energy Reviews* 34: 463–470. 2014.
16. Ekechukwe, O. V., Norton, B., Review of Solar Energy Drying Systems II: An Overview of Solar Drying Technology. *Energy Conversion and Manag.* 40, 616–655. 1999.
17. Kumar M, Sansaniwal S.K, Khatak P., Progress in solar dryers for drying various commodities. *Renew Sustain Energy Rev*; 55, 346–60, 2016. https://doi.org/10.1016/j.rser.2015.10.158.
18. Lingayat, A. B., Chandramohan, V.P. Raju, V.R.K., Meda, V., A review on indirect type solar dryers for agricultural crops – Dryer setup, its performance, energy storage and important highlights. *Applied Energy* 258, 2020.
19. Acar, C., Dincer, I. and Mujumdar, A., A comprehensive review of recent advances in renewable-based drying technologies for a sustainable future, *Drying Technol.*, 2020. DOI: 10.1080/07373937.
20. Lingayat, A., Chandramohan, V.P., Raju, V.R.K., Kumar, A., Development of indirect type solar dryer and experiments for estimation of drying parameters of apple and watermelon *Thermal Sci. and Eng. Progress*, 16, 2020.
21. Vijayan, S., Arjunan, T.V., Kumar, A., Exergo-environmental analysis of an indirect forced convection solar dryer for drying bitter gourd slices. *Renewable Energy* 146, 2210-2223, 2020.
22. Misra, N.R., Kenner, M.H., Roller, L.W., Solar heat for corn drying under Ohio conditions II. Summer storage of solar heat. *Transactions of the ASAE*, 459-464, 1982.
23. Miller WM. Energy storage via desiccants for food/agricultural applications. *Energy in Agric.*, 2(4), 341-354, 1983.
24. Goyal, R.K., Tiwari, G.N., Effect of thermal storage on a performance of a deep bed drying. *Ambient Energy*. 20(3), 125-136, 1999.
25. Maroulis, Z.B., Saravacos, G.D., Solar heating of air for drying agricultural products. *Solar & Wind Technology*; 3, 127-134, 1986.

26. Aboul-Enein, S., El-Sebaii, A.A., Ramadan M.R.I., El-Gohary, H.G., Parametric study of a solar air heater with and without thermal storage for solar drying applications. *Renewable Energy*; 21, 506-522, 2000.
27. Jain, D., Jain R.K., Performance evaluation of an inclined multi-pass solar air heater with in-built thermal storage on deep-bed drying application. *J. Food Eng.*, 65, 497-509, 2004.
28. Shalaby, S.M., Bek, M.A., Experimental investigation of a novel indirect solar dryer implementing PCM as energy storage medium, *Energy Convers. Manag.* 83, 1-8, 2019.
29. Yadav, S., Chandramohan, V.P., Performance comparison of thermal energy storage system for indirect solar dryer with and without finned copper tube, *Sust. Energy Technol.s and Assess.* 37, 1-14, 2020. https://doi.org/10.1016/j.seta.2019.100609.
30. Bal, L.M, Satya, S., Naik, S.N., Meda, V., Review of solar dryers with latent heat storage systems of agricultural products. *Renew Sust Energy Rev*, 15:876–880, 2011.
31. Atalay, H. Performance analysis of a solar dryer integrated with the packed bed thermal energy storage (TES) system, *Energy* 172, 1037-1052, 2019. 10.1016/j.energy.2019.02.023.
32. Bennamoun, L., Improving Solar Dryers' Performances Using Design and Thermal Heat Storage. *Food Eng. Reviews*, 5(4), 230-248, 2013.
33. Vlachos, NA, Karapantsios TD, Balouktsis, AI, Chassapis D., Design and testing of a new solar tray dryer. *Drying Tech*nol. 20(6), 1243-1271, 2002.
34. Amer, B.M.A., Hossain, M.A., Gottschalk, K. Design and performance evaluation of a new hybrid dryer for banana. *Energy Convers. Manag.*, *51*, 813–820. 2010.
35. Hossain, M.A., Amer, B.M.A., Gottschalk, K., Hybrid Solar Dryer for Quality Dried Tomato. *Drying Technol.*, 26, (12), 1591-1601, 2008.
36. Hossain, M.A., Amer, B.M.A., Gottschalk, K., Mathematical modelling of solar assisted hybrid dryer for banana., In *Proceedings of the 5th International Technical Symposium on Food Processing, Monitoring Technol. in Bioprocesses and Food Quality Management*, Potsdam, Germany, 748–755, 31 August–2 September 2009.
37. Amer, B.M.A., Gottschalk, K., Hossain, M.A., Integrated hybrid solar drying system and its drying kinetics of chamomile. *Renewable Energy*, 121 (3), **539-547**, 2018.
38. Reyes, A., Mahn, A., Cubillos, F., Huenulaf, P., Mushroom dehydration in a hybrid solar-dryer. *Energy Convers. Manag.* 70: 31-39, 2013.
39. Luna, D., Nadeau, J.P., Jannot, Y., Model and simulation of a solar kiln with energy storage. *Renew Energy*, 35: 2533–2542, 2010.
40. Nabnean, S., Janjai, S., Thepa, S., Sudaprasert, K., Songprakorp, R., Bala, B.K., Experimental performance of a new design of solar dryer for drying osmotically dehydrated cherry tomatoes. *Renew Energy*, 94, 147-156, 2016.

41. Prasad J, Vijay VK., Experimental studies on drying of Zingiber officinale, Curcuma longa l. and Tinospora cordifolia in solar-biomass hybrid drier. *Renewable Energy*; 30, 2097-2109, 2005.
42. Bennamoun, L., Belhamri, A., Design and simulation of a solar dryer for agricultural products. *J. Food Eng.*, 59, 259-266, 2003.
43. Janjai S., Praditwong P., Development of a solar fruit dryer for tropical areas. *2nd World Renewable Energy Congress.*, 2, 908-912, 1992.
44. Moussaoui, H., Bahammou, Y., Tagnamas, Z., Kouhila, M., Lamharrar, A., Idlimam, A., Application of solar drying on the apple peels using an indirect hybrid solar-electrical forced convection dryer, *Renewable Energy*, In Press, 2020.
45. Amer, B.M.A., Albaloushi, N.S., A developed design to improve the performance of solar dryer supported by sensible heat storage and photovoltaric module for drying cantaloupe. *Fresenius Environmental Bulletin*, 30, 2920-2930, 2021.
46. Chauhan PS, Kumar A., Nuntadusit, C., Heat transfer analysis of PV integrated modified greenhouse dryer. *Renewable Energy*. 121, 53-65, 2018.
47. Djebli, A., Haninia, S., Badaouia, O., Boumahdia, M. A new approach to the thermodynamics study of drying tomatoes in mixed solar dryer. *Solar Energy*. 193, 164–174, 2019.
48. Azam, M.M.. Eltawil, M.A., Amer, B.M.A.. Thermal analysis of PV system and solar collector integrated with greenhouse dryer for drying tomatoes. *Energy, 212*, 2020.
49. Adelaja, O., Babatope, B. I., Analysis and Testing of a Natural Convection Solar Dryer for the Tropics. *Journal of Energy* Article ID 479894: 8, 2013. doi:10.1155/2013/479894.
50. Agrawal, Ashish, and R. M. Sarviya. A Review of Research and Development Work on Solar Dryers with Heat Storage. *International Journal of Sustainable Energy* 35 (6): 583–605, 2014.
51. Sanghi, A., Kingsly Ambrose, R. P. and Maier, D., CFD simulation of corn drying in a natural convection solar dryer, *Drying Tech*nol., 36, 7, 859-870, 2018. DOI: 10.1080/07373937.
52. Wang, Dai-Chyi, Fon, Din-Sue, Fang, W., Sokhansanj, S., Development of a Visual Method to Test the Range of Applicability of Thin Layer Drying Equations Using MATLAB Tools, *Drying Tech*nol., 22,8, 1921-1948, 2004. DOI: 10.1081/DRT-200032878.
53. Kumar, A., Tiwari, G.N., Thermal modeling of a natural convection greenhouse drying system for jaggery: An experimental validation. *Solar Energy* 80, 1135–1144, 2006.
54. Iranmanesh, M., Akhijahani, H.S., Mohammad Saleh Barghi Jahromi Corrigendum to CFD modeling and evaluation the performance of a solar cabinet dryer equipped with evacuated tube solar collector and thermal storage system. *Renew Energy.* 145, 1192-1213, 2020.

55. Amer, B.M.A., Chouikhi, H., Smartphone Application Using a Visual Programming Language to Compute Drying/Solar Drying Characteristics of Agricultural Products. *Sustainability*, 2020. DOI: 10.3390/SU12198148.
56. Attkan, A. K., Kumar, N., & Yadav, Y. K. (2014). Performance evaluation of a dehumidifier assisted low temperature based food drying system. *Journal of Environmental Science, Toxicology and Food Technology*, 8(1), 43-49.
57. Attkan, A. K., Raleng, A., Kumar, N., Alam, M. S., & Yadav, Y. K. (2016). Performance Evaluation of Solid Desiccant Wheel Dehumidifier for Agricultural Crop Drying. *International Journal of Bio-resource and Stress Management*, 7(6), 1284-1289.
58. Attkan, A. K., & Kumar, N. (2016). Effect of drying on physico-chemical and nutritional properties of fenugreek leaves. *Poljoprivredna tehnika*, 41(3), 27-36.

7

Fluidized Bed Drying: Recent Developments and Applications

Praveen Saini*, **Nitin Kumar, Sunil Kumar and Anil Panghal**

Department of Processing and Food Engineering, College of Agricultural Engineering and Technology, CCS Haryana Agricultural University, Hisar, India

Abstract

Drying is an essential unit operation in various processing units. Out of various drying techniques, fluidized bed drying is one of the well-known and versatile methods. It is known for its uniform drying and high efficiency, hence is well adopted for industrial drying. However, drying agro-products using fluidized bed dryers is an emerging trend. It is proven to be energy efficient and causes minimal quality variation of dried products compared to other drying techniques. This chapter elaborates on the principle and working mechanism of dryers accompanying fluidization technique along with their design configurations. This chapter also discusses some modified fluidized bed dryers, proposed to overcome the limitations of conventional fluidized bed dryers.

Keywords: Fluidized bed drying, design, applications, food quality, hybrid drying

7.1 Introduction

The quality of stored grains is dominantly affected by the moisture content, and drying is the most common process for moisture reduction without deterioration. The moisture content of a suitable range (up to 14% wet basis or less) promotes safe storage of grains over a long duration and also reduces the handling costs. There are several methods for drying; however, short-term drying of high-moisture grains (paddy, parboiled rice, maize,

**Corresponding author*: praveensaini507@gmail.com

Nitin Kumar, Anil Panghal and M. K. Garg (eds.) Thermal Food Engineering Operations, (197–220)

and soybean) with a forced convection air drying such as cross-flow dryer, fluidized bed dryers, and the spouted bed has been presented as a successful and efficient drying method [1]. These types of dryers are suitable for heat-sensitive food particles.

Drying is one of the most energy-intensive processes and has negative impacts on the environment due to the reason that the primary source of energy is from fossil fuels. About 12–25% of industrial energy consumption is related to the dehydration process and the demands for energy for drying will further increase. Fluidized bed drying has been adopted widely due to its various advantages, including uniform moisture reduction with less drying time, higher drying rates, relatively high thermal efficiency, low energy consumption and cost of operation, etc. [2]. Consequently, the dryer size is compact compared to its drying capacity. Besides, this drying technique offers easy operation and dryer maintenance facilities, compatibility to automation, maintains the grain's physical quality, and also offers easy attachment of other processes like mixing, cooling, classification along with drying [3, 4]. Despite several advantages, a fluidized bed dryer has a few disadvantages like the high-pressure drop, lower fluidization velocity, attrition of solids and erosion of containing surfaces, etc. [5]. Still, it has great potential for commercialization.

Commercial use of fluidized bed dryers for drying granular materials, polymers, pharmaceuticals, chemicals, fertilizers, minerals, and crystalline products is common. For agricultural grains, literature studies using lab-scale fluidized bed dryer are for wheat grains [6]; green beans, potatoes, and peas [7]; macaroni beads [8]; paddy rice [9]; wheat and corn [10]; and with soybeans [11]. However, its large-scale drying is limited.

7.2 Principle and Design Considerations of Fluidized Bed Dryer

If airflow with sufficiently high velocity is used to overcome the gravitational forces of grains on a perforated plate or grid to uplift and support them without carrying them out in the flow, suspension of grains is called fluidization. If hot air is used as fluidizing air it causes dehydration of the wet product and thus achieves drying.

As air is made to pass across the bed, an increase in the superficial air velocity will also cause an increase in the pressure drop across the bed. As long as the bed particles are packed, the pressure drop is proportional to air velocity. When the frictional drag force will cause a pressure drop equal to the weight of particles any further increase in gas velocity will cause the upward movement of food particles [12]. At this stage, the grain bed expands and behaves like a fluid, and the state is known as a "fluidized-bed." When the particles became physically separated and free to move within the flow, it is known as the minimum fluidization condition. The minimum specific air velocity which will be needed for fluidization of bed is termed as "minimum fluidization velocity" and can be computed for spherical particles by the following given relationship:

$$v_f = \frac{(\rho_p - \rho)g}{\mu} \times \frac{d^2\varepsilon^3}{180(1-\varepsilon)}$$

where, v_f is the minimum fluidization velocity (m/s), ρ_p is the density of the food particles (kg/m^3), ρ is the density of drying air (kg/m^3), g is the acceleration due to gravity (m/s^2), μ is the viscosity of the air (N-s/m^2), d is the diameter of the particles (m), and ε is the voidage of the bed (the fraction of bed that is void).

As the air velocity passes the minimum fluidization condition, air bubbles will start to form that rise from the bottom of the perforated bed to the top layers of the bed. These bubbles after merging with others make large bubbles and cause mixing of bed particles. Further increase in air velocity causes the bed particles to entrain with the flowing air. This condition is called "entrainment velocity" and is calculated for the spherical particles by the following relationship:

$$v_e = \sqrt{\frac{4d(\rho_p - \rho)}{3C_d\rho}}$$

where v_e is the entrainment velocity (m/s), C_d is the drag coefficient (= 0.44 for Reynolds number of air 500 to 200000).

Fluidized bed drying is suitable for food particles with characteristic size ranging from 20 μm to 10 mm, uniform in size and shape, and not very sticky [13, 14]. Food materials such as cereal grains can easily fluidize at wide moisture ranges while some other food materials will only fluidize at low moisture conditions. Drying with fluidization is efficient because of its structure that enables the entire product surface to act like molten lava and aids in shuffling solid particles with the hot drying air, resulting in high rates of heat and mass transfer with shorter drying periods. A typical fluidized bed dryer is shown in Figure 7.1.

A fluidized bed dryer equipped with a mechanical agitator, located near the feed inlet is available which assists wet or sticky materials. The agitator

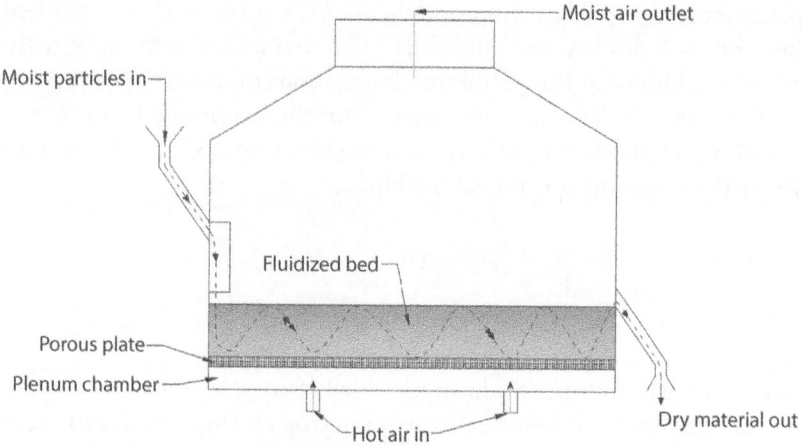

Figure 7.1 A typical fluidized bed dryer.

helps to disintegrate the feed and disperse into already dried material, and so facilitate fluidization. Air velocity needed for drying should be sufficient enough to promote rapid drying. Typical air velocities needed for fluidizing particles with a density of 1000 to 2000 kg/m^3 with varying sizes are shown in Table 7.1.

A fluidized bed drying system is more suitable for batch drying but could be applied for continuous drying operations. Drying conditions are easily controllable in batch operations and high drying rates with uniform product moisture can be achieved. However, batch units are preferred and used only for heat-sensitive materials, for small-scale operations. Based on the particles and airflow pattern there are two different continuous fluidized bed dryers, i.e., plug-flow type and well-mixed or backed-mixed type (Figure 7.2).

Table 7.1 Typical air velocity needed for fluidization of different sized particles and critical hot air temperature for different moisture; Adopted from [15] and [16].

Average size (μm)	Fluidization velocity (m/s)	Grain Moisture (%)	Air temperature (°F)
100 - 300	0.2 - 0.4	18-20	152 - 142
300 - 800	0.4 - 0.8	22-24	134 - 127
800 - 2000	0.8 - 1.2	28-30	114 - 110
2000 - 5000	1.2 - 3.0	-	-

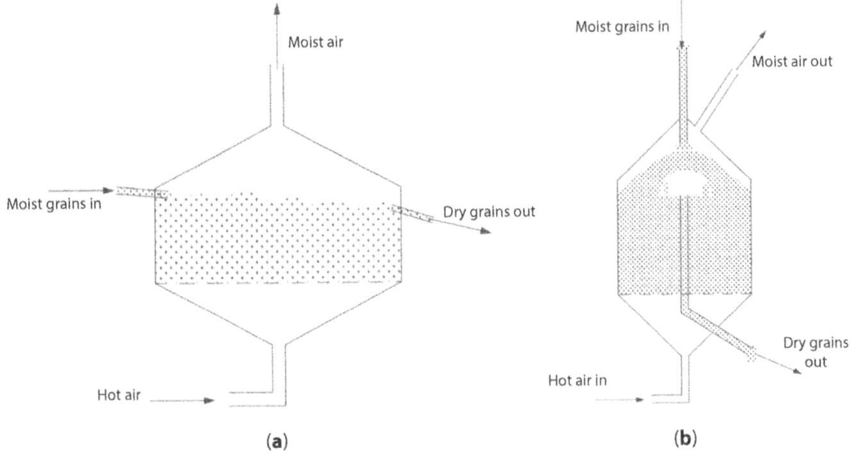

Figure 7.2 (a) Continuous plug flow and (b) Continuous well-mixed fluidized bed dryer.

The continuous plug-flow dryer has a relatively long and narrow bed, in which feed flows from the inlet to the outlet in a plug-flow pattern. In this dryer, uniform drying is achieved due to the narrow residence time of the particles. This type of dryer has limited use for readily fluidized food particles. In the other type, well-mixed continuous flow dryer, the bed length to width ratio is equal to or less than unity. The bed temperature is relatively uniform due to the vigorous mixing of the feed particles. In this type of dryer, the particle residence time is broadly distributed so the product moisture may be less uniform than plug-flow type continuous flow dryer [15].

A multistage fluidized bed dryer comprises a series of FBD that makes better control and maintenance of drying conditions at different stages, hence can yield a product with better uniformity of moisture than a single-stage FBD. Moreover, it may also lead to saving energy over a single unit system. One of many possible combinations is to use a well-mixed bed dryer to evaporate most of the surface moisture with a plug flow dryer to attain final drying with uniformity of moisture. Alternatively, for food drying applications, several plug flow dryers can be used in series. Both single- and multi-unit fluidized bed dryers have wide applications for drying different food materials just by varying a few design parameters and drying conditions for the specific food product.

7.2.1 Spouted Bed Dryer

Spouted bed dryer is a special particular design case of fluidized bed dryer developed [16]. The dryer is a cylindrical or rectangular vessel with a bottom of conical shape and fitted with a nozzle from which spouted air (drying air) is introduced into the dryer. Food material is fluidized and supported with the hot air of sufficient fluidization velocities and which results in the generation of a high-velocity region (a spout) of fast-moving food particles in the central column of the vessel. These food particles return to the base of the bed through the outer annulus surrounding the spout in a plug flow movement and get some reconditioning then again lifted by the upcoming air, at a relatively slow velocity. The material may be recirculated several movements by adjusting the discharge rate [13, 14]. In spout, high heat and mass transfer rate occurs and thus achieves a high rate of initial drying. Uniformity of moisture is achieved in SBD as the food material flows in a plug flow fashion, ensuring uniform residence time distribution. Spout bed drying is suitable for food particles that are too coarse to be readily fluidized generally, greater than 5 mm in diameter. Spouted bed dryers may also have a cylindrical draft tube at the center of the vessel. This vertical tall draft tube, some distance above the nozzle, eliminates the restriction of limited capacity and inability to scale up the dryer beyond 1 m diameter, by acting as a pneumatic conveyor. Figure 7.3 shows a draft tube spouted bed dryer. During startup, it is necessary to develop a high-pressure drop across the bed surface to form a spout. However, a spouted bed with a tangential air inlet features out the need for a high early pressure drop for the startup. Wheat, corn, and diced vegetables are among the various food materials which have been successfully dried using a spouted-bed drier.

Apart from various advantages of spouted bed dryers there are some problems, like spout features are greatly dependent on the flow rate of inlet air, spout generation is limited to a narrow range of airflow rates [16] and due to high flow rate requirements for spout generation, the contact times in the spout is reduced. It is also evident that other operating parameters like bed aspect ratio, spout dimension, inlet flow rate, and physical properties of the particles greatly affect the spout formation and a small change in these parameters may lead to a significant variation of spout and bed behavior. The formation of dead zones which cause reduced mixing is also a serious issue [17].

7.2.2 Spout Fluidized Bed Dryer

By considering the limitation of spout bed dryers a novel method was proposed which synthesise the features of both spout and fluidized bed, and is known as spout fluidized bed dryer or spout fluid bed dryer. In this type of dryer, an additional background gas (generally called fluidizing

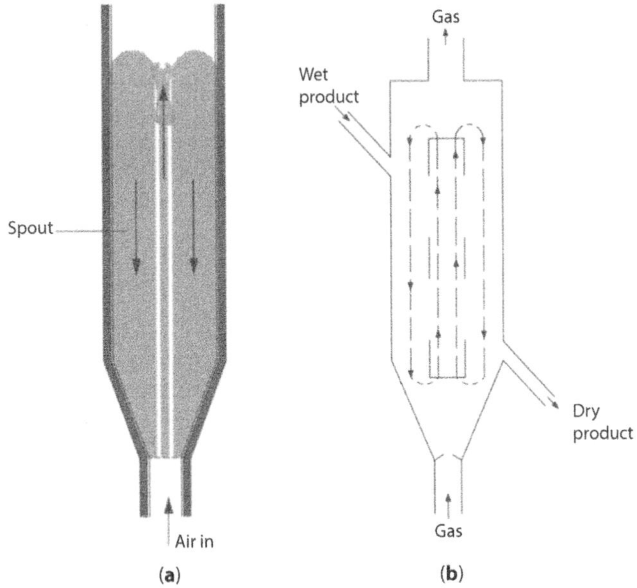

Figure 7.3 (a) A Conventional SBD, (b) a typical spouted bed dryer with draft tube.

or auxiliary gas) is also introduced to the drying chamber through the distributor plate, along with the spouting gas through the central nozzle. Auxiliary gas in the annular region leads to improved gas circulation and solid-gas mixing due to the generation of the bubble in the annulus. It also achieves a reduction in some of the common issues of SBD such as dead zones, particle agglomeration, and adhesions to the wall and base of the conventional spouted bed reactor. Additionally, it can also handle defluidization problems caused by sticky and irregular food particles [18].

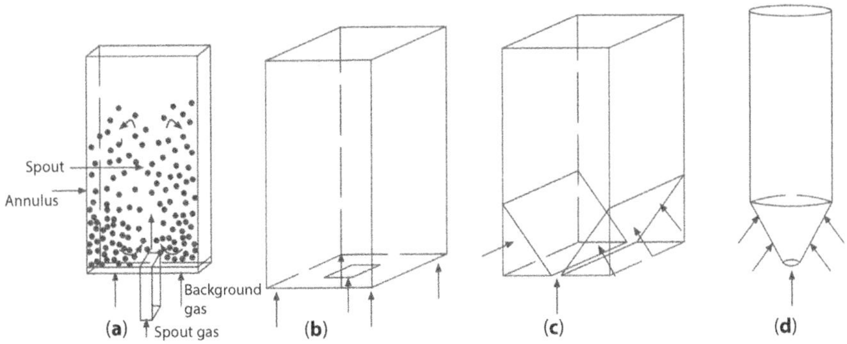

Figure 7.4 Different configuration of spout fluidized bed dryer: (a) pseudo 2D (b) rectangular (c) slotted rectangular and (d) cylindrical spout fluidized beds [17].

Table 7.2 Comparison of fluidized beds, spouted beds and spout fluidized beds.

Parameter	Fluidized bed dryer	Spouted bed dryer	Spout fluidized bed dryer
Usual particle size	< 1 mm	> 1 mm	> 1 mm
Usual particle size distribution	Broad	Narrow	Broad
Drying chamber section	Cylindrical shape	Cylindrical section with a Conical base	Cylindrical, square, or rectangular section with either conical or flat-bottom base
Pressure drop inside the bed in terms of weight of particles	96 to 100%	Less than 75%	Lower than both the beds mentioned
Axial pressure gradient	Independent of bed height	Significantly depends on bed height	Highly depends
Temperature gradient	Uniform over the entire bed	An axial and radial temperature gradient	The cross-section has a uniform gradient
Drying air motion	Non-uniform	Uniform and outwards of the spout	Uniform and outwards of spout and annuals region of flat bottom type
Particles separation	Very little	Significant, depending on particle size and density	Significant, depending on particle size and density

(*Continued*)

Table 7.2 Comparison of fluidized beds, spouted beds and spout fluidized beds. (*Continued*)

Parameter	Fluidized bed dryer	Spouted bed dryer	Spout fluidized bed dryer
Orifice diameter	Non-specific	Must be lower than 25 times of mean particle size	Must be lower than 25 times of mean particle size
Bed depth	A broad range, 0.1 to 20 m	Limited range, 0.2 to 2 m	A broad range, 0.1 to 50 m
Superficial air velocity	Broad range, 0.2 to 10 m/s	Limited range; 1-1.8u_{ms}	Depends on background air and ranges from 1-1.8u_{ms}
Use of internals	Common	Rare, except for the draft tube and heat coil	The use of draft tubes is common
Attrition	Little; in the cyclone and jet region	Significant; in spout and fountain area	Significant; in spout and fountain area

* Adopted from [17].

This combination can fluidize a wide range of particle sizes and density [17]. Different configurations of a spout fluid bed dryer are represented in Figure 7.4. Spout fluidized bed had been reported effective in the simultaneous drying of guava seeds [19]. Minimum spout velocity is not only dependent on the physical property of the product but also on bed configuration, so varies with the design. A structural comparison of fluidized beds, spouted beds and spout fluidized beds is presented in Table 7.2.

7.2.3 Hybrid Drying Techniques

For universal adaptability and to achieve a higher drying rate and hence efficiency, conventional FBD are modified with some attachments. Some of the recent trends are discussed below.

7.2.3.1 Microwave-Assisted FBD

In microwave drying, faster drying rates are achieved because microwaves raise the inside temperature of the product up to the boiling point of water

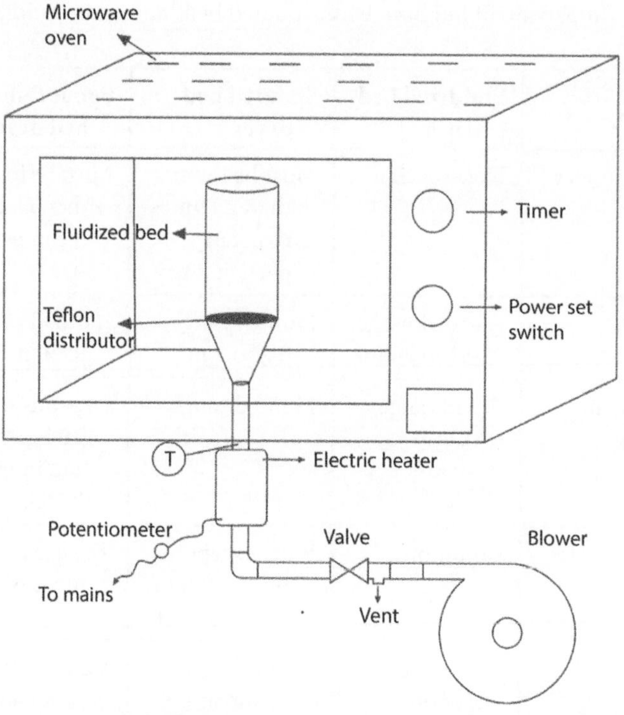

Figure 7.5 Microwave-assisted [29].

and generate a significant vapor pressure inside the product. This pressure force diffuses the moisture from the interior of the product to its surface, and so obtains higher evaporation rates. A fluidized bed dryer in combination with a microwave dryer (Figure 7.5) can lead to several desirable results; well mixing due to fluidization can provide uniformity of temperature and higher drying rates with the help of microwave energy [20, 21]. This combination dryer has successfully been used and found better than conventional FBD for drying of beetroot and kiwi fruits [22], for macaroni beads [23], for apple cubes [24], for corn [25], for macadamia nuts [26] and soybean [27, 28]. Overall microwave-assisted FBD drying improved the quality, energy efficiency, and drying rate of selective agro products.

7.2.3.2 *FIR-Assisted FBD*

Far infrared rays are the rays that are farthest from visible rays in the infrared region of the electromagnetic spectrum. The wavelength range of FIR is 2.5 to 30 μm. The rays accelerate the molecular vibration of absorbing material and thus cause a rise in the temperature of the material. The power and time combination of the process differs from commodity to commodity and also

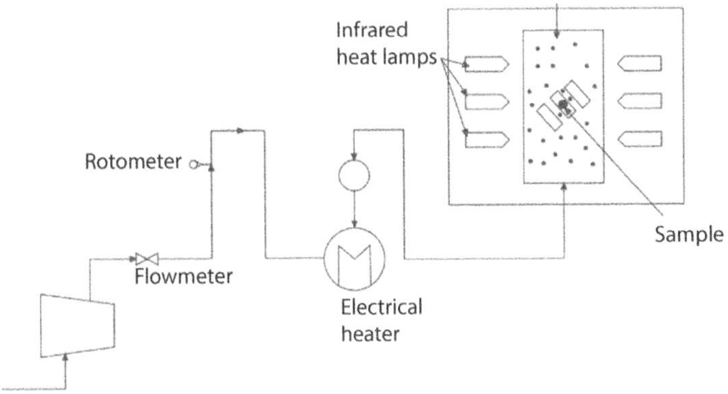

Figure 7.6 FIR-assisted FBD [29].

depends on drying air temperature, initial food moisture, etc. FBD in combination with FIR (Figure 7.6) rays was successfully used for drying paddy using multi-stage FBD and resulted in increased head rice yield, without variation in quality [30]. Drying of soybean using this combination caused minimum shrinkage, breakage, and cracking along with better quality due to low-stress development compared with other drying methods [31]. FIR-assisted fluidized bed drying resulted in improved product quality in terms of processing time, rehydration characteristics, and retention of vitamin C, compared with IR-assisted hot air drying [32]. Paddy dried with FIR-assisted FBD was found to be at 2% (d.b.) lower moisture rather than direct fluidized bed drying [30]. The response of individual food products is specific against FIR, so a detailed study is needed to further analyze the particular application.

7.2.3.3 Heat Pump–Assisted FBD

The principle behind the heat pump–assisted fluidized bed drying is to use the heat generated from a heat pump to assist in drying operation. The performance of heat pump–assisted FBD can be considered about 2 to 3 times of conventional FBD. If the quality of the dried product is of foremost importance then drying with heat pump–assisted FBD (Figure 7.7) can be considered as it offers enhanced product quality and reduced spoilage loss due to better control over the drying conditions [33]. A single-stage heat pump can supply hot air with a temperature range between 50 to 55°C, whereas double- or multi-stage pumps can be considered for higher temperature ranges [34].

7.2.3.4 Solar-Assisted FBD

Drying of agricultural commodities using a solar dryer is quite handy and useful but it becomes difficult to achieve the required drying capacity

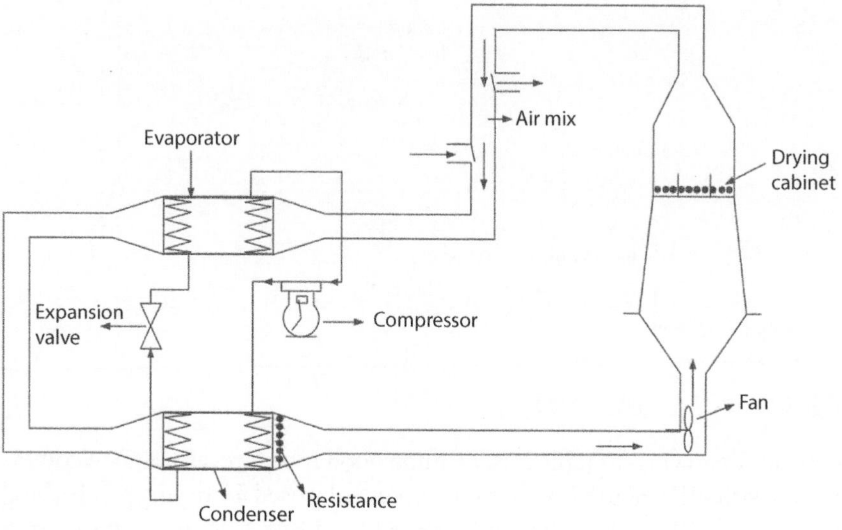

Figure 7.7 Heat pump–assisted FBD; reprint from [33].

during the season when enough sunlight is not available. Unconstrained temperature and air velocity is also a major problem of the solar dryer. Among various hybrids of the solar dryer, solar-assisted fluidized bed drying is a successful and energy-efficient method of drying. The combination consists of a solar collector covered with a glass sheet, a blower, and a spouting column (Figure 7.8). Solar dryer in the combination of fluidized bed has been used for drying of peas and found to be more efficient in terms of drying rate and rehydration capacity in comparison to open sun drying of peas [35]. A saving of 0.52 kg of natural gas for drying per kg of habanero chili in solar-assisted FBD is achievable [36].

Apart from the techniques discussed above, other hybrid techniques such as FBD dryer with immersed heater bed, FBD with recycling pipe, toroidal bed dryer, and ultrasonic transducer–assisted FBD dryer were also reviewed [29 and 37].

7.3 Design Alterations for Improved Fluidization Capacity

7.3.1 Vibrated Fluidized Bed

These beds are also called "Vibro-fluidizers". A vibrating base may be used to assist in easy fluidization in conventional fluidized bed dryers. Vibration

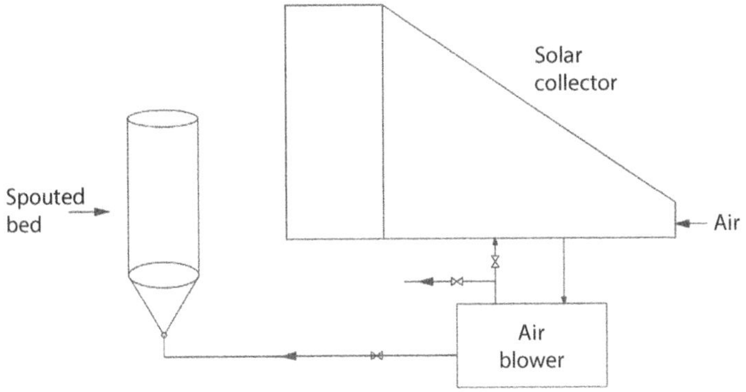

Figure 7.8 Solar-assisted FBD; reprint from [36].

along with the flow of air helps the product to move horizontally from the feed end to the dry end outlet. Due to the vibrating base, mixing of particles becomes easy and so the high rates of heat and mass transfer and overall increased drying efficiency. Compared to non-vibrating beds, they maintain lower residence time for the product per unit bed area. This technique requires lower gas velocity to achieve the condition of minimum fluidization and hence avoids the problem of fine particles entrainment [38]. Moreover, an appreciable decrease in attrition forces can also be seen, generated due to particle-particle and particle-wall interactions and so can be used for drying of abrasive, fragile, and heat-sensitive food materials [39]. A vibrating bed is more suitable for drying particles that are less uniform in size and sticky compared to conventional FBD. These dryers are generally shallow depth beds with plug flow structures (Figure 7.9).

7.3.2 Agitated Fluidized Bed

The strong cohesive force between the fine particles makes them difficult to fluidize in a conventional FBD. To improve their fluidization a

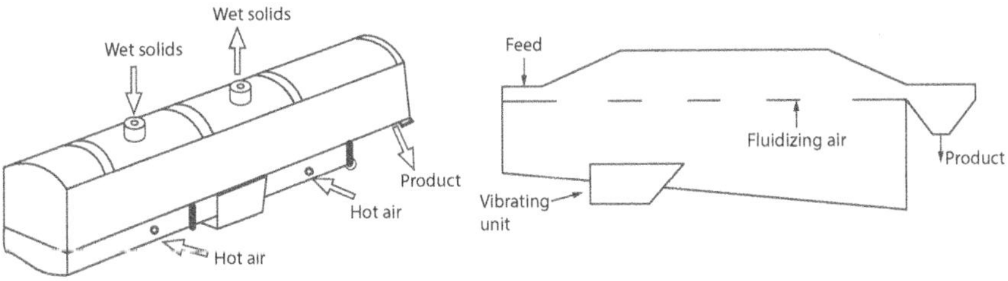

Figure 7.9 Different view of a typical vibrated FBD; courtesy of [38].

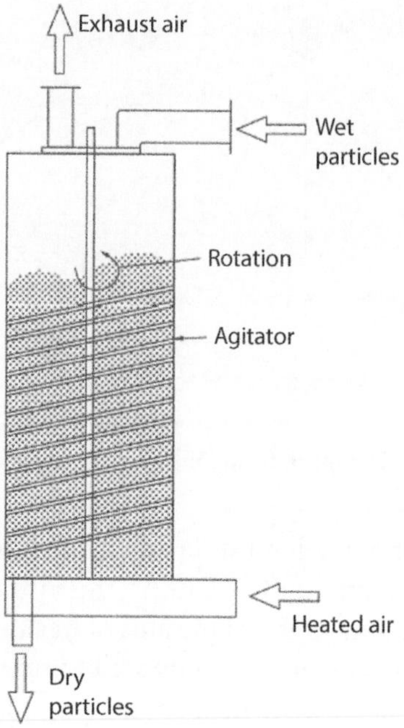

Figure 7.10 A typical agitated FBD [38].

mechanical agitator can assist FBD (Figure 7.10). Fine particles can be efficiently fluidized using this method due to homogeneous mixing and bed formation, without affecting the fluidization quality as a result of agitation [39]. Agitation reduces agglomeration and channeling as well as increases voidage and thus helps to achieve higher drying rates. Different agitation methods are straight blade; pitch blade; helical ribbon–type agitator; extension of helical ribbon agitator. Drying rates increase with the rate of agitation up to a maximum limit, then start decreasing [38, 40].

7.3.3 Centrifugal Fluidized Bed

In centrifugal FBD (Figure 7.11), a cylindrical chamber rotating on its horizontal axis generates suitable centrifugal force to balance the drag force of particles to be fluidized. The rapidly rotating chamber causes the food particles to form a bed in the annular space of the chamber. The walls of the cylindrical chamber are made perforated through which a high velocity (up to 15 m/s) drying air is injected that causes the fluidization of the bed. Feed enters at one end of the drying chamber occupying nearly 10 to 20% of chamber volume during the fluidization state and finally leaves at

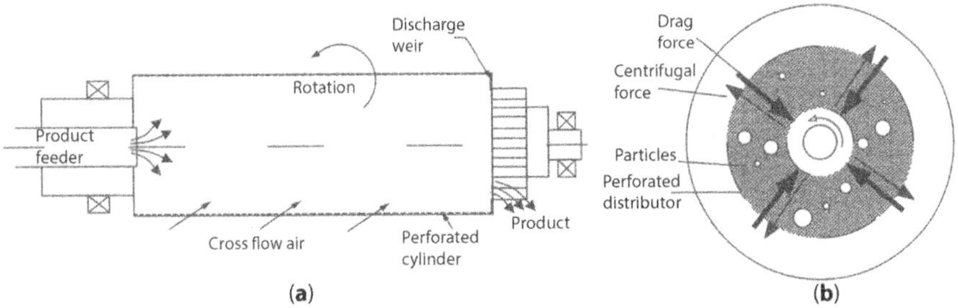

Figure 7.11 (a) Side view and (b) top view of a typical centrifugal fluidized bed dryer.

the other end as the dried product. By varying the rotating speed of the fluidization chamber the fluidization can be easily maintained. Under centrifugal force, good contact efficiency and lower agglomeration conditions are achieved along with easy fluidization [38].

7.4 Energy Consumption in Fluidized Bed Drying

For industrial drying, apart from the drying rates and product quality, the need to understand energy consumption, energy exergy relations is important by considering the future energy crisis. This calculation also helps to analyze the energy costs for industrial drying using fluidized beds and hence to promote scaling up. As a consequence, different studies had analyzed the energy demand of FBD. Industrial fluidized bed drying of paddy increases the SEC and energy demands with the rise in operating temperature and higher initial product moisture content. Only 31 to 37% exergy utilization was noted for drying with the remaining large amount of energy being the waste [41]. When the specific energy consumption (SEC) of FBD and SBD were compared, small values for a later case were found. However, drying using a higher drying temperature minimizes the difference between the SEC of both the dryers. It was also reported that SEC was higher for drying rice seeds than wheat seeds [42]. The energy consumption of FBD can significantly be improved by small modifications. A decrease in air velocity or an increase in drying air temperature can significantly increase the energy and exergy efficiency of FBD tested for the wheat [43]. Also, energy efficiencies of FBD can be increased up to 63% by partial recirculation of exhaust air and hence reduced heat requirement, been tested for drying of pre-treated soybean [44]. Use of inclined bed FBD with spirals inside the drying chamber for drying of paddy was reported with the highest drying efficiency and minimum energy consumption due to a decrease in blower energy consumption and heat input than conventional FBD [45]. Drying of paddy with ultrasound-assisted FBD successfully reduced the SEC by 22% in comparison to control, at selected drying conditions [46]. Further studies are needed to analyze the energy and exergy efficiencies of different hybrid fluidized beds.

7.5 Effect of Fluidized Bed Drying on the Quality

The selection of a drying method greatly depends on the quality aspect of the dried food product. Adaptation of inadequate drying conditions in fluidized bed drying may contribute towards serious changes in the product quality. Although the quality aspect of dried food products is not widely investigated, the literature summarises a few reports about quality changes during fluidized bed drying. In various previous investigations, color is reported as the primary quality indicator. Two different studies for drying white rice using FBD reported similar results that whiteness values of dried rice samples decreased compared to reference samples. And this decrease further increased with the increase in initial product moisture and drying time and temperature, resulted due to Maillard reaction [47, 48]. However, a study on purple rice reports that drying of purple rice using fluidized bed drying did not cause any substantial color change, even at increased initial moisture and drying temperatures, compared to sundried samples and reference samples. Other quality parameters such as anthocyanin, polyphenols, and antioxidant activity of purple rice were also not affected, when dried under FBD with recommended parameters [49]. Quality of wheat and rice seeds in terms of germination was highly dependent on dryer type (either FBD or SBD) along with other drying conditions [42]. Reduction in drying time (23%) and so the overheating, high power ultrasound-assisted FBD can improve the rice grain quality (cracked kernels and bending strengths), when operated under 20 kHz frequency, 50°C temperature and 18.7 kW/m^3 power density [46]. Fluidized bed drying of food has an important effect on its structure. A study investigated the effect of near-infrared radiations (from 4 to 8 kW) combined with FBD on quality parameters of soybean grains and reported that these radiations caused a significant decrease in cracking and breakage values to almost negligible (4.4% and 5.3%) compared to conventional fluidized bed drying of soybean with the reported values of 40-60% and 50-65% respectively. The total maximum color difference recorded for different intensities of radiations was in the range of 2.9–4.2 [31]. However, it had also been reported that during hot air fluidized bed drying of soybeans, cracking increases with the temperature, drying time, and velocity [11].

Table 7.3 Application of fluidized bed drying in different drying studies.

Product	Study	Findings
Wheat [6]	Effect of swirling flow on drying performance	Swirling flow significantly improved the drying rates, affected by drying air temperature, flow rate, and swirling flow field
Wheat and bulgur [50]	Compared the spouted bed dryer with microwave-assisted spouted bed dryer to compute appropriate operating conditions	Low temperature (50°C) and low microwave power (288 W) was recommended for improved drying
Wheat (Soft-Teal, Hard-condor and Very hard-eagle) [51]	Studied the heat effect on fluidized bed drying of different wheat varieties	An increase in moisture, increased the susceptibility to heat, and soft variety was found more susceptible to heat than hard
Paddy [45]	Investigated the performance of inclined spiral FBD	Inclined bed decreased the energy consumption and was more efficient than conventional FBD
Soybean [52]	The physical effect of air temperature and velocity variation in fluidized bed drying	Physical damage increased with the increase of air velocity and air temperature
Millet [53]	Studied the drying kinematics of millets in a bath FBD	Drying air temperature and velocity increased the drying rates and effective diffusivity, while it decreases with an increase in solids holdup
Macaroni beads [23]	Effect of application of microwave power in fluidized bed drying	Microwave power decreased the drying time. Temperature and microwave power increased the effective diffusivities correspondingly

(Continued)

Table 7.3 Application of fluidized bed drying in different drying studies. (*Continued*)

Product	**Study**	**Findings**
Apple cubes [54]	Evaluated new drying models for variation of air temperature, moisture ratio, and drying rate	Developed new empirical models estimating drying of apple cubes
Coconut [55]	Effect of temperature of drying air and air velocity on quality and oil content of coconut	Lower drying temperature decreased the surface oil content, while high inlet air velocity led to higher surface oil content; air temperature significantly affected the color
Tea [56]	Evaluated the performance of different FBD for tea drying	Use of multistage FBD with recirculation unit was more efficient
Maize and green peas [57]	Studied correlation for shrinkage, density, and diffusivity for drying of maize and green peas in FBD	Shrinkage and density was depended on moisture content while diffusivity depended on both temperature and moisture content
Mushrooms [58]	Compared FBD with different dryers: hot air cabinet dryer, vacuum dryer, and microwave oven drying	Considering the drying time and product quality fluidized bed drying was recommended over other dryers
Carrot cubes [59]	Effect of drying air temperature on the kinematics of spout fluidized bed drying of carrot	The drying curve shape was not affected by the drying air temperature
Carrot [60]	Evaluated the solar cabinet dryer, microwave oven, and fluidized bed dryer for blanched carrots drying	Recommended FBD for better rehydration properties, color, and overall sensory acceptability

7.6 Applications of Fluidized Bed Drying

Drying of agricultural commodities using fluidized bed dryers is quite evident. They are suitable for drying different fruits, vegetables, and grains except for high-moisture products, which are difficult to fluidize. However, the limitations of conventional FBD can be handled by adopting hybrid drying techniques. Table 7.3 shows the performance studies of past work for fluidized bed drying of different agro commodities.

7.7 Concluding Remarks

Drying agro products with a fluidized bed dryer is a versatile and emerging technology. Although there is very limited literature for large-scale industrial utilization of fluidized bed drying of grains, it nonetheless displays significant potential. Its use gives a good-quality product and is proven to be an energy-efficient method of drying with a short and uniform rate of drying compared to many other drying techniques. The drying performance of a fluidized bed dryer can be assessed by drying time, capacity, the temperature of dry air and product along with the product quality; all vary from product to product. Moreover, there are product-dependent design variations needed in a fluidized bed dryer. So a product-specific detailed study is needed to scale it up for industrial-level drying.

References

1. Mujumdar, A. S., & Devahastin, S. (2003). Applications for fluidized bed drying. *Chemical Industries*, New York-Marcel Dekker, 469-484.
2. Kassem, A. S., Shokr, A. Z., El-Mahdy, A. R., Aboukarima, A. M., & Hamed, E. Y. (2011). Comparison of drying characteristics of Thompson seedless grapes using combined microwave oven and hot air drying. *Journal of the Saudi Society of Agricultural Sciences*, *10*(1), 33-40.
3. Tirawanichakul, S. (2004). Influence of fluidized-bed drying temperature on chemical and physical properties of paddy, Doctoral dissertation, PhD. Dissertation, King Mongkut's University of Technology Thonburi, Thailand.
4. Chandran, A. N., Rao, S. S., & Varma, Y. B. G. (1990). Fluidized bed drying of solids. *AIChE Journal*, *36*(1), 29-38.
5. Setty, Y. P., & Murthy, J. V. R. (2003). Development of a model for drying of solids in a continuous fluidized bed dryer. *Indian Journal of Chemical Technology*, *10*(5), 477-482.

6. Ozbey, M., & Soylemez, M. S. (2005). Effect of swirling flow on fluidized bed drying of wheat grains. *Energy Conversion and Management, 46*(9-10), 1495-1512.
7. Senadeera, W., Bhandari, B. R., Young, G., & Wijesinghe, B. (2003). Influence of shapes of selected vegetable materials on drying kinetics during fluidized bed drying. *Journal of Food Engineering, 58*(3), 277-283.
8. Goksu, E. I., Sumnu, G., & Esin, A. (2005). Effect of microwave on fluidized bed drying of macaroni beads. *Journal of Food Engineering, 66*(4), 463-468.
9. Izadifar, M., & Mowla, D. (2003). Simulation of a cross-flow continuous fluidized bed dryer for paddy rice. *Journal of Food Engineering, 58*(4), 325-329.
10. Syahrul, S., Dincer, I., & Hamdullahpur, F. (2003). Thermodynamic modeling of fluidized bed drying of moist particles. *International Journal of Thermal Sciences, 42*(7), 691-701.
11. Soponronnarit, S., Swasdisevi, T., Wetchacama, S., & Wutiwiwatchai, W. (2001). Fluidised bed drying of soybeans. *Journal of Stored Products Research, 37*(2), 133-151.
12. Perry, H., & Green, D. (1997). *Perry's Chemical Engineers Handbook*, McGraw Hill Inc. *New, York*.
13. Mujumdar, A. S. (2007). Book Review: *Handbook of Industrial Drying*: A Review of: Publisher: CRC Press, Boca Raton, FL, 2007.
14. Mujumdar, A. S. (2014). 15 Impingement Drying. *Handbook of Industrial Drying*, 371.
15. Hovmand, S. (1995). Fluidized bed drying. *Handbook of Industrial Drying, 1*, 195-248.
16. Mathur, K. B., & Gishler, P. E. (1955). A technique for contacting gases with coarse solid particles. *AIChE Journal, 1*(2), 157-164.
17. Sutkar, V. S., Deen, N. G., & Kuipers, J. A. M. (2013). Spout fluidized beds: Recent advances in experimental and numerical studies. *Chemical Engineering Science, 86*, 124-136.
18. Aguado, R., Prieto, R., San Jose, M. J., Alvarez, S., Olazar, M., & Bilbao, J. (2005). Defluidization modelling of pyrolysis of plastics in a conical spouted bed reactor. *Chemical Engineering and Processing: Process Intensification, 44*(2), 231-235.
19. Osorio-Revilla, G., Lopez-Suarez, T., & Gallardo-Velazquez, T. (2004). Simultaneous Drying and Cleaning of Guava Seeds in a Spout-Fluid Bed with Draft Tube. *Canadian Journal of Chemical Engineering, 82*(1), 148-153.
20. Jumah, R. Y., & Raghavan, G. S. V. (2001). Analysis of heat and mass transfer during combined microwave convective spouted-bed drying. *Drying Technology, 19*(3-4), 485-506.
21. Wang, W., Thorat, B. N., Chen, G., & Mujumdar, A. S. (2002). Simulation of fluidized-bed drying of carrot with microwave heating. *Drying Technology, 20*(9), 1855-1867.
22. Chandrasekaran, S., Ramanathan, S., & Basak, T. (2013). Microwave food processing—A review. *Food Research International, 52*(1), 243-261.

23. Goksu, E. I., Sumnu, G., & Esin, A. (2005). Effect of microwave on fluidized bed drying of macaroni beads. *Journal of Food Engineering*, *66*(4), 463-468.
24. Askari, G. R., Emam-Djomeh, Z., & Mousavi, S. M. (2013). Heat and mass transfer in apple cubes in a microwave-assisted fluidized bed drier. *Food and Bioproducts Processing*, *91*(3), 207-215.
25. Momenzadeh, L., & Zomorodian, A. (2011). Study of shelled corn shrinkage in a microwave-assisted fluidized bed dryer using artificial neural network. *International Journal of Agriculture Sciences*, *3*(3), 150.
26. Silva, F. A., Marsaioli Jr, A., Maximo, G. J., Silva, M. A. A. P., & Goncalves, L. A. G. (2006). Microwave assisted drying of macadamia nuts. *Journal of Food Engineering*, *77*(3), 550-558.
27. Ranjbaran, M., & Zare, D. (2013). Simulation of energetic-and exergetic performance of microwave-assisted fluidized bed drying of soybeans. *Energy*, *59*, 484-493.
28. Anand, A., Gareipy, Y., & Raghavan, V. (2019). Fluidized bed and microwave-assisted fluidized bed drying of seed grade soybean. *Drying Technology*, 1-21.
29. Sivakumar, R., Saravanan, R., Perumal, A. E., & Iniyan, S. (2016). Fluidized bed drying of some agro products–A review. *Renewable and Sustainable Energy Reviews*, *61*, 280-301.
30. Meeso, N., Nathakaranakule, A., Madhiyanon, T., & Soponronnarit, S. (2004). Influence of FIR irradiation on paddy moisture reduction and milling quality after fluidized bed drying. *Journal of Food Engineering*, *65*(2), 293-301.
31. Dondee, S., Meeso, N., Soponronnarit, S., & Siriamornpun, S. (2011). Reducing cracking and breakage of soybean grains under combined near-infrared radiation and fluidized-bed drying. *Journal of Food Engineering*, *104*(1), 6-13.
32. Vishwanathan, K. H., Giwari, G. K., & Hebbar, H. U. (2013). Infrared assisted dry-blanching and hybrid drying of carrot. *Food and Bioproducts Processing*, *91*(2), 89-94.
33. Patel, K. K., & Kar, A. (2012). Heat pump assisted drying of agricultural produce—an overview. *Journal of Food Science and Technology*, *49*(2), 142-160.
34. Le Lostec, B., Galanis, N., Baribeault, J., & Millette, J. (2008). Wood chip drying with an absorption heat pump. *Energy*, *33*(3), 500-512.
35. Sahin, S., Sumnu, G., & Tunaboyu, F. (2013). Usage of solar-assisted spouted bed drier in drying of pea. *Food and Bioproducts Processing*, *91*(3), 271-278.
36. Rodriguez, E. C., Figueroa, I. P., & Mercado, C. A. R. (2013). Feasibility analysis of drying process habanero chili using a hybrid-solar-fluidized bed dryer in Yucatan, Mexico. *Journal of Energy and Power Engineering*, *7*(10), 1898-1908.
37. Tirawanichakul, S., Prachayawarakorn, S., Varanyanond, W., & Soponronnarit, S. (2009). Drying strategies for fluidized-bed drying of paddy. *International Journal of Food Engineering*, *5*(2).

38. Daud, W. R. W. (2008). Fluidized bed dryers—Recent advances. *Advanced Powder Technology*, *19*(5), 403-418.
39. Law, C. L., & Mujumdar, A. S. (2006). Fluidized bed dryers. *Handbook of Industrial Drying*, *3*.
40. Reyes, A., Diaz, G., & Marquardt, F. H. (2001). Analysis of mechanically agitated fluid-particle contact dryers. *Drying Technology*, *19*(9), 2235-2259.
41. Sarker, M. S. H., Ibrahim, M. N., Aziz, N. A., & Punan, M. S. (2015). Energy and exergy analysis of industrial fluidized bed drying of paddy. *Energy*, *84*, 131-138.
42. Jittanit, W., Srzednicki, G., & Driscoll, R. H. (2013). Comparison between fluidized bed and spouted bed drying for seeds. *Drying Technology*, *31*(1), 52-56.
43. Syahrul, S., Hamdullahpur, F., & Dincer, I. (2002). Exergy analysis of fluidized bed drying of moist particles. *Exergy, an International Journal*, *2*(2), 87-98.
44. Irigoyen, R. M. T., & Giner, S. A. (2016). Drying–toasting of presoaked soybean in fluidised bed. Modeling, validation and simulation of operational variants for reducing energy consumption. *Journal of Food Engineering*, *171*, 78-86.
45. Thant, P. P., Mahanta, P., & Robi, P. S. (2015). Performance Enhancement of Inclined Bubbling Fluidized Bed Paddy Dryer by Design Modification. *International Journal of Engineering and Applied Sciences*, *2*(8), 257847.
46. Jafari, A., & Zare, D. (2017). Ultrasound-assisted fluidized bed drying of paddy: Energy consumption and rice quality aspects. *Drying Technology*, *35*(7), 893-902.
47. Soponronnarit, S., Chiawwet, M., Prachayawarakorn, S., Tungtrakul, P., & Taechapairoj, C. (2008). Comparative study of physicochemical properties of accelerated and naturally aged rice. *Journal of Food Engineering*, *85*(2), 268-276.
48. Swasdisevi, T., Sriariyakula, W., Tia, W., & Soponronnarit, S. (2010). Effect of pre-steaming on production of partially-parboiled rice using hot-air fluidization technique. *Journal of Food Engineering*, *96*(3), 455-462.
49. Rattanamechaiskul, C., Junka, N., Wongs-Aree, C., Prachayawarakorn, S., & Soponronnarit, S. (2016). Influence of hot air fluidized bed drying on quality changes of purple rice. *Drying Technology*, *34*(12), 1462-1470.
50. Kahyaoglu, L. N., Sahin, S., & Sumnu, G. (2010). Physical properties of parboiled wheat and bulgur produced using spouted bed and microwave assisted spouted bed drying. *Journal of Food Engineering*, *98*(2), 159-169.
51. Ghaly, T. F., & Van Der Touw, J. W. (1982). Heat damage studies in relation to high temperature disinfestation of wheat. *Journal of Agricultural Engineering Research*, *27*(4), 329-336.
52. Darvishi, H., Khoshtaghaza, M. H., & Minaei, S. (2015). Effects of fluidized bed drying on the quality of soybean kernels. *Journal of the Saudi Society of Agricultural Sciences*, *14*(2), 134-139.

53. Srinivasakannan, C., & Balasubramanian, N. (2009). An investigation on drying of millet in fluidized beds. *Advanced Powder Technology, 20*(4), 298-302.
54. Kaleta, A., Gornicki, K., Winiczenko, R., & Chojnacka, A. (2013). Evaluation of drying models of apple (var. Ligol) dried in a fluidized bed dryer. *Energy Conversion and Management, 67*, 179-185.
55. Niamnuy, C., & Devahastin, S. (2005). Drying kinetics and quality of coconut dried in a fluidized bed dryer. *Journal of Food Engineering, 66*(2), 267-271.
56. Temple, S. J., Tambala, S. T., & Van Boxtel, A. J. B. (2000). Monitoring and control of fluid-bed drying of tea. *Control Engineering Practice, 8*(2), 165-173.
57. Hatamipour, M. S., & Mowla, D. (2003). Correlations for shrinkage, density and diffusivity for drying of maize and green peas in a fluidized bed with energy carrier. *Journal of Food Engineering, 59*(2-3), 221-227.
58. Walde, S. G., Velu, V., Jyothirmayi, T., & Math, R. G. (2006). Effects of pre-treatments and drying methods on dehydration of mushroom. *Journal of Food Engineering, 74*(1), 108-115.
59. Zielinska, M., & Markowski, M. (2010). Air drying characteristics and moisture diffusivity of carrots. *Chemical Engineering and Processing: Process Intensification, 49*(2), 212-218.
60. Prakash, S., Jha, S. K., & Datta, N. (2004). Performance evaluation of blanched carrots dried by three different driers. *Journal of Food Engineering, 62*(3), 305-313.

8

Dehumidifier Assisted Drying: Recent Developments

Vaishali Wankhade[1]*, Vaishali Pande[2], Monalisa Sahoo[3] and Chirasmita Panigrahi[4]

[1]*Mandsaur University, Mandsaur, India*
[2]*Maharashtra Institute of Technology, Aurangabad, India*
[3]*Center for Rural Development and Technology, Indian Institute of Technology, Delhi, India*
[4]*Department of Agriculture and Food Engineering, Indian Institute of Technology, Kharagpur, India*

Abstracts

The rapid development and changes in modern societies and their lifestyles coupled with advancement in science and technology has laid enormous stress on nature and ecology. Scientific research and technology have woken up and shifted towards developing harmony and ecological balance with sustainable development. Drying is a vital unit operation in the agro-food processing industries to store, transport, enhance shelf life, ease packaging, and add value to food products. But conventional drying methods are energy inefficient, time-consuming, and laborious. Therefore, new and advanced drying systems have been designed and developed in recent years. Dehumidifier-assisted drying is more advantageous compared to conventional drying methods, being energy efficient and environmentally friendly. Different dehumidifier drying systems have been designed for different applications in food and agricultural commodities. Recent advancements in dehumidifier dryings with their working principles and mechanisms are elaborated in this chapter.

Keywords: Heat pumps, desiccated dryer, adsorption air dryer, food processing

8.1 Introduction

The process of removal of moisture from a solid surface by use of heat as energy input is referred to as dehumidification drying. In almost all

**Corresponding author*: vaishalirpatil12@gmail.com

Nitin Kumar, Anil Panghal and M. K. Garg (eds.) Thermal Food Engineering Operations, (221–236)

agricultural countries, a large number of agricultural commodities and food products are dried to enhance shelf life, reduce transportation cost and weight, lower packaging cost, and retain flavor, color, and nutritional values [1, 2]. But drying is an intensive energy-consuming process where latent heat is supplied to evaporate the moisture in the drying materials. Therefore, conventional dryers require high-temperature air and a large amount of energy which is not environment friendly [1].

Researchers have reported that the drying process in the industries consumes approximately 10-15% of the total energy consumption [3]. Therefore, to reduce energy expenditure, it is essential to design, develop and select an efficient drying system to overcome the problems of conventional drying systems. Dehumidified dryers are efficient, low energy consumption, and environmentally friendly technologies [1, 4, 5]. Different types of dehumidified dryers are available based on the mechanism and principles which are described in this chapter. Integration of a heat pump to a dryer is known as heat pump dehumidifier. Dehumidification of the hot air in the drying process is the main feature that is attained when the evaporator surface temperature of the process air is lower than the dew point temperature at the evaporator inlet. Heat pump dehumidifiers have many advantages in comparison with conventional dryers which include high energy efficiency, better product quality, high coefficient of performance, and easy handling and control of drying conditions [5]. Moreover, this dryer can be operated in ambient weather independently as compared to solar drying systems [6].

8.2 Absorbent Air Dryer

Since ancient times, food including vegetables and fruits has traditionally been sun-dried. Solar drying is the most conventional and free of cost method for food drying. But it also tends to result in a few drawbacks such as loss of original texture and color, loss of vitamin content, etc. Also, it is not always possible to use this method for drying the food. For example, in seasons like monsoon and winter, the drying time gets affected, thus this causes loss in the quality of the product. A few recent methods use sun drying as the source [7]. A few more researchers have also used the sun as a source of drying [8, 9]. However, these have drawbacks, as mentioned previously. Some drying methods using heat have also been proposed [10]. But removal of moisture from food items using heat energy might result in lower vitamin levels, and loss of texture and color at times.

Low-temperature food drying can aid in preserving the color of food items. Desiccant-based food driers are found to be effective [11–13] where only temperature control is applied. But this can result in slow speed of drying and loss of product quality. A few other methods [14] use pulsating airflows for drying grains, while a few have mentioned the use of microwaves [15]. An attempt was made to maintain the color and expansion ratio of food using vacuum control [16]. All of the mentioned methods were effective in some or other ways but none of them satisfied all the five major requirements to maintain color, shape, texture, vitamin content, and fast speed of drying. Low-temperature adsorbent/desiccant drying allows food to be dried without color and texture loss.

8.2.1 Working Principle of Adsorption Air Dryer

An adsorption air dryer works on the adhesion principle. When compressed air flows through a desiccant/drying agent, desiccant adsorbs the moisture from the air by the adhesion principle. This is called dispersive adhesion (Figure 8.1). This results in dry air formation after it leaves the desiccant chamber. This adsorption is a reversible process.

8.2.2 Design Considerations and Components of the Absorbent Air Drier

8.2.2.1 Desiccant Drying System

8.2.2.1.1 Dehumidifier

A dehumidifier is attached to a desiccant rotor. This rotor is filled up with silica gel. There is a total of three areas or parts of this rotor:

- Operating area: This area contains silica gel. Here silica captures moisture from the air and supplies the drying chamber with dry air. This chamber is opposite to the operating area.

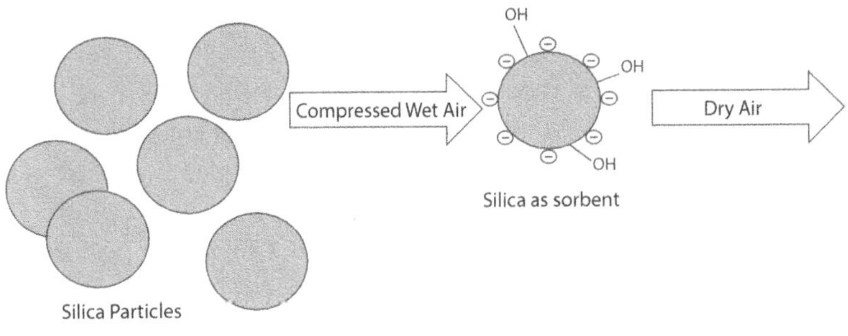

Figure 8.1 Dispersive adhesion.

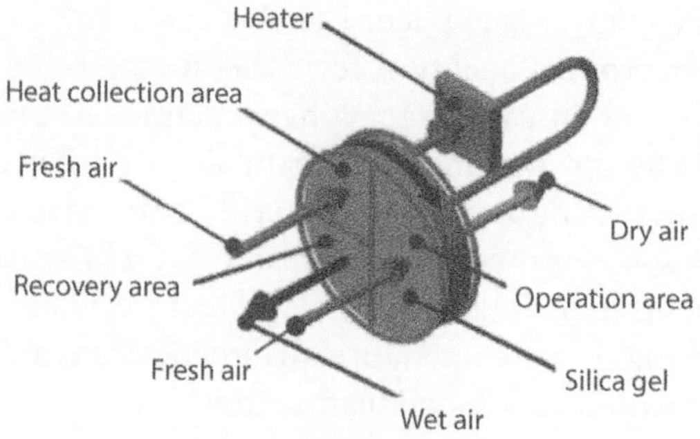

Figure 8.2 Desiccant drying motor, source: [17].

- Recovery area: Here damp air is heated and finally expelled out.
- Area of heat collection: This is the area where heat is released to the unprocessed external air by silica gel.

The cycle repeats slowly as the rotor rotates (Figure 8.2). As a result, dry air is produced each time without the need to replace silica gel every time. The blower can bear airflow of up to 50 m^3/h and the heater is adjusted at 1450 W.

8.2.2.1.2 Food Drying System

The drying chamber involved in the current system aids in retaining the original shape of food items being dried (Figure 8.3). It is composed of clear acrylic plates of size (0.9 x 0.9 x 1.8 cm) [17]. The product to be dried is kept in chamber 3 which is attached with the netting rack 4. This arrangement makes it possible for the desiccated air to blow from the base of the chamber through food and finally to the outside via the upper outlet. Thus, dry air blows from the bottom to the top of the drying system. The temperature of the whole chamber is set below 49°C. The range of temperature is set such that it maintains the texture, color, and vitamin content of the food. If the speed needs to be increased to ensure the drying conditions are uniform, this can be achieved by controlling airflow and temperature of the chamber. The dehydrated air obtained from desiccators is warmed up to 49°C accurately [17]. This process can be done by a heater 8, which is a 2-kW box heater. This heater can be turned on and off with the help of computer control. Thermal transducers disclose the temperature which governs the computer control. These transducers are mounted at both the upper and lower side of the drying chamber to

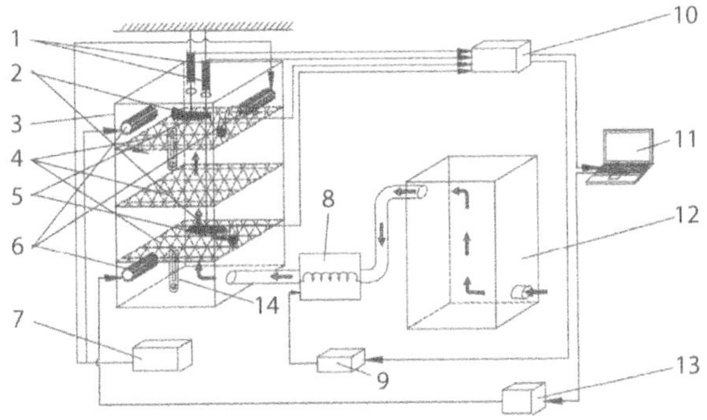

Figure 8.3 Desiccant food drying system with temperature and airflow control
1. linear potentiometer; 2. food drying net; 3. food drying chamber; 4. netting with rack; 5. thermal transducer; 6. fans; 7. DC power supply; 8. heater; 9. control box; 10. DSP; 11. computer; 12. dehumidifier; 13. amplifier; 14. thermometer, source: [17].

ensure precise detection. To obtain the airflow, temperature control is combined with fans 6 of 25V having 50 W capacity. These fans are also fitted at both the head and base chamber and are also controlled by computer 11 and digital signal processor 10 (DSP) [17]. The drying chamber also involves attached thermometers 14 at the head and base to ensure quick and easy visualization of temperature fluctuations, if any. Two linear slide potentiometers 1 monitor and detect the moisture removal and reduction in weight of food items present on the nets. These potentiometers are suspended with the help of strands to the supporting nets 2. Both of these are present in opposite directions [17].

8.2.2.1.3 Temperature and Airflow Control

Temperature and Airflow control are two major steps that need crucial surveillance in the desiccating dryer system. Figure 8.4 is a flow chart diagram of both airflow and temperature control [17]. Here, V_1 is voltage analogous to 49°C. V_0 is the preferred control voltage to activate the heater relay of 9V and V_c is the actual voltage exerted on the heater relay for command. t_0 is the mean temperature of thermal transducers located on top and bottom [17]. The heater is supervised in such a way that temperature t_0 is maintained about V_1. All the fans are run continuously excluding one base fan which contributes very little. One of the bottom fans is applied with control with the help of feedback algorithm given as Equation (8.1).

$$V_f = g(V_u - V_T) \tag{8.1}$$

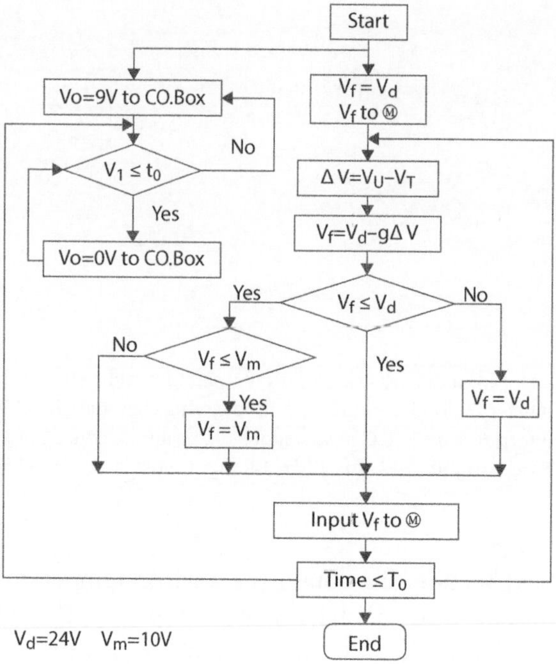

Figure 8.4 Flowchart of control algorithm, source: [17].

Where V_f = Fan control voltage

V_u and V_T = Linear potentiometer voltages for head and base nets.

g = Feedback gain

Initially, the fan is run at an optimal voltage (V_d = 25 V). In this case, V_f can be calculated as follows (Equation 8.2)

$$V_f = V_d - g\,\Delta V \tag{8.2}$$

(minimum operating voltage V_m = 12V and maximum is V_d)

$$\text{Where, } \Delta V = V_u - V_T$$

This control aids in arranging the weights of food on both the bottom and top to come in close proximity, inferring a uniform and high-speed drying at a constant temperature.

8.2.3 Performance Indicators of Desiccant Air Dryer System

Performance of desiccant air dryer systems with and without temperature control as well as air circulation systems is determined by considering their efficiency [17]. The facts about the effectiveness of the desiccant air dryer in both the above-mentioned cases are explained in the followed sections of this chapter.

8.2.3.1 Low Temperature Drying With No Temperature Control and Air Circulation System

To evaluate the efficacy of the current system, food dehumidification with no basic airflow and heat control was carried out. Here, the texture of food items was maintained using a thermostat device to confirm that system temperature does not exceed 49°C. Now, the capacity of water evaporation was tested using water dishes. These water-containing dishes were kept in the head and base net racks and the initial temperature was maintained at 18°C for water with the total weight of dish-containing water was 500g [17]. The temperature was found to increase very rapidly in the initial one hour. After that, the temperature decreased gradually. In the case of humidity, it decreased rapidly in the first hour only. This process was carried out for six hours and at the end of the time, the top dish was found to lose 19 grams of weight while the bottom dish lost 16 grams. The temperature and humidity fluctuations are shown in Figures 8.5 and 8.6. This experiment clearly indicates that without temperature and airflow control, the drying tends to be uneven with a slower rate than the basic system explained [17].

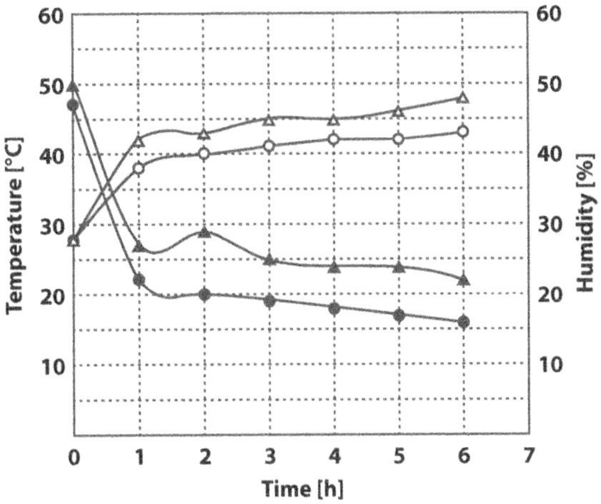

Figure 8.5 Humidity and temperature variation with no temperature and airflow control with respect to time, source: [17].

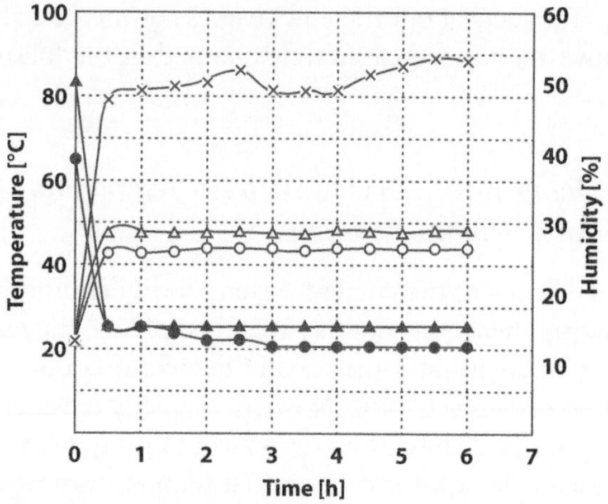

Figure 8.6 Humidity and temperature change with respect to time under controlled conditions, source: [17].

8.2.3.2 Low Temperature Drying With Air Circulation and Temperature Control

The success of controlled airflow and temperature was evaluated by carrying out fundamental drying without temperature control, with two fans of the bottom functioning for the two water-containing dishes. 30g of water was observed to evaporated from the top dish and 84.5g from the bottom [17]. This corresponds to five times achievement without circulation. So, by using fans, the drying speed can be controlled.

The above Figure 8.5 indicates investigational results using heat and airflow control. At the end of six hours, both dishes show identical loss of weight which corresponds to six times of basic experiment. Within the first 30 minutes, the temperature reached 49°C and relative humidity reduced to 15%. After this, both temperature and humidity remained steady for the rest of the drying process [17].

8.3 Heat Pump–Assisted Dehumidifier Dryer

Heat pump–assisted drying is a well-known technology for drying agricultural commodities at controlled temperature and humidity with energy-efficient operation at the lowest cost and in this technology, 90 to 95% of heat is recoverable [18]. A heat pump–assisted dehumidifier drier works based on combined and hybrid application of different modes of heat transfer with two or more stages with similar or dissimilar kinds of dryers which are assisted with a heat pump for the drying process [19]. Drying with a heat pump–assisted drier is carried out at a low temperature and with better control results in lower consumption of energy, relative humidity, and temperature along with several advantages over the other conventional driers [20]. The utilization of heat pumps for drying became popular in a short time as it has a high coefficient of performance and has shown promising enhancements in the quality of dehydrated foods as it is operated at lower temperatures and works better even at high humidity weather conditions, causing less pollution [21]. It is mentioned in the literature that heat pump drying has shown more effective results on the performance evaluation of the R-114 heat pump dehumidification system [22]. Heat pumps are basically a refrigeration system raising the energy by cooling from a low-temperature energy carrier with additional help from external energy to a high temperature, further transmitting it to an energy-conducting medium. In the heat pump, both the cooling and heating actions of refrigerators are useful [21]. Initially, heat pumps were being used as accompanying system components for heating, but subsequent research and developments bought advancements of drying systems and processes exclusively involving heat pumps. Its commercial application was also attested to in various regions of Europe, Australia, and Asia [23]. In the past years, a heat pump–assisted dehumidifier drier is recognized by many researchers due to its special features and improved food quality with less energy consumption. Working principles, design considerations, and performance indicators of a heat pump–assisted drier along with recent developments are described briefly in the following sections of this chapter.

8.3.1 Working Principles of a Heat Pump–Assisted Dehumidifier Dryer

The heat pump is mainly comprised of two subassemblies, *viz.*, a heat pump and a drying chamber. Heat pumps work on the principles of a thermodynamic cycle similar to the refrigeration technique. It works like a typical heat pump and transfers low-temperature heat from the surrounding heat sources, which may be atmospheric air, river water or lake water which are also referred as "ambient energies" [18] and upgrades it to the higher temperatures. The heat pump also functions with waste heat progressed coming from industrial processes. Absorption heat pumps and chemical heat pumps are the variants of conventional heat pumps. The absorption heat pump consists of a chemical absorbent which works as a mechanical compressor in compressor-driven systems. The upgradation of thermal energy resulting from the heat of chemical reaction is the principle behind the working of a Chemical heat pump [24, 25]. The heat pump–assisted dehumidifier drier consists of four primary components: a compressor, a condenser, an evaporator, and a throttle valve, establishing a closed cycle through a processed fluid (refrigerant freon) is moved. Processed fluid evaporates into the evaporator and gets condensed in a condenser, raising the level of temperature of the evaporator and condenser. In the case of an open cycle heat pump water vapor is a processing fluid and makes a part of the heat retrieval system in the dryer. After mechanical compression, condensation of this water vapor is carried out in a heat exchanger where water is further evaporated, whereas the condensate is hauled out of the heat pump system.

The working of a heat pump–assisted dehumidifier dryer is explained in this section. Figure 8.7 demonstrates the schematic view of the entire system, i.e., a heat pump–assisted dehumidifier dryer in which latent heat arising from the drier exhaust air is recovered. A is the fraction of moist air departing from the dryer that gets cooled in the evaporator resulting in its lower temperature T_B (B) and bringing it adjacent to dew point temperature T_{ev}. Accordingly, condensation of a fraction of moisture out of this airstream occurs and thus with the influence of heat recovered boiling of processing fluid in the evaporator occurs. This dehumidified air is further mixed with the residual fraction of moist air from the dryer,

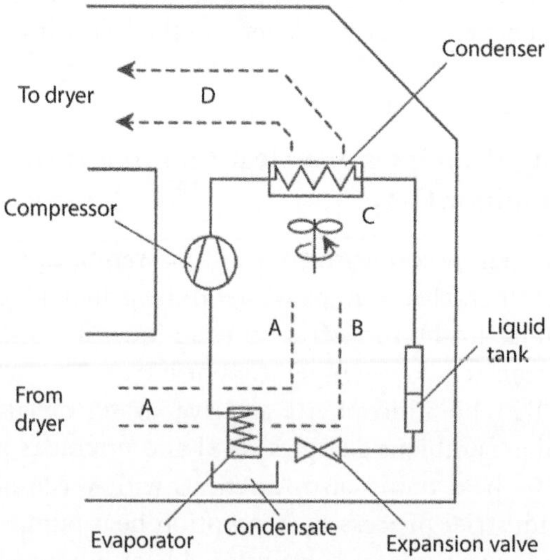

Figure 8.7 Schematic illustration of heat pump dehumidifier dryer, source: [26].

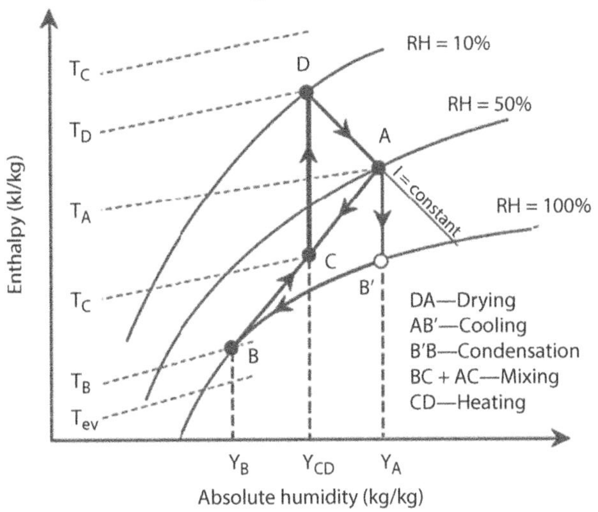

Figure 8.8 Mollier chart representing thermodynamic cycle for the air stream, source: [26].

C. The pressure and temperature of the evaporated processing fluid are increased by accumulation of external work dispensed by the compressor, followed by condensation of it into the condenser resulting in heat transfer to the mixture of air (C) to raise its temperature to T_D (D), lower to condensation temperature T_C. At the end, hot and dehumidified air is directed towards the drying chamber. The Mollier chart representing the thermodynamic cycle for the air stream is shown in Figure 8.8 [26].

8.3.2 Performance Indicators of Heat Pump-Assisted Dehumidifier Dryer

Calculation of energy consumption is an important topic in today's scenario of the increasing population, as energy consumption is increasing day by day with the increase of world population. Drying in food processing needs a larger volume of energy for its industrial process and thus should be efficient. Therefore, the efficiency of the drying systems should be considered for its selection especially when the source of energy used is high grade (Fuel or electricity). For heat pump dehumidifier assisted drier efficiency is determined by evaluation of coefficient of performance (COP) of the heat pump as following equation (8.3) [26].

$$COP = \frac{Usefull\ heat\ ouput}{Power\ input} \tag{8.3}$$

Maximum theoretical efficiency is evaluated by the Carnot efficiency as following (8.4)

$$COP_{Carnot} = \frac{T_{Condenser}}{T_{Condenser} - T_{Evaporator}} \tag{8.4}$$

Heat pump efficiency ranging between 40–50% of the theoretical Carnot efficiency can be attained in actual exercise [26].

One more parameter for drying to get suitable efficiency is the specific moisture evaporation (extraction) rate (SMER) determined as following Equation (8.5) [26, 27].

$$SMER = \frac{Amount\ of\ water\ evaporated}{Energy\ used} \quad (\text{kg/kWh}) \tag{8.5}$$

The SMER is influenced by the maximum air temperature within the dryer, the relative humidity of the air, thc temperature of evaporation and condensation, and the efficiency of a refrigeration system. Heat pump dryer efficiency (HPDE) is defined as the energy required to remove 1 kg of water and hence is stated as the reciprocal of the SMER of heat pump dryer (Strommen and Kramer, 1994). A refrigeration system with low COP results in an observable drop of temperature and thus decreases the amount of removed water per unit of input energy at the lower temperatures; therefore drying operation should be carried out at higher temperatures, if there is no need for temperatures below freezing point due to some quality issues.

The classical definition of SMER does not include the energy required to force the air in the drying chamber with the help of a fan. In the case of low temperature drying this required energy is a fraction of overall consumption. In this case the following definition (Equation 8.6) of COP is suggested [28].

$$COP^{*} = \frac{Q_o}{E + E_F} \tag{8.6}$$

Thus, the HPDE, i.e., Heat pump dryer efficiency can be determined as follows (Equation 8.7).

$$HDPE = \frac{\Delta H / \Delta Y}{COP^{*}} = \frac{1}{SMER^{*}} \tag{8.7}$$

Some advanced techniques can be incorporated to improve the efficiency of the heat pump dryer. In a two-stage heat pump, two heat pumps are connected in parallel in such a way that the evaporator of the first heat pump operates at a higher pressure than the second one. Theis type of heat pump has shown much more recovery of latent energy from the atmosphere as compared to a single heat pump [29]. The studies revealed that theoretical SMER of heat pump dehumidifier drier has shown favorable results compared to conventional drying methods and concluded that heat pump dryers are 10 times more efficient than old-style convective driers.

8.4 Applications of Dehumidifier-Assisted Dryers in Agriculture and Food Processing

There has been a growing demand for dehumidifier-assisted drying technology for food and agricultural material where well-controlled and low drying conditions are required to obtain an enhanced quality product. Extremely heat-sensitive high-value products are always freeze-dried. But this is an expensive process [30]. Therefore, a heat pump dehumidifier is a low-cost substitute for the freeze dryer. In recent years, a few researchers have experimented on food products, as explained in this section. Many researchers have investigated the different applications of dehumidifier-assisted driers to agricultural commodities. The effects of different operating parameters on the drying characteristics, nutritional and functional properties of the different food products have been examined by several research groups across the world. Commercial use of dehumidifying dryers has been reported in several parts of Asia, Australia, and Europe (France, the Netherlands, and Norway). The technology is not limited to the processing of marine products only but also covers the drying of vegetables and fruits.

Work has been carried out to design and develop a heat pump dehumidifier dryer to study the drying characteristics of horse mackerel with different process parameters. Results found that the drying characteristics could be influenced by drying air temperature, air velocity, evaporator bypass ratio, and surface load. It was also concluded by the authors that heat pump dryers are the best methods to produce intermediate moisture foods [1].

Experiments were made on drying of instant foods in an Adiabatic Fluidized Bed Heat Pump Dryer and a Carbon Dioxide Heat Pump Dryer. It was observed that the dried powder or granule retained natural color with porous and free-flowing nature [31]. Research has been done for drying of apple in a heat pump dehumidifier-assisted system (2.3 KW heat pump dehumidifier with external water-cooled condenser) and compared with the hot air-dried products. A difference in the product quality was observed. Results demonstrated that the heat pump dehumidifier dryer could be used successfully for drying agricultural materials [32, 33].

8.5 Concluding Remarks

Drying and dehydration of agricultural commodities is an important unit operation to enhance their quality and shelf life. The drying operation is carried out on a large scale with qualitative and shelf-stable products, but after consuming a huge amount of energy in various forms, as well as time, it is costly, with added hazards to the environment in some cases. In the present chapter, light has been thrown on recent developments in dehumidifier-assisted drying. It can be concluded that different types of dehumidifier-assisted driers have shown very good performances when experimented with for drying of various agricultural commodities. Special features of the dehumidifier-assisted drier, working even at higher humidity regions, and at low temperature in the case of heat pump drying, have given it special attention in the fields of agricultural process engineering. It is a good alternative to conventional drying methods and dryers for producing superior quality dried foods with a judicious reduction in time and energy consumption.

References

1. Shi, Q. L., Xue, C. H., Zhao, Y., Li, Z. J., & Wang, X. Y. Drying characteristics of horse mackerel (Trachurus japonicus) dried in a heat pump dehumidifier. *Journal of Food Engineering*, 84, 1, 12-20, 2008.
2. Sokhansanj, S., & Jayas, D. S. Drying of foodstuffs. *Handbook of industrial drying*, Marcel Dekker, New York, A.S. Mujumdar (ed.), 517-554, 1987.
3. Chua, K. J., Mujumdar, A. S., Chou, S. K., Hawlader, M. N. A., & Ho, J. C. Convective drying of banana, guava and potato pieces: effect of cyclical variations of air temperature on drying kinetics and color change. *Drying Technology*, 18, 4-5, 907-936, 2000.
4. Sosle, V., Raghavan, G. S. V., & Kittler, R. Low-temperature drying using a versatile heat pump dehumidifier. *Drying Technology*, 21, 3, 539-554, 2003.
5. Strommen, I., Bredesen, A. M., Eikevik, T., Neska, P., Pettersen, J., & Aarlien, R. Refrigeration, air conditioning and heat pump systems for the 21st century. *Bulletin of the International Institute of Refrigeration*, 2, 3-18, 2000.

6. Jain, D. Determination of convective heat and mass transfer coefficients for solar drying of fish. *Biosystems Engineering*, 94, 3, 429-435, 2006.
7. Tomita, M., Bautista, R. C., & Toji, E. Fissure produced in drying process of rough rice: comparison of non-waxy and waxy rice. *Journal of Japanese Society of Agricultural Machinery*, Tohoku Branch, 43, 61–64, 1996.
8. Tiris, C., Tiris, N. O., & Dincer, I. Thermal performance of a new solar air heater. *International Communications in Heat and Mass Transfer*, 22, 3, 411–423, 1995.
9. Salsilmaz, C., Yildiz, C., & Pehlivan, D. Drying of apricots in a rotary column cylindrical dryer (RCCD) supported with solar energy. *Renewable Energy*, 21, 2, 117–127, 2000.
10. Taira, E., Kasajkma, H., & Kohachi, S. Development of food dry system using of thermal radiation panel. *Report of Miyagi Industrial Technology Center*, 38, 63–68, 1994.
11. Sato, K., Katahira, M., & Toji, E. Drying characteristics of raw bulb and dehumidified air and moisture content distribution of different bulb parts. *Journal of Japanese Society of Agricultural Machinery*, Tohoku Branch, 44, 43–46, 1997.
12. Okano, H. Honeycomb rotor type dehumidifiers: comparison between honeycomb rotor type dehumidifiers and various dehumidifying systems and explanation on their outlines. *Clean Technology*, 3, 33–37, 1998.
13. Miller, W. M. Energy storage via desiccant for food agricultural applications. *Energy in Agriculture*, 2, 341–354, 1983.
14. Dias, S. R. S., Futata, F. P. L., Garvalho, J. A., Couto, H. S., Jr., & Ferreira, M. A. Investigation of food drain drying with pulsating air flows. *International Communications in Heat and Mass Transfer*, 31, 3, 387–395, 2004.
15. Khraisheh, M. A. A., McMinn, W. A. M., & Magee, T. R. A. (2004). Quality and structural changes in starchy foods during microwave and convective drying. *Food Research International*, 37, 5, 497–503.
16. Louka, N., & Allaf, K. Expansion ratio and color improvement of dried vegetables texturized by a new process (controlled sudden decompression to the vacuum application to potatoes, carrots and onions). *Journal of Food Engineering*, 65, 2, 233–243, 2004.
17. Nagaya, Kosuke, Ying Li, Zhehong Jin, Masahiro Fukumuro, Yoshinori Ando, and Atsutoshi Akaishi, Low-temperature desiccant-based food drying system with airflow and temperature control. *Journal of Food Engineering*, 75, no. 1, 71-77, 2006.
18. Patel, Krishna Kumar, and Abhijit Kar. Heat pump assisted drying of agricultural produce—an overview. *Journal of Food Science and Technology*, 49, 2, 142-160, 2012.
19. Lazzarin, Renato M. Heat pumps in industry—I. Equipment. *Heat Recovery Systems and CHP*, 14, 6, 581-597, 1994.
20. Best, R., J. M. Cruz, J. Gutierrez, and W. Soto. Experimental results of a solar assisted heat pump rice drying system. *Renewable Energy*, 9, 1-4, 690-694, 1996.

21. Daghigh, Ronak, Mohd Hafidz Ruslan, Mohamad Yusof Sulaiman, and Kamaruzzaman Sopian. Review of solar assisted heat pump drying systems for agricultural and marine products. *Renewable and Sustainable Energy Reviews*, 14, 9, 2564-2579, 2010.
22. Tai KW, Devotta S, Watson FA, Holland FA. The potential for heat pumps in drying and dehumidification systems- III: an experimental assessment of the heat pump characteristics of a heat pump dehumidification system using R114. *Int J Energy Res.* 6,4, 333–340, 1982.
23. Sosle, V., Raghavan, G. S. V., & Kittler, R. Low-temperature drying using a versatile heat pump dehumidifier. *Drying Technology*, 21, 3, 539-554, 2003.
24. Furutera, M., T. Origane, T. Sawada, Y. Kunugi, T. Kashiwagi, M. Arizawa, and H. Mori. "Advanced absorption heat pump cycles", 1996.
25. Labidi, J., Nikanpour, D. and De-Parolis, L. Hybrid absorption/compression heat pump for space application. *Proc. Int. Ab-Sorption Heat Pump Conference. Natural Resources Canada, Montreal, Canada*, 489–496, 1996.
26. Kudra, Tadeusz, and Arun S. Mujumdar. Advanced drying technologies. CRC press, 2009.
27. Hawlader MNA, Chou SK, Jahangeer KA, Rahman SMA, Lau KWE. Solar-assisted heat-pump dryer and water heater. *Appli Energy*, 7, 1, 185–193, 2003.
28. Kudra, T. and Strumillo. C. (eds.) *Thermal Processing of Bio-Materials*. Gordon and Breach Science Publisher, Amsterdam, 669, 1998.
29. Chou, S. K., Hawlader, M. N. A., Ho, J. C. and Chua, K. J. On the study of a two-stage heat pump cycle for drying of agricultural products. *Proc. ASEAN Seminar and Workshop on Drying Technology. Phitsanulok, Thailand*, 3, 5, 1–4, 1998.
30. Baker CGJ. *Industrial drying of foods*, 1st ed., New York: Springer-Verlag:1995
31. Alves-Filho, Odilio. Combined innovative heat pump drying technologies and new cold extrusion techniques for production of instant foods. *Drying Technology*, 20, 8, 1541-1557, 2002.
32. Chua, K. J., S. K. Chou, J. C. Ho, and M. N. A. Hawlader. Heat pump drying: Recent developments and future trends. *Drying Technology*, 20, 8 ,1579-1610, 2002.
33. Strommen DI and Kramer MS. New applications of heat pumps in drying processes. *Drying technology*. 1994.

9

Refractance Window Drying: Principles and Applications

Peter Waboi Mwaurah[1]*, Modiri Dirisca Setlhoka[2] and Tanu Malik[3]

[1]*Department of Agricultural and Biosystems Engineering, Jomo Kenyatta University of Agriculture and Technology, Nairobi, Kenya*
[2]*Department of Food Science and Technology, Botswana University of Agriculture and Natural Resources, Gaborone, Botswana*
[3]*Centre of Food Science and Technology, Chaudhary Charan Singh Haryana Agricultural University, Hisar, India*

Abstract
Refractance window (RW) drying is a modern drying technology that yields superior quality products. The dryer incorporates a technology that can be used to dry or evaporate both fluids and semi-fluid-like products; hence, the dryer applies to a wide range of products, such as agricultural products, herbal extracts, spices, and various food ingredients. RW technology is a low temperature (60-70 °C) short time (2-6 min) drying process that allows for the retention of quality attributes, such as color, flavor, and nutrients, as well as the preservation of bioactive compounds, such as beta carotene, ascorbic acid, and anthocyanins. The low operating temperature makes the RW dryer an appropriate drying technique for heat-sensitive products. Compared to other conventional dryers, the RW dryer has high thermal efficiency because the hot water is recovered through recirculation. RW drying technique expends approximately one-third of the freeze-drying cost and uses less than half of the energy consumed by a freeze dryer for a unit quantity of the wet material. The chapter explores the potential of this novel technology in the drying and preservation of food products.

Keywords: Refractance window drying, food preservation, heat-sensitive food, quality, technology, infrared radiation

**Corresponding author*: petermwaurah16@gmail.com, petermwaurah@hau.ac.in

Nitin Kumar, Anil Panghal and M. K. Garg (eds.) Thermal Food Engineering Operations, (237–260)

9.1 Introduction

Drying is a vital unit operation as far as food processing is concerned. Drying involves removing water from a product to halt the growth of spoilage microorganisms and slow down the rate of chemical reactions in the food product. According to USDA standards, dried products are those with over 2.5% water on a dry basis, while products with water content lower than 2.5% are considered dehydrated products [1]. With the rapid advancements in technology, a diverse range of drying techniques are in place to date. The techniques can be categorized into first, second, third, and fourth generations [1, 2]. First-generation techniques focused on drying food grains using hot air and include cabinet and bed type of dryers. On the contrary, second-generation dryers such as drum and spray dryers are used to dry pastes, purees, and slurries. Third-generation dryers, for example, osmotic and freeze-drying, are designed to dry fruits and vegetables. Refractance window drying is a fourth-generation drying technology used for drying heat-sensitive products. Other novel technologies in this category include radio-frequency drying, microwave drying, and fluidized bed drying [2]. The shortcomings of the different dryer types have resulted in continuous improvement with new models coming up each day. The limitations of the different generations of dryers can be traced as follows.

In convective dryers, for example, tray dryers, the product is in direct contact with hot air, causing cross-contamination and, consequently, quality degradation. Drum dryers operate at a temperature range of between 120-170 °C, and this results in enormous quality loss. To reduce these quality losses, the drum dryer is sometimes embedded in a vacuum chamber to achieve drying at lower temperatures [3]. Similarly, spray dryers operate at 150-300 °C. Apart from the high temperatures, spray drying requires that the product be atomized, making the technique unsuitable for solids and other products sensitive to mechanical damage. Among the third-generation dryers, freeze-drying is excellent as far as product quality (color, flavor, shape of the crystals) is concerned and is perhaps the best drying method when high-quality products are required. However, freeze-drying is 3-7 times more expensive than air drying techniques [3, 4]. Further, freeze-drying yields porous products that tend to rehydrate when exposed to a humid environment [5].

Although product discoloration can be caused by non-enzymatic browning, discoloration is also a critical physical factor that indicates quality deterioration due to heat treatment during drying. Apart from color, some minerals and nutrients are also sensitive to heat, and their quality depreciates considerably. Over the years, customers' preference and choice of food products are more inclined towards wholesome foods with high retention of nutrients, color, and other nutritional ingredients. Different products require different drying methods; fruit and vegetables have different drying requirements from cereals and grains [6]. Consequently, researchers are continuously exploring new innovative drying methods to deliver quality products to consumers. Research is driven by the need to develop an efficient drying technique that is fit for heat-sensitive products.

Drying and evaporation remain two important and demanding unit operations vital in extending the shelf-life of agricultural products. In developing countries, particularly, the lack of appropriate drying technologies is the major cause of the deterioration of food that is intended to provide the required nutrition to millions of families. Lack of appropriate drying technologies results in inferior products. Regardless of the level of technology, there is a dire need to properly adopt the available techniques to improve product quality because even some of the highly advanced devices are hardly put to good use. As mentioned above, we are in the fourth generation of dryers, yet some of the existing dryers have not been explored exhaustively. As part of the same endeavor, this study intends to shed more light on refractance window drying, its heating and drying kinetics, industrial applications, and product quality effects. The chapter will explore the potential of this novel technology in the drying and preservation of both dried and value-added products.

9.2 Refractance Window Drying System

9.2.1 History and Origin

RW dryer was developed in 2000 by MCD Technologies Inc. following a patent by Richard Magoon in 1986. The dryer incorporates a technology that can be used to dry or evaporate both fluids and semi-fluid-like products. The development came as a result of the need to dehydrate heat-sensitive food products and was based on two decades of research on novel techniques of removing water from food products. RW dryer is a contact drying method and compares well with other types of dryers in this category, e.g., drum dryer [4]. RW dryer is designed to convert the dried products into high-quality concentrates and powders for extension of shelf-life. Accordingly, the dryer applies to a wide range of products, such as agricultural products, herbal extracts, spices, and various food ingredients. Commonly dried agricultural products include mangoes, carrots, avocadoes, asparagus, strawberries, and squash, among others. Studies have shown that RW not only successfully dries these products but allows for the retention of quality attributes, such as color, flavor, and nutrients [7]. This, therefore, means that the RW dryer has the potential of being used for diverse applications in the drying of agricultural products.

9.2.2 Components and Working of the Dryer

RW dryer applies a simple design with the major components being plastic belt conveyor, water heating unit, steam injection system, hot water pump, water troughs (for both hot and cooling water), and an exhaust vent to carry away the vapor from the drying product (Figure 9.1). The plastic belt is transparent to infrared radiations and floats on the hot recirculating water. The plastic belt (film) has special refractive properties such as Mylar to enhance radiation heat transfer to the product being dried. The hot circulating water is heated directly by steam using a steam injector [8–10]. The steam gives off its latent heat to the circulating water increasing its temperature. RW drying technology uses circulating hot water, normally at 90-95° C, at atmospheric pressure to drive off moisture from products [11]. The hot water's thermal energy is transferred to the product by being in contact with the plastic conveyor. The product to be dried is spread on an infrared-transparent plastic (impervious) conveyor belt that is in contact with hot circulating water contained in a shallow trough.

The belt is so thin (0.2-1 mm) that the product comes into thermal equilibrium with the hot water almost immediately, initiating the drying process. The water is continuously reheated to maintain a constant temperature throughout the drying time. The three modes of heat transfer, that is, conduction, convection, and radiation, come into play during drying. The net energy reaching the belt's surface is transferred to the drying product through conduction and radiation. Conduction heat transfer occurs as a result of the product being in contact with the belt, while radiation transfer occurs when heat energy is radiated from the hot plastic belt into the drying product. The heat from the steam is, on the other hand, transferred to the bottom of the hot circulating water through convection.

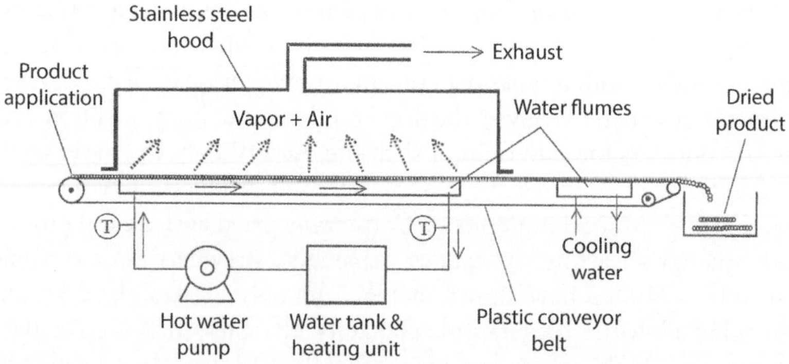

Figure 9.1 Schematic drawing of refractance window dryer. Adapted from reference [12].

The uniqueness of RW drying is that it uses water just below the boiling point of water (90-95 °C), which results in actual product temperature in the range of 60-70 °C. This makes RW dryer an appropriate drying technique for heat-sensitive products. To obtain the right quality of the product being dried, it is crucial that a uniform spread be made on the conveyor belt. For flakes, thin slices should be maintained (3-5 mm) for fast and uniform drying [4, 8, 12]. The belt could be moving or stationary depending on the product and the drying requirements. Typical belt speeds are in the range of 0.6-3 m/min [2].

Studies indicate that radiative heat transfer contributes approximately 3% of the total heat, with conduction being the most dominant [13]. Accordingly, the RW dryer is a conductive drying process. Before the product exits the dryer, it is brought back to atmospheric temperature by cooling water contained in a cooling water trough. On reaching the end of the dryer, the product is scraped off the belt using a blade that spans the conveyor's entire width. The industrial dryer is presented in Plate 9.1 below.

The thickness and nature of the plastic belt is an important aspect of the RW dryer. The more transparent it is, the more the heat is radiated from its surface in the form of infrared radiations. The thicker it is, the more resistance it offers to conduction heat transfer, and the higher will be the temperature difference between the circulating hot water and the upper surface of the belt [2]. Consequently, this would lower the thermal efficiency of the dryer since a lot of heat would be wasted heating the material

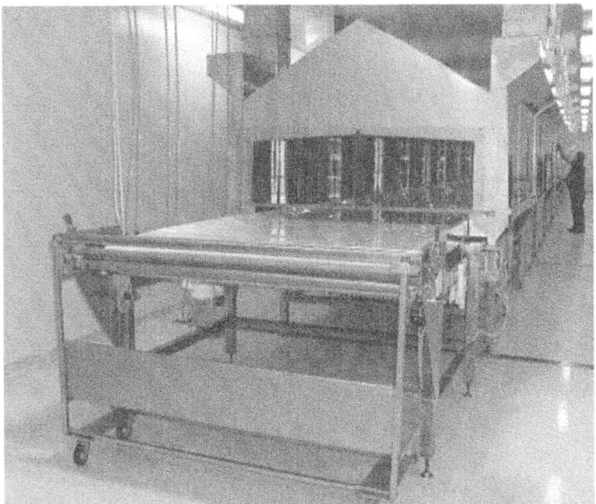

Plate 9.1 Refractance window dryer. Adapted from MCD Technologies Inc.

than heating the product to be dried. A thin material allows the system to reach thermal equilibrium almost immediately.

RW dryer has high thermal efficiency because the hot water is recovered through recirculation; the water is recycled and reused. By recirculating the hot water, less energy (steam) is required to bring back the water to the required temperature. The high thermal efficiency is the major strength of the RW dryer over other conventional dryers. Consequently, the process results in very rapid drying of the wet material with an evaporation capacity of up 10 kg $m^{-2} h^{-1}$ [13]. The rapid drying (2-6 minutes) in RW dryer is a result of a combination of three effects, which includes:

1. The use of water at a temperature just below the boiling point of water reduces the time required to evaporate the moisture from the wet product.
2. The thin plastic film increases its thermal conductivity and thus a faster rate of heat transfer.
3. The transparent conveyor has infrared properties whose transmission wavelength matches the absorption spectrum for water: 3.0, 4.7, 6.0, and 15.3 mm.

RW dryer applies a low-temperature short-time continuous process to dry the product. Consequently, the drying method applies to delicate and heat-sensitive products with high product color retention, aromatic pigments, flavor, bioactivity, and nutritional contents. Studies indicate that the quality of RW dried products compare well with those obtained from freeze-drying and superior to other conventional technologies such as sun drying, drum drying, and spray drying. Further, the product obtained has a high bulk density [14]. RW dryer falls in the category of contact, indirect and film-drying technique. This is because the wet material is deposited in contact with the plastic conveyor as a thin film.

9.2.3 Principle of Operation

RW dryer is based on the principle of refractance. According to the inventor, the working fluid, water, has unique properties in that it is good at absorbing infrared radiations under wavelengths in the range of 3.0-15.3 mm [15]. Further, the transmission of infrared radiation is strongest when the plastic conveyor is in intimate contact with water on one side and a moist material on the other side. To better illustrate the principle of operation of refractance window technology, the process can be divided into four stages, as shown in Figure 9.2(a-d). The four stages are discussed below.

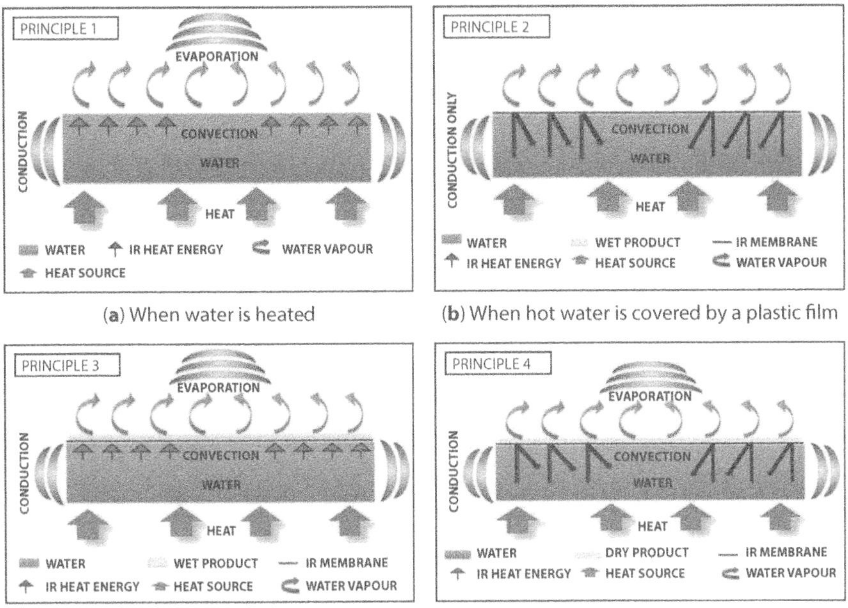

(a) When water is heated

(b) When hot water is covered by a plastic film

(c) When the product is loaded on membrane

(d) When the dried product is obtained

Figure 9.2 (a-d) Illustration of the principle of operation of refractance window technology. Adapted from reference [6].

1. Considering a situation where water in a trough is placed over a heat source, the infrared energy is transmitted to the water by conduction through the trough. The infrared energy is further transmitted by convection within the water mass. If the trough is open from the top, the heat energy is lost from the hot water through evaporation.
2. If the trough is covered by a transparent membrane that is in contact with the surface of the hot water, heat loss through evaporation is blocked, and heat transfer is only possible through conduction. The transparent membrane serves as a mirror that reflects the infrared energy back into the hot water.
3. However, if a moist material is placed on top of the plastic-transparent membrane, the water present in the material creates a "window" that provides a passage for the flow of infrared energy into the product. At this point where the moist material is in contact with the membrane and the membrane is in contact with the hot water, all the

three modes of heat transfer (conduction, convection, and radiation) act exceptionally to provide an effective transfer of heat into the product resulting in rapid evaporation of moisture. Based on the principle behind the operation of this dryer, the wet product receives the drying heat from hot circulating water mainly through infrared radiation, which is the reason why the dryer is termed as a refractance dryer.

4. As drying continues, the water content in the material reduces, and the infrared "window" closes, limiting the heat transfer to conduction only. The infrared energy is reflected back into the hot water, and the product is no longer exposed to heat. The limited transfer of heat prevents the product from overheating and thereby maintains the quality attributes of the product. Accordingly, the RW drying process is referred to as a "self-limiting drying process" because the process requires less operator intervention [6, 14].

From the above discussion, it is clear that RW drying is a novel, thin-film, and self-limiting drying technique that dominantly relies on infrared radiations together with the conductivity of water as opposed to the direct reliance on extreme temperatures to drive moisture from a product. The system uses water to drive off water from the wet product.

9.3 Heat Transfer and Drying Kinetics

RW dryer is unique in that during the drying process, all three modes of heat transfer are active to provide rapid drying of the wet product. Thermal energy from the hot water is transferred to the wet product causing rapid evaporation of the moisture. Heat transfer by radiation in RW technology relies on reducing the mismatch in refractive index between the water-plastic and the plastic-food interface. If the refractive index between these two interfaces is close to each other, reflection is minimized, and the transmission of infrared energy into the food material is maximized. The refractance "window" opens when the refractive index mismatch is reduced, and radiations are allowed to pass into the material (Figure 9.3b). The window closes as the mismatch increases. As the product dries, its refractive index increases, resulting in a greater mismatch in the refractive index at the water-plastic-food interface forcing the infrared energy to be refracted into the hot water, limiting heat transfer [6, 14, 16]. Figure 9.3a represents the transmission of infrared radiations when there is no product

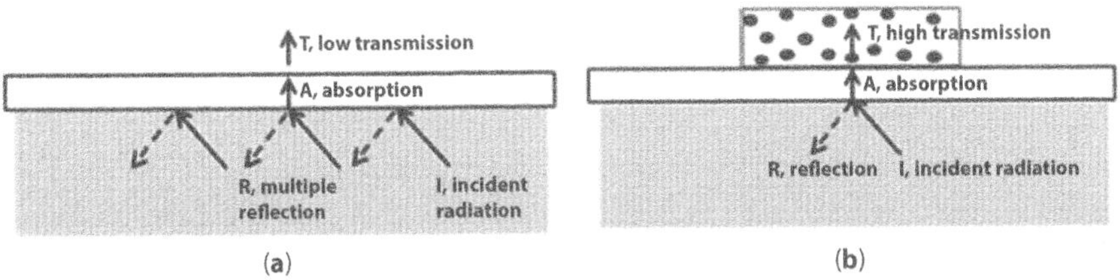

Figure 9.3 Drying kinetics (a) water-plastic-air represents the dryer without the product to be dried or a very dry product, and (b) water-plastic-wet food product. Adapted from reference [16].

on the plastic membrane, while Figure 9.3b represents the scenario when a food product is placed on the membrane. Due to the presence of water in the product, a high transmission rate of the radiations is expected compared to when there is no product.

9.3.1 Drying Rate and Moisture Reduction Rate

Studies indicate that it takes approximately 6 minutes to remove 1 kg of water from a carrot puree using the RW drying technique, while 120 minutes are required to remove the same quantity of water for carrot puree using a tray dryer [17]. As far as moisture reduction is concerned, the RW technique can deliver products with moisture content lower than that of a freeze dryer [18]. From one of the studies, it took approximately 30 min for the RW drying technique to dry mango slices (1 mm thickness) to below 5% wb and about 60 min to attain the same moisture content using 2-mm thick slices. In contrast, a tray drier operating at 62 °C took 240 min to reach the same moisture content [19]. Further studies indicate that 0.074 and 3.5 h are required to dry asparagus puree to a moisture content of <0.1 DB using RW and tray dryer, respectively [20].

9.4 Effect of Process Parameters on Drying

9.4.1 Effect of Temperature of the Hot Circulating Water

Proper control of the circulating water is particularly very crucial for the smooth running of the drying process. The temperature of the hot circulating water affects the dryer throughput, the drying time, and the retention of heat-sensitive compounds contained in the wet product. If the hot water boils, then vapor will form between the water-plastic interface, which will

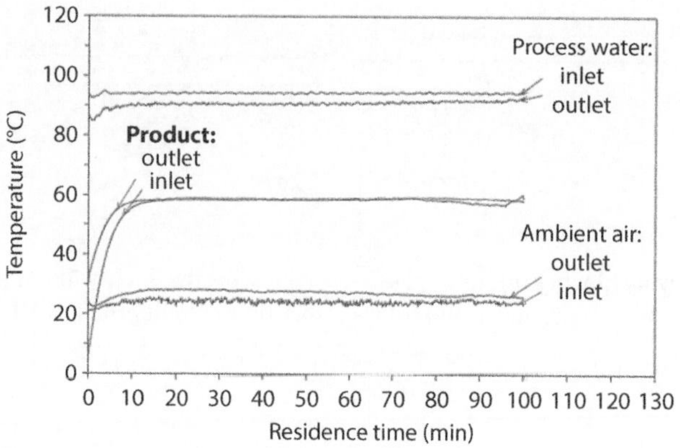

Figure 9.4 Typical temperature profiles in a refractance. Adapted from reference [9].

interfere with heat transfer processes. If the temperature of the circulating water is so low, the rate of drying and evaporation will significantly be affected, resulting in longer drying time and low throughput. Fluctuations in the temperature of the circulating water, on the other hand, lower the equipment's thermal efficiency [4, 9, 14]. Accordingly, the temperature of the circulating water should, at all times, be maintained within the range of 95-98 °C. Typical process parameters are illustrated in Figure 9.4 below. Figure 9.4 shows that it takes about 10 min for the product at 10 °C to achieve thermal equilibrium (60 °C) with the hot circulating water.

9.4.2 Effect of Product Inlet Temperature and Thickness

For a fast drying process, the product should reach an equilibrium with the circulating hot water as soon as possible. Very low product inlet temperature results in an extended time to reach this equilibrium, as is the case in Figure 9.4 above. Where possible, the product should be pre-heated in the holding tank before introduction into the dryer to minimize temperature depression. Pre-heating the product results in rapid drying and a high evaporation rate. Similarly, a thin layer of the wet material allows for faster drying rates because less time is required for heat to traverse through the layer. It should be noted that RW drier is a thin film type of dryer. Relatively thick samples (about 1.25 cm thick) also dry at the same rate as thin samples [16]. However, a thick bed could result in different layers drying at different rates because as the product dries, a thermally resistive layer is formed on the bottom side where the product contacts the plastic membrane. This reduces the heat flux from the hot circulating water [16]. Regardless of the nature of the product, a thick layer lowers the drying rate by impending the rate of transfer of heat through both conduction and convection. It is worth noting that radiation heat transfer is hardly affected by the thickness of the product because radiations are rapid and travel at the speed of light.

9.4.3 Effect of Residence Time

For a constantly moving conveyor, the speed determines how long the product is exposed to the hot circulating water for drying. Studies show that 2-6 minutes is typical for an RW dryer at a typical speed of 0.6 -3 m/min. Higher than this speed implies that the product will be exposed to drying for a shorter time, and the process may not achieve the required moisture content/concentration levels [21]. Unlike conventional dryers, RW dryer is a self-limiting type of dryer. When the infrared window closes, heat transfer to the product is limited to conduction, and the product can no longer be overheated.

9.4.4 Effect of Ambient Air Temperature (Air Convection)

RW dryer can run on either natural or forced convection. During the drying process, the wet product exhibits a higher average temperature when subjected to natural convection, whereas forced convection results in lower average temperatures. Natural convection results in lower moisture kinetics compared to drying under forced convection [21, 22]. In the latter scenario, the air carries with it the vapor creating a high partial pressure for more water to evaporate and, consequently, a high moisture loss. Based on these studies, therefore, the RW dryer is best suited for forced convection over natural convection to achieve higher evaporation rates.

9.5 Comparison of Refractance Window Dryer with Other Types of Dryers

Compared to other contact/thin-film dryers such as drum dryer and solid steel belt dryers, RW dryer not only has a higher thermal efficiency due to recirculation of hot water but also prevents cross-contamination of the product since the product is never in contact with the heat transfer media. Also, most of the contact type of dryers use hot water, saturated steam, and glycol solution, among other readily available fluids.

Conventional dryers take hours to dry the product. Freeze drying, for example, requires approximately 12-72 hours depending on the product to be dried, while sun drying may even take days. However, the RW dryer

Table 9.1 Energy consumption of refractance window in comparison with other dryers [9].

Dryer type	Product temperature (°C)	Thermal efficiency (%)
Rotary dryer	About 175	50-25
Spray dryer	80-120	51-20
Drum dryer	120-130	78-35
RW Dryer	60-70	77-52

Table 9.2 Energy efficiency of refractance dryer when compared to other dryers [24].

Dryer type	Calculated energy needed for drying 1kg sample (kWh)	Energy consumption for drying 1kg sample (kWh)	Overall energy efficiency (%)
Freeze dryer	1.46	130.65	1.12
RW dryer	1.36	4.31	31.56
Spray dryer	1.42	11.01	12.92

requires 2-6 minutes to dry a majority of the products. This is attributed to the high heat and mass transfer rates when compared to the other methods. For the unit quantity of the wet material, the RW drying technique costs approximately one-third the cost of freeze-drying and uses less than half of the energy consumed by a freeze dryer. Further, the RW dryer has a high retention of product color, vitamins, phytochemicals, and heat-sensitive compounds that compare well to a freeze dryer but are superior to other conventional dryers [22, 23]. In terms of energy consumption and thermal efficiency, different studies indicate that RW drying is an energy-efficient alternative when compared to other types of dryers (see Tables 9.1 and 9.2).

9.6 Effect of Refractance Window Drying on Quality of Food Products

Food quality is one crucial food and safety aspect. During processing, raw materials, such as fruits and vegetables, undergo pre-heat treatment (blanching or pasteurization) to reduce microbial load and halt the rate of product deterioration. Pre-heat treatment procedures are not always preferred when drying heat-sensitive materials since the food item may lose important quality attributes, such as color, nutrients (e.g., vitamins), and flavor in the course of these procedures [6]. Losing such nutritional components is a common challenge that concerns the nutritional quality of dried food products. Accordingly, a drying method capable of reducing microbial load and, at the same time, preserving the heat-sensitive compounds is desired for optimum quality of dried food products. RW drying is one such technique. In a study to evaluate the capacity of RW drying in reducing the microbial load from pumpkin puree, it was noted that Coliforms, *E.coli* and *Listeria innocua* exhibited a 6.1, 6.0, and 5.5 log reduction, respectively. The RW dryer operated at 95 °C for 5 minutes. The study also indicates that the microbial load was within the acceptable safety bracket as the counts were below the minimum detectable limit of <5 CFU/ml [8].

9.6.1 Effects on Food Color

Color is the foremost quality attribute used by customers to determine product quality. Consumers prefer food products that adhere to a specific color [25]. Although the color is a primary quality parameter, it is highly affected by thermal processing and, at a minimum, describes the chemical changes brought about by these processes [26]. Drying is one unit operation that exposes a product to heat, resulting in loss of color [27]. Non-enzymatic browning (Millard reactions) is the primary cause of discoloration in food products [28]. Among the different drying techniques, RW drying and freeze-drying are commonly utilized in drying heat-sensitive materials. Comparison of different methods of drying of various products (See Plate 9.2) suggests that the RW drying technique has better retention of the color of the dried product than other conventional drying techniques [12, 29–31]. In Plate 9.2a below, the RW dried products portray higher retention of fresh product color than drum and tray dried products (Plate 9.2b).

RW drying technique yields color akin to a raw product [29]. Analysis of color for dried asparagus using RW, tray drying, and freeze-drying methods (Table 9.3) indicated that RW-dried asparagus had color closer to that of the fresh product [8]. Table 9.3 below shows the results of color measurements for both fresh and dried asparagus using the different method as well as their operating temperatures.

From Table 9.3, the hue angle and chroma for RW dried asparagus (106.0 and 16.2) were closest to the fresh asparagus (104.7 and 12.1) and

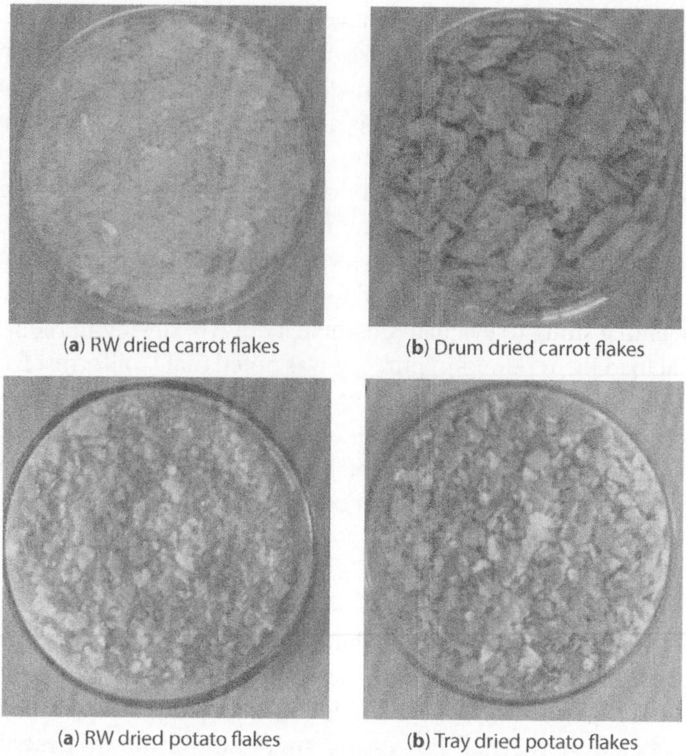

Plate 9.2 Comparison of color preservation on potato and carrot flakes by RW drying and other drying techniques. Adapted from reference [30, 31].

very close to those of freeze-dried asparagus. Similarly, RW dried mango powder had characteristics similar to freeze-dried powder [29]. A recent study on the drying of Mammee apple pulp indicates that the RW drying technique results in minimal loss of color (See Table 9.4) [32]. The before and after drying readings show a very close resemblance in the dry and fresh product color.

9.6.2 Effects on Bioactive Compounds

Most food products contain bioactive compounds at varying concentrations. Fruits and vegetables are essential to human health and are the primary sources of these compounds. Among these bioactive compounds, beta carotene, ascorbic acid, and antioxidants are good indicators of heating effects brought about by drying [33]. In this section, we discuss different bioactive compounds and how they are affected by RW drying technology.

Table 9.3 Color measurements of asparagus dried using five drying techniques [8].

Drying method	Temperature (°C)	L	a*	b*	Hue angle	Chroma
Tray Drying	50	24.8 ± 0.7	-2.6 ± 0.7	20.6 ±1.4	92.7	20.8
	60	14.4 ± 1.7	-0.6 ± 0.1	12.2± 1.0	92.6	12.2
MSWB	50	17.9 ± 3.8	-3.3 ± 0.7	14.7± 3.2	102.8	15.1
	60	26.8 ± 2.0	-1.9 ± 0.5	17.6±1.8	96.1	17.7
RW	95	18.3 ± 3.4	-4.5 ± 0.4	15.6±1.6	106.0	16.2
Freeze Drying	20	37.7 ±4.5	-4.9 ± 0.7	20.7±1.4	103.4	21.3
Fresh Asparagus	-	40.2 ± 2.9	3.1 ± 0.7	11.7±2.6	104.7	12.1

*MSWB-Microwave Spouted Bed, RW-Refractance Window.

Table 9.4 Color analysis for mammee apple pulp after drying using RW drying technique [32].

Color parameter	Before drying	After drying
L*	43.88±0.23	39.85±0.59
a *	7.31±0.09	0.69±0.13
b*	29.37±0.17	25.15±0.24
C	30.26±0.20	25.16±0.24
h	76.01±0.10	88.42±0.30

9.6.2.1 Carotene Retention

Carotenes are prone to heat and oxidation, and to prevent the effects of the two, it is necessary to choose the right drying method. Carotenes are vital to human health, and their preservation is needed to achieve quality products [34]. Drying carrot puree resulted in total carotene loss of

Table 9.5 Carotene loss for the different drying methods [12, 17].

Drying method	Total carotene loss (%)	Total carotene loss (%)
Refractance window	9	8.7
Freeze drying	4	4
Drum drying	56	56
Food Dehydrator	18	-

56%, 8.7%, and 4.0% from drum drying, RW, and freeze-drying, respectively (See Table 9.5) [12, 17]. RW drying technique was slightly behind freeze-drying in the retention of carotene. The lower the loss, the better the method of drying in retaining the carotenes. Similar studies on the RW drying method indicated high carotenoid retention in carrots [12]. In a recent study, approximately 64% of total carotenoids were retained after drying *mammee* apple pulp using the RW technique [32].

From Table 9.5, RWD had a 9% carotene loss, second after freeze-drying. Other methods recorded a high loss of carotene, implying that RWD provides an opportunity to dehydrate colored food products.

9.6.2.2 *Ascorbic Acid Retention*

Apart from carotenes, ascorbic acid is the other essential nutrient that is highly affected by heat and oxygen. Due to its abundance of fruits and vegetables, optimum drying conditions are needed to avoid its loss. Research suggests that the RW drying technique provides such conditions [6, 8, 12]. Studies on drying of strawberry purees indicated that 94% of ascorbic was retained using this technique compared to 93.6% obtained using freeze-drying [12, 35]. High retention of ascorbic acid was also recorded on the drying of asparagus using RW drying [8].

9.6.2.3 *Anthocyanin Retention*

Fruits and vegetables remain to be a good source of bioactive compounds. Anthocyanins, like other bioactive compounds, are affected by thermal treatment and drying processes [36, 37]. Fortunately, research indicates that drying methods such as RW result in minimal loss of these compounds. A study carried out on colored potato flakes reported that RW drying resulted in 23% loss in anthocyanins compared to 45 and 41% in freeze and drum drying, respectively [36, 37]. In another study, RW drying resulted in a 92% retention of anthocyanins in Haskap berry puree [38].

9.7 Applications of Refractance Window Drying in Food and Agriculture

Over the years, there has been intense research on the uses of the RW drying technique in food and Agriculture. The majority of studies focus on comparing RW drying against other drying techniques on different food products and, particularly, fruits and vegetables. In comparison to other drying methods, the RW technique exhibits superior product attributes, making it a method of choice in drying a variety of food products [18, 33]. Besides food items, RW technology has also been applied to skincare products. This section focuses on the applications of the RW technique in the drying of various food products in food and agriculture. It also focuses on the application of RW technology to food safety.

9.7.1 Applications of Refractance Window Drying in Preservation of Heat-Sensitive and Bioactive Compounds

RW drying is a technique of choice in the drying of fruits and vegetables. This is due to its ability to retain and preserve heat-sensitive compounds [8]. In the previous section of this chapter, heat sensitive and bioactive compounds such as carotene and vitamin C were discussed. Drying of fruits and vegetables significantly affects these compounds, and depending on the severity of the heat treatment, the compounds may be reduced to undesirable levels. Compared to other drying methods, RW drying exhibits promising results from a wide range of food products, including carrot, potato, strawberry, and mango purees. In the drying of carrots using RW drying, approximately 90% of total carotene is retained (See Table 9.5). For this reason, a more enhanced fresh carrot color is observed in dried carrot flakes (See Plate 9.2). In equal measures, bioactive compounds such as vitamin C can highly be retained using RW drying technology.

Besides heat sensitive and bioactive compounds, other important physical attributes such as color and chlorophyll are susceptible to drying heat. In comparison to other drying methods, the extent to which RW drying affects color is minimal. In the drying of green asparagus, for instance, the color of the RW dried product was very close to that of the fresh product [8, 12]. Although freeze-drying is known for providing superior quality products, it faces several limitations in terms of cost and energy consumption and produces a porous product that is susceptible to absorption of moisture if placed in a humid environment. The ability of RW drying to preserve the nutritional attributes of a product could offset the limitations of freeze-drying, making RW the method of choice.

RW drying technique offers a novel solution to the pharmaceutical industry. In recent years, the demand for skincare products has grown drastically. Plants such as aloe Vera have made significant strides in this regard due to their therapeutic properties [39]. Aloe Vera is a heat-sensitive plant subject to the high levels of antioxidants in it. Despite being sensitive to heat, aloe Vera requires that dehydration be done using the right technique for its safe storage. To avoid the loss of heat-sensitive antioxidants, a suitable method/technique of drying should be employed. In this regard, various techniques such as freeze-drying, drum drying, tray drying, spray, and RW drying have been tested. The latter is ideal and more promising as far as the preservation of antioxidants is concerned [40]. RW drying technique yields Aloe Vera with stable antioxidant activity and, consequently, allows for its prolonged storage.

9.7.2 Applications of Refractance Window Drying on Food Safety

Drying is the oldest food preservation method utilized in extending the shelf life of food products by reducing their water activity [41]. It has been proven that water activity below 1 inhibits the proliferation of microorganisms in the food matrix. Furthermore, with reduced water activity, food products can be stored easily for quite a long time [42]. Over the years, several drying techniques, for example, RW, freeze-drying, drum, and spray drying, have been applied in extending the shelf life of various products. Regardless of the effectiveness of the method, each of the techniques ensures food safety to a varying extent. Different microbial organisms thrive at different water activity levels; for instance, bacteria, yeast, and molds are inhibited at a water activity of 0.87, 0.88, and 0.80, respectively [41]. RW drying technique has a high potential to ensure microbial food safety because a water activity below the stated values can be achieved in a relatively shorter time.

One of the studies noted that 60 minutes were sufficient to achieve a water activity of 0.5. In contrast, drum drying took 240 min to reach the same target [43, 44]. A combination of low water activity and heat treatment is powerful in reducing the microbial load to acceptable levels. Heat denatures proteins that make up the bacteria. Food pathogens such as *E.coli* and *Listeria innoca* can be reduced to acceptable levels of <5 CFU/ml using RW drying technology [44].

9.8 Advantages and Limitations of Refractance Window Dryer

RW drying technique has multiple advantages, as highlighted in the different sections of this chapter. Firstly, the dryer offers a low-cost drying technology capable of delivering superior products. Secondly, the dryer operates at lower temperatures and at the same time provides rapid drying of the wet product. Further, besides a simple design, RW dyer prevents cross-contamination since the product being dried does not come in contact with the heating media, a scenario common with other types of contact dryers. All these aspects of the dryer work together to deliver quality products. On the contrary, RW technology is modest at drying fluid and semi-fluid-like products. Based on its principle of operation, the dryer is not suitable for solid or granular material, and this limits its scope of operation. Also, despite being suited for fluids and semi-fluids, the drier is not convenient for fluids with high sugar content because such products stick on the plastic film inhibiting their free flow. Lastly, as a thin film type of dryer, drying large quantities of a product calls for a large surface area for heat-exchange and drying. This particular limitation acts as the major bottleneck for the scale-up of the technology into a large industrial dryer (1-7).

9.9 Recent Developments in Refractance Window Drying

Recently, there has been an attempt to develop an evaporator using RW technology. The evaporator, used to concentrate fluids and fluid-like food products, utilizes circulating hot water approximately at 95-98 °C and an infrared-transparent plastic sheet to dry the fluid. To concentrate the product, the fluid makes several passes on the plastic sheet. The evaporator employs the same principles as the RW dryer only that the evaporator is inclined at an angle to allow for gravitational flow (Figure 9.5) [45]. The configuration eliminates the need for a powered conveyor, as is the original design (Figure 9.1). The proprietary company, MCD Technologies, has developed a similar model of the evaporator. The evaporator can be used as stand-alone equipment, or it can be part of the product's drying process. This equipment allows a product to be concentrated to higher total solids at the same time, maintaining a liquid state [33].

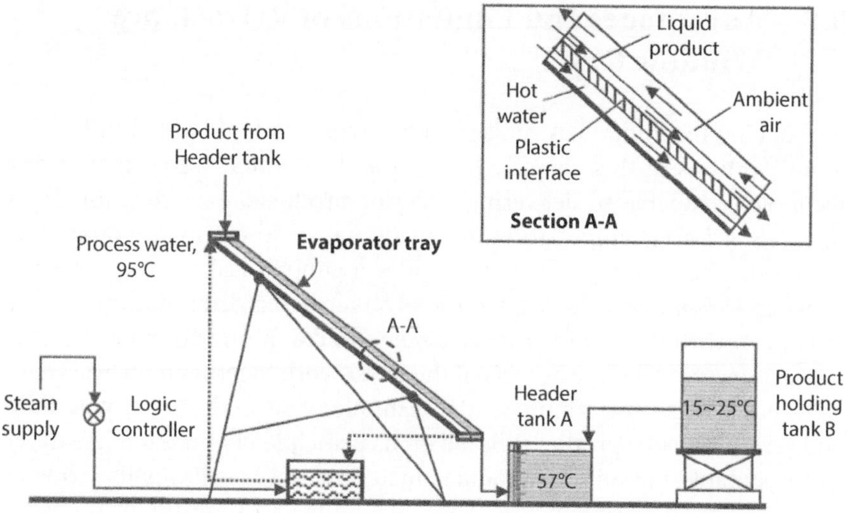

Figure 9.5 Layout of a refractance window evaporator. Adapted from reference [45].

9.10 Conclusion and Future Prospects

As a fourth-generation dryer, RW dryer is not only a simple and low-cost type dryer but is also a technology that applies to a diverse range of products from the farm level to the industrial level. Current understanding maintains that radiation heat transfer is the most dominant mode resulting in rapid drying. However, proponents suggest that quantitative studies are needed to substantiate the claim that conduction heat transfer is the predominant mode of heat transfer. While the technology is on a pilot scale, large-scale, continuous industrial dryers and evaporators could be developed adhering to the same principle. Such equipment is vital in providing high-quality and high-nutritive products for healthy living and to boost the immunity subject to the ongoing Covid-19 pandemic. Although RW drying is still a novel technology, there is a need to investigate its effectiveness in drying and concentrating different food materials and the effect on microorganisms such as yeast and probiotics. It is worth examining the effect of drying on the granules and microstructure of the product being dried. Even though the role of each of the process parameters is known, modeling the RW drying could lead to more improvements as well as optimization of the drying process in terms of design and energy consumption. This could be an avenue to RW industrial dryers. Lastly, the current equipment uses convectional energy sources to heat the water and to run the conveyor; however, studies should explore how these energy sources can be replaced with renewable and inexhaustible sources for sustainability purposes.

References

1. Vega-Mercado, H., Góngora-Nieto, M. M., & Barbosa-Cánovas, G. V. (2001). Advances in dehydration of foods. *Journal of Food Engineering*, *49*(4), 271-289.
2. Raghavi, L. M., Moses, J. A., & Anandharamakrishnan, C. (2018). Refractance window drying of foods: A review. *Journal of Food Engineering*, *222*, 267-275.
3. Nindo, C. I., Tang, J., Cakir, E., & Powers, J. R. (2006). Potential of Refractance Window technology for value added processing of fruits and vegetables in developing countries. In *2006 ASAE Annual Meeting* (p. 1). *American Society of Agricultural and Biological Engineers.*
4. Nindo, C. (2008). Novel drying method for vegetables, fruits, herbs, and aquatic resources. In *Proceedings of the CSBE/SCGGA 2008 Annum Conference.*
5. Abascal, K., Ganora, L., and Yarnell, E. (2005).The effect of freeze-drying and its implications for botanical medicine - A review. *Phytotherapy Res.*, 12: 660-665.
6. Trivedia, M., D'costaa, V., Shituta, J and Srivastava. S. (2017). Refractance Window Technology – A promising drying technique for the food industry. *International Journal of Innovative and Emerging Research in Engineering*, *4*(1): 2394-5494.
7. Bolland, K. 2005. President, MCD Technologies Inc. Personal communication.
8. Nindo, C., Sun, T., Wang, S. W., Tang, J., & Powers, J. R. (2003). Evaluation of drying technologies for retention of physical quality and antioxidants in asparagus (*Asparagus officinalis* L.). *LWT-Food Science and Technology*, *36*(5), 507-516.
9. Nindo, C. I., & Tang, J. (2007). Refractance window dehydration technology: a novel contact drying method. *Drying Technology*, *25*(1), 37-48.
10. Kudra, T., & Mujumdar, A. S. (2009). *Advanced Drying Technologies.* CRC Press.
11. Moses, J. A., Norton, T., Alagusundaram, K., & Tiwari, B. K. (2014). Novel drying techniques for the food industry. *Food Engineering Reviews*, *6*(3), 43-55.
12. Abonyi, B.I., H. Feng, J. Tang, C.G. Edwards, D.S. Mattinson, and J.K. Fellman. (2002). Quality retention in strawberry and carrot purees dried with Refractance Window® System. *Journal of Food Science*, 67: 1051-1056.
13. Zotarelli, M. F., Carciofi, B. A. M., & Laurindo, J. B. (2015). Effect of process variables on the drying rate of mango pulp by Refractance Window. *Food Research International*, *69*, 410-417.

14. Bolland, K. M. (2017). Refractance Window food drying system delivers quality product efficiently. Retrieved from https://www.foodonline.com/doc/refractance-window-food-drying-system-deliver
15. Sandu, C. (1986). Infrared radiative drying in food engineering: a process analysis. *Biotechnology Progress, 2*(3), 109-119.
16. Jimena Ortiz-Jerez, M., Gulati, T., Datta, A. K., & Isabel Ochoa-Martinez, C. (2015). Quantitative understanding of Refractance Window (TM) drying. *Food and Bioproducts Processing, 95*, 237-253.
17. Abonyi, B.I., Tang, J, and Edwards, C.G. (1999), Evaluation of energy efficiency and quality retention for the Refractance Window™ drying system. *Research Report*, Washington State University, Pullman WA.
18. Baeghbali V, Niakosari M, Kiani M. (2010). Design, manufacture and investigating functionality of a new batch Refractance Window system. In: *Proceedings of 5th International Conference on Innovations in Food and Bioprocess Technology*, 7(9).
19. Ochoa-Martínez, C. I., Quintero, P. T., Ayala, A. A., & Ortiz, M. J. (2012). Drying characteristics of mango slices using the Refractance Window™ technique. *Journal of Food Engineering, 109*(1), 69-75.
20. Abonyi, B. I., Feng, H., Tang, J., Edwards, C. G., Chew, B. P., Mattinson, D. S., & Fellman, J. K. (2002). Quality retention in strawberry and carrot purees dried with Refractance WindowTM system. *Journal of Food Science, 67*(3), 1051-1056.
21. Ortiz-Jerez, M. J., & Ochoa-Martínez, C. I. (2015). Heat transfer mechanisms in conductive hydro-drying of pumpkin (Cucurbita maxima) pieces. *Drying Technology, 33*(8), 965-972.
22. Baeghbali, V., & Niakousari, M. (2018). A review on mechanism, quality preservation and energy efficiency in Refractance Window drying: A conductive hydro-drying technique. *J. Nutr. Food Res. Technol, 1*, 50-54.
23. Castoldi, M., Zotarelli, M. F., Durigon, A., Carciofi, B. A. M., & Laurindo, J. B. (2015). Production of tomato powder by refractance window drying. *Drying Technology, 33*(12), 1463-1473.
24. Baeghbali, V., Niakousari, M., and Farahnaky, A. (2016). Refractance Window drying of pomegranate juice: Quality retention and energy efficiency. *LWT-Food Science and Technology*, 66: 34–40. doi:10.1016/j.lwt.2015.10.017.
25. Pathare, P. B., Opara, U. L. and Al-Said, F. A. (2013). Colour measurement and analysis in fresh and processed foods: A Review. *Food Bioprocess Technol.*, 6: 36–60.
26. Ahmed, J., U.S. Shivhare and K.S. Sandhu. (2002).Thermal degradation kinetics of carotenoids and visual color of papaya puree. *Journal of Food Science*, 72(67): 26922695.
27. Ruchika, Z., Preetinder, K and Mukul, S. (2020). Refractive window drying-A better approach to preserve the visual appearance of dried products. *Pantnagar Journal of Research*, 18(1).

28. Severini, C., Baiano, A., Pilli, T. D., Romaniello, R. and Derossi, A. (2003). Prevention of enzymatic browning in sliced potatoes by blanching in boiling saline solutions. *Journal Food Science and Technology*, 36: 657–65.
29. Caparino, O. A., Tang, J., Nindo, C. I., Sablani, S. S., Powers, J. R. and Fellman, J. K.(2012). Effect of drying methods on the physical properties and microstructures of mango (Philippine '*Carabao'var.*) powder. *Journal of Food Engineering*, 111 (1): 135-148.
30. Shende, D. and Datta, A. K. (2019). Optimization study for refractance window drying process of Langra variety mango. *Journal Food Science and Technology.* doi: 10.1007/s13197-019-04101-0.
31. Zalpouri, R. (2018). Development and Evaluation of Refraction Based System for Dehydration of potato. Master Thesis, Punjab Agricultural University, Ludhiana, India.
32. Nascimento, C., Rodrigues, A. M and Silva, L.M. (2019). Development of a dehydrated product with edible film characteristics from mammee apple (*Mammea americana* L.) using Refractance Window drying. *Food Science and Technology, Campinas*, 40(1): 245-249.
33. Clarke, P. T. (2004). Refractance Window—Down Under. In *Proceedings of the 14th International Drying Symposium, IDS* (pp. 813-820).
34. Pénicaud, C., Achir, N., Dhuique-Mayer, C & Dornier., M. (2011). Degradation of β-carotene during fruit and vegetable processing or storage: Reaction mechanisms and kinetic aspects: A review. *Fruits.* 66. 417-440. 10.1051/fruits/2011058.
35. Sablani, S.S. (2006). Drying of fruits and vegetables: Retention of nutritional/ functional quality. *Dry Technology*, 24:123–135.
36. Tonon, R.V., Brabet C and Hubinger,M, D. (2010). Anthocyanin stability and antioxidant activity of spray-dried açai (*Euterpe oleracea Mart.*) juice produced with different carrier agents. *Food Research International*, 43(3):907–914.
37. Nayak, B., Liu, R.H and Tang, J. (2015). Effect of processing on phenolic antioxidants of fruits, vegetables, and grains—A review, *Critical Reviews in Food Science and Nutrition*, 55(7), 887-918.
38. Celli, G. B., Khattab, R., Ghanem, A., & Brooks, M. S. L. (2016). Refractance Window™ drying of haskap berry–preliminary results on anthocyanin retention and physicochemical properties. *Food chemistry, 194*, 218-221.
39. Hu, Y., Xu, J., & Hu, Q. (2003). Evaluation of antioxidant potential of Aloe vera (*Aloe barbadensis Miller*) extracts. *Journal of agricultural and food chemistry, 51*(26), 7788-7791.
40. Femenia, A., García-Pascual, P., Simal, S., & Rosselló, C. (2003). Effects of heat treatment and dehydration on bioactive polysaccharide acemannan and cell wall polymers from Aloe barbadensis Miller. *Carbohydrate Polymers, 51*(4), 397-405.
41. Beuchat, L. R., Komitopoulou, E., Beckers, H., Betts, R. P., Bourdichon, F., Fanning, S., ... & Ter Kuile, B. H. (2013). Low-water activity foods: Increased

concern as vehicles of foodborne pathogens. *Journal of Food Protection, 76*(1), 150-172.

42. Brown, Z. K., Fryer, P. J., Norton, I. T., Bakalis, S., & Bridson, R. H. (2008). Drying of foods using supercritical carbon dioxide—Investigations with carrot. *Innovative Food Science & Emerging Technologies, 9*(3), 280-289.
43. Tontul, I., & Topuz, A. (2017). Effects of different drying methods on the physicochemical properties of pomegranate leather (pestil). *LWT-Food Science and Technology*, 80, 294-303.
44. Bourdoux, S., Li, D., Rajkovic, A., Devlieghere, F., & Uyttendaele, M. (2016). Performance of drying technologies to ensure microbial safety of dried fruits and vegetables. *Comprehensive Reviews in Food Science and Food Safety, 15*(6), 1056-1066.
45. Nindo, C. I., Powers, J. R., & Tang, J. (2007). Influence of Refractance Window evaporation on quality of juices from small fruits. *LWT-Food Science and Technology, 40*(6), 1000-1007.

10

Ohmic Heating: Principles and Applications

Sourav Misra, Shubham Mandliya and Chirasmita Panigrahi*

Agricultural and Food Engineering Department, Indian Institute of Technology, Kharagpur, West Bengal, India

Abstract
Ohmic heating (OH) has evolved as an efficient processing method over the last two decades. The implementation of OH in the food industry has grown considerably due to various associated advantages. This technology is currently being applied for blanching, preheating, pasteurization, and sterilization of food products. Studies on process modelling and prediction, analysis of food properties, detailed characterization of treated product and determination of the heating pattern of complex foods are subjects of the future scope of research. These are required to assist the design of food sterilization or pasteurization ohmic-based processes and for the successful commercialization of developed products. This chapter presents an overview of the principles and mechanism of OH, its applications in various food processing sectors and the influence of treatment on nutritional and sensory characteristics of food products.

***Keywords*:** Ohmic heating, principles, applications, processing effects

10.1 Introduction

Globally, industrial manufacturing operations related to different sectors such as food, chemicals, paper, refining, mining, etc., are utilizing more than 30% of energy usage in the world [1] and there is a need to minimize the consumption of global energy mainly in the above energy-intensive sectors. There are great opportunities and many possible ways to control as well as improve the quality of energy, which are still underexploited.

**Corresponding author*: chirasmitapanigrahi8@gmail.com

Nitin Kumar, Anil Panghal and M. K. Garg (eds.) Thermal Food Engineering Operations, (261–300)

Thermal energy may be produced by the combustion of fuels (generally solid, liquid, or gas) based on industrial applicability and employed to the material either directly or indirectly.

Many of the grocery items that people buy today are packaged foods. Food production's key objectives are to ensure microbiological protection and increase the storability of food by destroying toxins, enzymes, etc. A heating process's efficiency has to be improved to attain a desirable food product that covers three basic processing parameters such as high throughput, low processing time, and low cost. A practical and energy-efficient process creates a high-quality product with less input energy and minimal degradation of sensory and nutritional properties [1]. The conventional heating methods have been proved to provide microbiological stability to food products but are inefficient to preserve the qualitative parameters. Though food fortification can compensate for the certain nutritional loss in conventional processes, the organoleptic parameters such as aroma, flavor, appearance, and texture are difficult to retain. Traditional heating techniques involve the generation of heat externally and then transfer through conduction, convection, or radiation to the food material. To achieve microbial lethality, the center of food material should be heated sufficiently, which is also known as the coldest point. Time taken to reach the desired temperature at the center of the product may cause over-processing in the surrounding region that ultimately leads to inherent flavor and nutrient destruction. This is a common problem during the conventional processing of products containing large particulates.

To overcome the above drawbacks, different non-thermal and novel thermal technologies have emerged to process and preserve food products, which can satisfy consumer demands in terms of higher quality, being free of additives and having a prolonged shelf life. The main objective of these technologies is mainly to inactivate the pathogens as well as the enzymes and allergens that are causing spoilage without any loss in nutritional and organoleptic properties. So there is a demand for quick, clean, and uniform heating methods that can fulfill desired microbial stability without or with minimal alteration of the overall quality of processed food products.

Ohmic heating has evolved as an alternative processing method over the past 20 years. It is also known as electroconductive heating, electrical resistance heating, or Joule heating. When electricity passes through a material, it produces heat. The heating takes place within the product in the form of internal energy transformation (from electrical to thermal) [2]. Unlike inductive and microwave heating, the food products come in contact with electrodes, but these are absent in the above methods. Compared to the radio or microwave frequency range, the applied frequency is smaller, and the waveform is therefore unrestricted but usually sinusoidal. Ohmic heating is currently used for blanching, preheating, pasteurization, and sterilization of fruit juices, vegetable products, and meat products [3].

10.2 Basic Principles

The general idea of Joule's effect is the dissipation of electrical energy into heat via an electric conductor. The energy generated is proportional to the square power of the confined electrical field strength and electrical conductivity (σ). The power produced (ΔP) perpendicular to the electric field (E) in a homogeneous element conductor of specified length Δz in the z-direction and area ΔA is given by [4]

$$\Delta P = \frac{\Delta U^2}{\Delta \Omega} = \sigma E^2 \Delta A \Delta z = \sigma E^2 \Delta V$$

Where $\Delta U = E\Delta z$ is the potential difference in the z-direction; $\Delta \Omega = {}^{\Delta z}/_{\sigma \Delta A}$ is the ohmic resistance. The equation above demonstrates the volumetric nature of the OH process. Inside the product, the heat power per unit volume is $G = \sigma E^2$. The food item is placed between two electrodes where the electric field is generated for heating. The development of current flow depends on the field strength, product conductivity, and electrode configuration.

When an electric potential is applied to a non-homogeneous material having local electrical conductivity, then the potential field is derived by using Laplace's Law as follows:

$$\nabla(\sigma \nabla U) = 0$$

The electrical field E is generated and produces local heat output per unit volume G within the material ($E = \nabla U$ and $G = \sigma E^2$).

Ohm's law defines a relationship between the current density and the electrical field strength. For a homogeneous element, Ohm's law states:

$$J = \frac{E}{\sigma}$$

The terms E and G are valid only locally in a medium consisting of various phases and electrical conductivity, and their values are dependent on physical and electrical homogeneity. Thermal conductivity k and volumetric heat capacity G decide the local temperature of the element; hence, in the conductive medium:

$$\frac{\partial \rho C T}{\partial t} = \nabla(k\nabla T) + G$$

Where T is the local temperature at time t, C is specific heat, and ρ is density.

According to Ohm's law, the flow of current contributes to the energy input that can be characterized by the more or less complete conversion of electrical energy into heat, high energy density, and fast heating cycles. The voltage range is between 400 and 4000 V with a field strength of a range of 20-400 V/cm, which occurs when the gap between the electrode is between 10-50 cm. The heating rates obtained depend on the power supply output, the configuration of the treatment equipment, and the properties of the product (viscosity, conductivity, and specific heat capacity).

The energy (P) of the ohmic heating system at the prescribed temperature is determined using the current (I) and the voltage (ΔV) values during the heating time (Δt):

$$P = \sum VI\Delta t$$

Based on the component and taking into account the product properties, the required electrical output can be measured based on mass flow, temperature rise, and specific heat capacity. Sensible heat is produced due to the passage of electrical current through the heating sample, which causes the sample temperature to rise from T_i to T_f, and the amount of heat provided to the system can be determined from the following equation:

$$Q = mC_p(T_f - T_i)$$

Where Q is the total sensible heat (kJ); m is the weight of the sample (kg); T is the temperature (°C); the subscripts i and f stand for initial and final, respectively.

The energy efficiency is measured and analyzed to determine the heating process's performance by using the ohmic heating method. Energy efficiency is defined as:

$$\text{Energy efficiency} = \frac{\text{Energy utilized to heat the sample}}{\text{Total input energy}} = \frac{mC_p(T_f - T_i)}{\sum VI\Delta t}$$

Direct or alternating current may be used to produce ohmic heating; however, alternating current is typically used to avoid electrochemical and electrolytic effects and to suppress unwanted reaction products. Current densities exist at the electrodes in the 0.5-20 A/cm^2 range. These current densities make it possible for the products involved to have unspecific electrochemical reactions. Alternating current at frequencies greater than 20 kHz is used to decrease oxidation reactions and metallic contamination of the product. When alternating current is used, electrochemical reactions occur less often, owing to the reversed field effect. Due to inhibiting Faraday reactions, electrode erosion is minimized at frequency values greater than 20 kHz. The electrode material is ordinarily stainless steel due to its low propensity to oxidize [5].

10.3 Process Parameters

The essential parameters to be considered during ohmic heating are:

Processing parameters: Electrical field strength, temperature, time, frequency of the electric current
Product parameters: Viscosity, electrical conductivity, specific heat, density, homogeneous or solid-liquid systems, tendency to form fouling
Equipment parameters: Size and shape of the electrodes, size of the ohmic cell, composition of electrode, configuration of the electrode

10.3.1 Electrical Conductivity

In ohmic heating, electrical conductivity is one of the significant parameters which impact the product heating rate. It is defined as the material's ability to transfer electric current per unit area, per unit time across a unit potential gradient. The Siemens per meter (S/m) is its SI derived unit, for any material electrical conductivity can be determined from the following equation:

$$\sigma = \frac{L}{A} \times \frac{I}{V}$$

Where σ is the electric conductivity (S/m); L is the distance between the two electrodes (m); A is the area of cross-section of the material in the heating cell (m^2); I is an AC passing through the material (A); V is the voltage across the material (V).

The electrical conductivity is a temperature-dependent constraint and it varies with different food materials. It usually shows linearity with temperature and increases as the temperature of the material increases. Using linear regression analysis, the relationship between electrical conductivity and electrical field strengths as well as concentrations is obtained using the following equation:

$$\sigma_T = \sigma_i + M$$

Where M is the temperature factor (S/m °C); the subscript T stands for temperature.

Biological materials have been categorized as poor conductors of electricity. Broadly, these materials are classified based on the electrical conductivity levels, which can be represented as [4]:

- $\sigma > 0.05$ S/m: materials having decent conductivity. Ex. milk desserts, eggs, condiments, gelatin, fruit juice, yogurt, hydrocolloids, wine, etc.
- $0.005 < \sigma < 0.05$ S/m: materials having low conductivity and require high electrical field strength. Ex. powders, marmalade, margarine, etc.
- $\sigma < 0.005$ S/m: materials having very low conductivity and require very high electrical field strength as well as possess difficulties during ohmic processing. Ex. foam, syrup, frozen foods, liquor, fat, etc.

The food products having an electrical conductivity of 0.1-10 S/m are generally processed by ohmic heating. With voltage gradient, water content, frequency, and temperature, food products' electrical conductivity changes. The heating of ohmically treated products occurs due to the presence of ion components and insulators in the complex food matrix.

10.3.2 Electrical Field Strength

An increase in the ohmic heating rate is attributed to a rise in electrical field strength and electrical conductivity. The use of field strength leads to improved capillary fluid motion and exhibits a linear relationship with electrical conductivity [6]. Different researchers have suggested different electrical field strength levels for various food applications of ohmic heating, such as the inactivation of microbes has been achieved effectively at a higher electrical field strength of 104-105 V/cm [7]. In contrast, a lower level of field strength, i.e., below 100 V/cm, has proved to be sufficient for the expression and extraction process in various food constituents [8–11].

10.3.3 Frequency and Waveform

Generally, the frequency in the range of 50-60 Hz is employed for OH of food materials [12]. The waveform and frequency of the voltage applied to affect the sample's electrical conductivity and its heating process.

10.3.4 Product Size, Viscosity, and Heat Capacity

The orientation relative to the electrical field has less effect on electrical conductivity in emulsions or colloids (usually smaller particles of less than 5 mm), but the electrical properties and heating rate are greatly affected by the effect of orientation in the case of large particulates (15-25 mm) [13]. There was a decrease in electrical conductivity with an increase in particle size that resulted in the lowering of the heating rate [14]. Since solid particles are also suspended in the liquid medium and have comparable electrical conductivity, the lower heat capacity portion between them will tend to heat faster. Foods with high density and high specific heat values are favorable for slower heating [14].

10.3.5 Particle Concentration

It has been observed that an increase in the concentration of strawberry pulp results in decreasing its electrical conductivity [15]. Increasing the solid concentration of carrot cubes also caused an increase in the heating time [16].

10.3.6 Ionic Concentration

Heating of biological materials causes an increase in the ionic mobility resulted from the structural changes in cellular level such as deterioration of cell wall, softening of tissue, and lowering of the viscosity in the aqueous phase that alternately occurs due to the increase in the electrical conductivity [17]. Faster heating rate results from the increase in ionic concentration.

10.3.7 Electrodes

The type of product and the electrode material significantly affect the heat loss from the food products that result in an unacceptably high-temperature gradient in the food materials. Electrodes during ohm heating are a possible source of heat loss. It has been studied that the temperature rises at a lower rate in the case of a thicker electrode [18]. Different materials such as platinized titanium, stainless steel, titanium, and aluminum were tested by [18] as electrode materials. They revealed that the titanium electrode surface showed a higher temperature than the same thickness stainless steel. The highest and lowest temperatures were achieved at thinner aluminum and platinized aluminum of a similar thickness, respectively.

10.4 Equipment Design

Ohmic heating setup mainly consists of a heating cell with the data logger, electrodes, power supply source, thermocouple, and voltage control unit (Figure 10.1). The function of a generator or power supply is to generate electricity. The electrodes are attached to the power supply source and pass the electric current by keeping in direct contact with the food materials. Depending upon the system's size, the electrical field strength can be varied by adjusting the distance between the electrodes (also known as the electrode gap). A data-logger is used in the system to record temperature, the intensity of the current, and voltage from time to time. Current and voltage transducers are connected to the system to measure current and voltage.

There are different generic configurations:

- **Batch configuration:** In this configuration, the electrodes have been placed parallel either in a plane or cylindrical geometry (coaxial pattern) in an electrically and thermally insulated batch system (Figure 10.2). The fundamental parameters such as heating time, the product's electrical conductivity, and process homogeneity can be calculated in this system. This is an efficient system. The processing effect on the final food quality can be monitored as well as initial optimum product composition can be tested. For this reason, it is easier to find the optimum condition by controlling

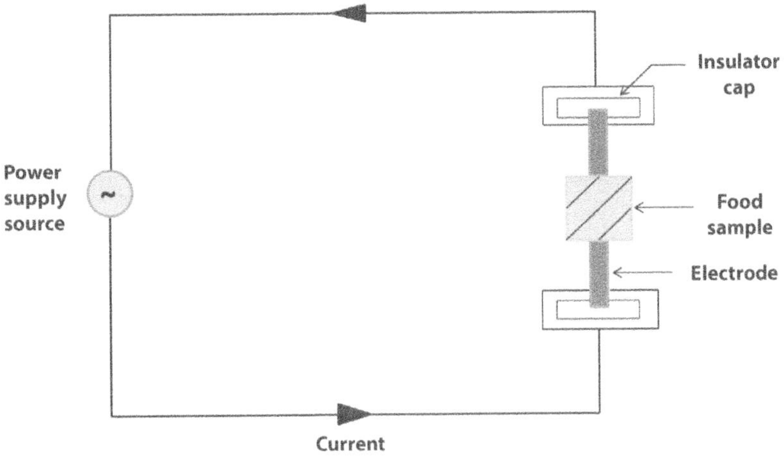

Figure 10.1 Schematic diagram of the principle of ohmic heating.

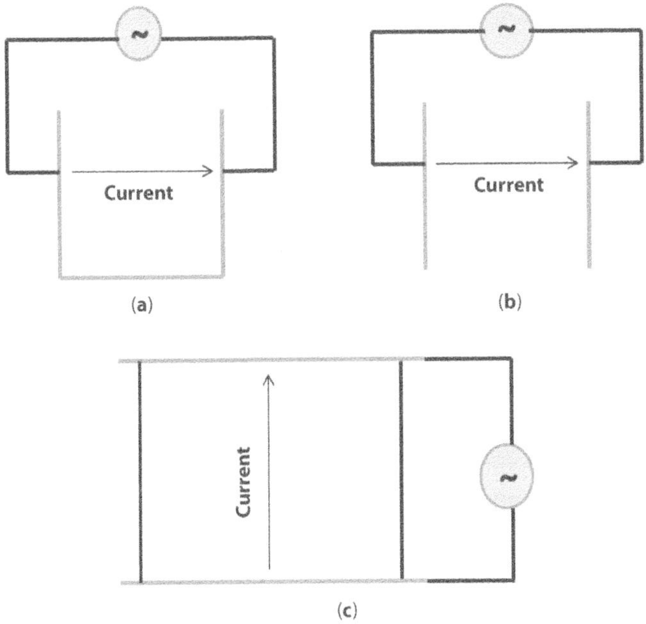

Figure 10.2 Different electrode configurations: (a) Batch, (b) Transverse, (c) Collinear.

the three stages of the continuous ohmic heating process, i.e., heating, holding, and cooling [19].

- **Transverse configuration:** In this configuration, there is a parallel-flow of the product with the slightly spaced electrodes that are generally flat or coaxial and perpendicular flow to the electric field (Figure 10.2). The current intensity is higher due to the low voltage (< 96 V) between the electrodes and large product-surface contact. This type of configuration is unsuitable for treating non-particulate fluids, such as milk beverages, due to localized boiling, overheating, and erosion of electrodes.
- **Collinear configuration:** The electrodes are widely spaced and the flow of the product is parallel to the electric field or perpendicular to the electrodes with the product from one electrode to another (Figure 10.2). The product-contact surface area is small and the applied voltage is high (up to 4500 V). For horizontal, longitudinal, or inclined cylinders, this configuration is used.
- **In-line field system:** The electrodes are mounted at different positions along the direction of product flow. Due to a drop in voltage gradient in the system, the material upstream exhibits greater field strength than the downstream section.
- **Cross-field system:** The product flows perpendicular to the electrodes with a constant electrical field strength throughout the process.

10.5 Application

Over the past two decades, the implementation of ohmic heating in the food industry has grown considerably, and the lack of these accomplishments has been linked to solving problems of electrode design, such as fouling and electrode polarization. In contrast, ohmic heating allows food to be heated at an extremely rapid rate. The wide application of ohmic heating for different food products has been illustrated in Table 10.1.

Table 10.1 Application and effects of ohmic heating on quality attributes of different food products.

Food products	Application	Processing conditions	Salient findings/ effects	References
Fruit juices/products				
Orange juice	Pasteurization	18.2 V/cm; 65-90 °C	Similar heating profile and decomposition of vitamin C as for conventional heating	[20]
Pomegranate juice	Pasteurization	10-40 V/cm; 90 °C for 3-12 min	No electrical effect on the quality of the product though there was a change in phenolic content, color, and rheological properties during the warm-up phase	[21]
Apple juice	Pasteurisation	20-60 V/cm	Change in the electrical conductivity with the voltage gradient, temperature, and concentration of juice	[22, 23]

(*Continued*)

Table 10.1 Application and effects of ohmic heating on quality attributes of different food products. (*Continued*)

Food products	Application	Processing conditions	Salient findings/ effects	References
Strawberry products	Pasteurization	25-100 V/cm, 100 °C	No effect of field strength on the electrical conductivity of the product except for strawberry pulp; minimal degradation of vitamin C due to the lower electrical field strength	[15]
Apricot pieces in syrup	Pasteurization	Continuous ohmic heater (30 kW); 60 V/cm, 200 kHz, 90 °C for 113 s	Improved storability, microbiological stability, and retention of quality attributes	[24]
Acerola	Pasteurization	120-200 V, electrode spacing not specified	Occurrence of electrochemical reaction; lower degradation of vitamin C as compared to conventional heating	[25, 26]
Blueberry pulp	Pasteurisation	Up to 240 V; up to 90 °C	Better retention of anthocyanins due to low voltage gradient than that of conventional heating method	[27]

(*Continued*)

Table 10.1 Application and effects of ohmic heating on quality attributes of different food products. (*Continued*)

Food products	Application	Processing conditions	Salient findings/ effects	References
Vegetables				
Cauliflower	Sterilisation	Continuous system (10 kW, 130 kg/h, 130-131°C)	Precooking at low temperature retained the stability and structure of cauliflower	[28]
Pea puree	Blanching	20-50 V/cm; 100 °C	Faster inactivation of peroxidase enzyme and better retention of color than with the conventional process	[29]
Carrot pieces	Blanching	60-90 °C, 50 Hz, 1-40 min	No significant difference in pectin methylesterase, peroxidase activity as compared to conventional and microwave heating	[30]
Meat/fish products				
Pork cuts	Heating/ thawing	5-7 V/cm	No difference in texture and sensory test with a slight difference in elasticity and color with the conventional process	[31]

(*Continued*)

Table 10.1 Application and effects of ohmic heating on quality attributes of different food products. (*Continued*)

Food products	Application	Processing conditions	Salient findings/ effects	References
Surimi	Heating/ thawing	6.7-16.7 V/ cm; 90 °C for 40-180 s	Higher concentration of sulfanyl compounds with better preservation of the color and higher water retention as compared to the conventional heating process	[32]
Beef	Thawing	10, 20 and 30 V/cm, 25 °C	Fewer histological and structural changes than through conventional warming	[33]
Sausage meat	Cooking	230 V; 50 Hz; 3.5-7 V/ cm	Reduction in electrical conductivity compared to steam cooking, no difference in texture, taste, and microscopic structure with a higher fat content	[34]

(*Continued*)

Table 10.1 Application and effects of ohmic heating on quality attributes of different food products. (*Continued*)

Food products	Application	Processing conditions	Salient findings/ effects	References
Other food products				
Milk	Pasteurisation	50 Hz-10 kHz	Less surface heat than with conventional heating, lowering of corrosion and fouling with increasing the frequency	[35]
Rice bran	Cell disintegration	100 V/cm, 1-60 Hz	Higher oil yield with the lowering of frequency due to the electroporation effect and non-thermal effect on lipase activity during ohmic heating	[36]
Whey solution	Pasteurisation	20-40 V/cm; 30-80 °C	Less sensitive to temperature and temperature changes, variation of electrical conductivity with temperature and concentration	[37]
Liquid whole egg	Heating/ thawing	20 V/cm; 20-60 °C	No significant difference in flow properties, apparent viscosity, and activation energy	[38]

10.5.1 Blanching

In food processing, blanching is a significant operation. This operation has numerous functions: enzymatic or microbial degradation, enhanced rehydration, gas elimination within tissues, bad taste or cloudiness correction, sugar reduction replacement, and so on. Water, time, and energy-consuming traditional blanching is by water or steam. This results in the depletion of soluble solids that need to be treated in plants' wastewater. As an alternative blanching process for vegetable purees, ohmic treatment can be used efficiently, resulting in high color attribute preservation.

Application of ohmic heating in blanching operation has led to numerous benefits such as shortening the deep-frying time of sliced potatoes by 10-15% with improved color and crispier texture; retaining higher solid content in mushroom caps with a reduced treatment time and same loss in weight/volume as that of the conventional blanching and rapid inactivation of peroxidase enzyme with a better color value in pea puree as compared to traditional water blanching [39, 40].

10.5.2 Pasteurisation/Sterilization

Conventionally, pasteurization is done by heat exchange, where heat is transferred from hot water or steam to the product by conduction. The heat exchange efficiency depends on the pasteurizer configuration and technical choices [41]. As ohmic heating (OH) guarantees higher thermal efficiency, this process is gaining popularity and is now being extensively used in the food industry [42]. The industry has implemented this technology to process liquids and solid-liquid blends [43]. In the pasteurization/sterilization of food products, ohmic heating is very widely used, resulting in excellent efficiency. In the early 20th century, the resistance heating method was used for milk pasteurization [44]. Ohmic heating can be applied for the sterilization of food at ultra-high temperatures (UHT), especially those containing large particles (up to 2.5 cm), which are challenging to be sterilized by other means. Due to vitamin destruction and flavor component damage, heat treatments of food are associated with quality deterioration. For example, carotenoids are susceptible to light, oxygen, temperature, metal, and chemical exposure during processing [45]. As heat can change nutritional and sensory attributes, improvements are required to mitigate food exposure to heat in process technologies. It was observed that ohmic heating minimized the loss of xanthophyll and carotene in grapefruit and blood orange juice, as well as the attainment of the high temperature of pasteurization during ohmic heating, resulted in a substantial increase in the organoleptic and nutritional properties of citrus juices [46]. Similarly, flavor compounds were preserved in orange juice when pasteurized with ultra-high temperature continuous ohmic heating, and storability was improved nearly two times that of the conventionally pasteurized sample [47].

10.5.3 Extraction

Conventional extraction is operated at medium to high temperature, i.e., 50 °C to 60 °C for a longer duration from 3 h to 20 h [48]. If an appeal for a more extended period, heat can degrade thermo-labile compounds such as polyphenols, flavonoids, antioxidants, and other bioactive diminishing their functionalities. There are constraints of low extraction percentage and relative instability in some extraction processes used at present. Most of the food materials are composed of ionic elements, such as acids and salts, which allow the electric current to be performed [49]. OH may generate heat within the product, converting electrical energy into thermal energy and thereby, heat materials at an extremely fast rate, with no requirement of any heating surface or medium. This process avoids quality degradation, especially sensitive food components, including vitamins, pigments, and other nutrients [50]. This treatment is widely applied to extract the essence of aromatic herbs and offers numerous benefits, like, minimizing energy consumption, reducing extraction time, and increasing the concentration of extract's components [51].

Colorant powder was prepared from rice bran using OH-assisted solvent extraction. It was found to have more significant amounts of bioactive compounds, including α-tocopherol and γ-oryzanol, and anthocyanins, like, cyanidin-3-O-rutinoside, cyanidin-3-O-glucoside, delphinidin, and malvidin [49]. It has been observed that OH could extract biomaterials from plant cell walls more efficiently than conventional heating [52]. They attempted to extract pectin from orange juice and found that higher voltage gradients could dramatically increase the efficiency of the pectin extraction yield. An approximately 75% reduction in extraction time was achieved when extracting the maximum amount of oil from ohmically processed rice bran relative to untreated bran [53]. Pulsed ohmic heating (POH) was found to be an effective method of merging electrical and thermal treatments to extract valuable cell components using medium electric fields and medium temperatures [54]. OH's effect on cell membrane damage and polyphenols extraction enhancement from red grape pomace has been investigated by [55]. Many related studies have been reported; for example, OH has been revealed to enhance the extraction yields of sucrose from sugar beets [56], rice bran oil, and some bioactive substances from rice bran [57], beet dye from beetroot [20, 58], apple juice from apples [59], and many more.

10.5.4 Dehydration

The influence of OH on mass transfer can also be utilized to increase the removal of water, reduce moisture content, and aid in the dehydration of foods. The thermal dehydration processes are energy-intensive and time-consuming, and hence, OH can be best investigated as a future pre-treatment accelerator. Grape drying was greatly improved by ohmic pre-treatment with an electric field strength of 14 V/cm having temperature and frequency 60°C and 30 Hz respectively, due to breakage of the skin of berry [60]. With an augmentation in electric field strength and electrical energy input, an increase in effective moisture diffusivity and drying rate of potato tissue was observed [61]. For potato tissue, ohmic pre-treatment lowered the drying temperature by around 20 °C. They demonstrated that through an electroporation mechanism, OH could impart visible damage to potato tissue. For ohmic pretreated samples, the structure permeabilization and water redistribution inside the vegetables are considered critical reasons for a shift in sorption isotherms [62]. After 60 min of osmotic dehydration, ohmically pretreated apple cubes produced 25% water loss over 240 min for controls [63]. In strawberries, ohmically blanched samples contain 68% dry matter, which is significantly higher than untreated samples having only 20.3% dry matter [64]. According to these researchers, even the short period of ohmic treatment led to substantial alterations in the fruit's cellular structure, which during osmotic dehydration, led to the improvement of mass transfer. Vacuum impregnation in syrup with or without citric acid was done prior to osmotic dehydration to increase homogeneity during ohmic treatment [65]. The pulsed vacuum and OH could best aid the osmotic dehydration treatment at 40 °C for 240 min and subsequent drying at 60 °C for dehydration of blueberries, considering the promising results, like, mass transfer improvement, lower losses of phenolics, anthocyanins, and flavonols, and reduced drying time [66]. They acted as pre-treatments to improve the efficiency of dehydration processes, focusing on the retention of bioactive compounds. The application of vacuum impregnation and ohmic heating accelerated the mass transfer during osmotic dehydration of apples [67], pears [68], and strawberries [69]. The authors observed that OH-induced cellular breakage due to its electro-thermal effect and vacuum application had a favorable impact on the maintenance of firmness of fruit samples.

10.5.5 Fermentation

Nowadays, the food fermentation industry employs a few innovative techniques to decrease fermentation time, increase production yield, and improve product quality. The feasibility of using novel processing technologies such as ohmic heating and moderate electric field (MEF) to improve the fermentation bioprocess has been explored by [70]. They found that volumetric ohmic heating could speed up fermentation by rapidly supplying optimum fermentation temperatures. By varying the substrate, such as releasing its micronutrients, electric fields can influence the fermentation process. It is possible to view these methods as future industrial fermentation tools.

To this date, *Lactobacillus acidophilus* is the most researched microorganism concerning the application of MEF/OH-assisted fermentation concerning the number of conducted studies. According to [71], in a low-intensity ohmic process, i.e., E = 1.1 V/cm or less, the reduction in the lag period for fermentation at 30 °C was found to be 18 times, compared to the conventional heating method. As a consequence of membrane electroporation, this finding may be due to enhanced nutrient transport through the cell's inner part. Also, the beneficial effects of OH on the early phases of lactic fermentation were reported by [71]. In later fermentative stages, OH induced a productivity decrease due to the passage of inhibitory metabolites. Ohmic treatment caused the development of lacidin A (a bacteriocin synthesized by *L. acidophilus*) to be postponed for 3 hours and reduced its activity. These findings reinforced the need to implement OH as a potential tool for the early-stage industrial fermentation process. Furthermore, it appears that the use of OH in the stationary fermentation stage should also be prevented [72].

The ohmic-assisted proofing of bread dough improved the heating rate, enabling the yeasts to smoothly reach the optimal temperature for their activities [73]. According to the authors, in the bread proofing process, OH could reduce the time required to achieve the desired expansion ratio by around 50%. At the beginning of the bread fermentation period, the lag phase was shortened from 1 h for conventional fermentation to 20 min for the ohmic-assisted fermentation method. For gluten-free bread dough testing, an ohmic-assisted fermentation system was developed recently, which was found to be a beneficial tool to evaluate the crumb structure formation and monitor the fermentation temperature [74]. Besides, the role of OH during the fermentation of fresh Arabica coffee beans on the minimization of coffee acidity was analyzed by [75]. The authors found that after 18 h of fermentation at a constant temperature of 30 °C, it reduced the total acidity of coffee beans from 0.53 to 0.18% lactic acid.

10.5.6 Ohmic Thawing

Frozen foods are kept between two electrodes and an alternating current is applied to them for thawing using ohmic heating. This method's benefits are that water and wastewater are not produced because of volume heating, thawing can be relatively uniform and the process can be easily managed. In the literature, a possible issue regarding ohmic heating has also been reported. As the frozen item thaws, the current flows more quickly into the thawed part of the block as frozen shrimp have a conductivity 100 times lower than that of the thawed shrimp [76]. Shrimp in that section will be cooked with the current passing through the thawed part of the block while the rest of the block is still in frozen form. This is called runaway heating or hot spot forming. If this dilemma is solved, there will be a new approach for thawing beef.

10.6 Effect of Ohmic Heating on Quality Characteristics of Food Products

10.6.1 Starch and Flours

10.6.1.1 Water Absorption Index (WAI) and Water Solubility Index (WSI)

Water absorption index (WAI) and water solubility index (WSI) characterize the conversion of the degree of starch from granule during processing due to the interaction with water [77]. During ohmic heating, the rise in these properties of flour may be due to the greater accessibility of granules to the absorption of water, the effects of the length of the amylopectin chains, the proportion of long or short chains, and the molecular weight of the starch. By conveniently dispersing in cold water, ohmic treated samples show desired WAI and low WSI and form a stable suspension.

10.6.1.2 Pasting Properties

Gelatinization is caused by the heating of starch in excess of water which is the most significant phenomenon since it results in food products possessing physical and chemical properties. Viscosity is an important feature that measures starch granule fragmentation and the operation's sternness [78]. Ohmic treatment is likely to induce changes that decrease the re-accommodation of amylose linear molecules or amylopectin parts. A negative linear interaction between electrical conductivity and the temperature of a starch suspension before and after gelatinization has been observed. In the gelatinization zone, the declining pattern of electrical conductivity (changes between initial and final viscosity) is attributed to the swelling of the starch granules and an increase in viscosity, resulting in a decrease in the field of motion of the starch particles and an increase in the resistance to the activity of swollen particles. A higher reduction in viscosity and complete retrogradation of cassava and jicama starches was observed by OH (107-115 V and 10-20 min) [79]. Both treatments demonstrated lower total retrogradation and decreased final viscosity compared to local treatments.

10.6.1.3 Thermal Properties

Starches are used as moisture retainers, stabilizing agents, and thickeners in food systems. Extreme deterioration of starch granules is caused by drum driers and various extrusion methods currently used in the food industry to prepare pre-gelatinized starches, increasing the soluble solid volume. Thermal properties at varying frequencies and voltages of traditionally and OH processed rice flours and starch was examined by [80]. The gelatinization temperature rose as the samples were pre-gelatinized due to traditional and ohmic heating, and due to starch-chain interactions, they became stiffer. Furthermore, enthalpy during gelatinization decreased for ohmically and conventionally heated starch. Thus during the differential scanning calorimeter (DSC) study, less energy for gelatinization was needed. The highest decrease in enthalpy was found in ohmic heating, possibly due to the maximum extent of pre-gelatinization. Minimum enthalpy was found at an electric field strength of 20 V/cm, whereas at higher temperatures up to 100 °C and longer heating time resulted in absolute pre-gelatinization at the lower voltage. OH at 70 V/cm electric field strength lowered the initial gelatinization temperature of white flour, resulting in a faster swelling of rice flour output, whereas the traditionally heated samples demonstrated higher resistance to heat. It was found that with temperature, electrical conductivity increased but decreased with the degree of starch's gelatinization [81, 82].

10.6.2 Meat Products

In recent years, numerous studies have been conducted on meat product's thermophysical properties and their behavior during the OH process. The heating of meat emulsion and meat batters has been effective in ohmic heating [83, 84]. The addition of lean to fat, the alignment of larger particles (15-25 mm), the direction of muscle fiber and dielectric properties, and the addition of phosphate and sodium chloride are affected by the electrical properties and heating rate of meat products [85, 86]. The effect of ohmic treatment on meats' electrical properties was extensively studied by [87]. The addition of phosphate and sodium chloride to processed beef triggered a significant rise in electrical conductivity, which would, in turn, result in an increase in OH rates. In addition, high electrical conductivity and decreased OH cycle were seen in whole meats having low fat or processed meats with the least added fat. When the current flow aligns the beef fibers with them, the optimum electrical conductivity was investigated.

10.6.3 Fruits and Vegetable Products

10.6.3.1 Electrical Properties

During traditional thermal processing, the nutritional content of most of the fruit and vegetable products is modified. This includes the quest for alternative manufacturing processes that result in end products of better quality. The economic feasibility of OH technology relies on the opportunity to apply it to the best of the goods produced by the food industry [15]. The authors have measured the appropriateness of ohmic heating for strawberries with varying amounts of solids. In a particle mixture, electrical conductivity was found to fall with a rise in solid material but a significant decrease for larger particles. The findings have indicated that electrical conductivity was too limited for higher solids (> 20 percent w/w) and sugar content with a Brix value (> 40 B) to be used in traditional ohmic heaters. The results of ohmic pre-treatment were analyzed by [81, 82], and no significant improvements in the moisture content of final products were noticed.

10.6.3.2 Soluble Solids Content and Acidity

Due to the combination of thermal and electrical effects, ohmic treatments result in successful damage of cells, resulting in the substantial diffusion of soluble nutrients during OH. When electropermeabilization happens in fruit/vegetable pulps or purées, when exposed to electric fields, pores created in cell membranes induce a decrease in the resistance when ions are permitted to travel through the membrane [9]. Thus during ohmic heating, a small rise in soluble solids and pH of liquid foods is likely. The potential fouling of protein fluids during OH will, however, alter their concentration and pH.

10.6.3.3 Vitamins

It has been observed that the difference in the percentage degradation of ascorbic acid was not statistically different during ohmic and traditional heating [88]. During traditional and ohmic heating of pre-pasteurized orange juice with a temperature range of 65-90 °C, the activation value for degradation of ascorbic acid was close to 52.8 kJ/mol. Similarly, degradation of ascorbic acid in the strawberry pulp was not affected by the occurrence of the field strength of 20 V/cm having a temperature range of 60-97 °C [50]. The vitamin C content of orange juice during the traditional and ohmic pasteurization process having a temperature from 90 °C to 150 °C and time from 0.68 s to 1.13 s is 15% less than that of fresh orange juice [89]. OH facilitated more remarkable nutrient preservation at all temperatures in orange juice (50-90 °C) relative to any other method such as conventional, microwave, and infrared heating [90]. High activation energy is required for vitamin C destruction during OH rather than by conventional heating. It was enlightened by the high dielectric properties of the orange juice and instantaneous heat production. The difference in buffer pH, salt concentration, and electrical energy used during OH alters the ascorbic acid reaction kinetics [91]. The reaction environment's pH level can be considered one of the most significant factors affecting the degradation rate. In electrode degradation and electrolytic reactions, however, pH is an essential element. The effect of corrosion and electrolysis on ascorbic acid degradation at pH 3.5 was described by [92]. pH can be increased by increasing the concentration of NaCl or by voltage rise.

Purees made from vegetables can be heated ohmically, leading to higher retention of the vitamin. Vitamin C deficiency in the range of 48.02-65.80% and 71.19% were found respectively, for ohmic and water blanching at 85 °C for artichoke by-product puree [93]. In an analysis by [94], ohmic heating (in the range of 10-40 V/cm) resulted in high β-carotene retention in a puree made from spinach. They determined no statistically significant influence of the voltage gradient on the β-carotene content ($p < 0.05$). However, based on the applied voltage gradient, β-carotene content was raised in a range of 6.8-28.4% ($p < 0.05$) due to the constant temperature for 600 s retention time. The rise in β-carotene is due to the potential enhancing effect of ohmic heating.

10.6.3.4 Flavor Compounds

An essential contribution to the unique fragrance of citrus juices is flavor compounds. The flavor compounds are highly sensitive to heat. Thus, they are not stable at a higher temperature compared to lower temperatures [95]. Significant findings on the preservation of flavor compounds during OH have been published. The relative concentrations (pinene, limonene, octanal, decanal, and myrcene) during 90 and 120 °C ohmic-heating treatments were observed by [89] as above 100%. Flavor compound concentrations in ohmic heated treatments were more outstanding than conventionally pasteurized at 150 °C. Two corresponding phenomena clarified these results: bonded components can be released from the medium due to thermal treatment and/or that the released flavor compounds in ohmic heating are not deteriorated as rapidly as in traditional pasteurization due to its shorter retention time in OH. During storage at 4 °C, the preservation of flavor compounds (myrcene, limonene, decanal, and octanal) in OH orange juice was greater than that in traditionally pasteurized juice [95].

10.6.3.5 Phenolic Compounds

During ohmic heating, the total phenolic content (TPC) of vegetable purees and fruit juices may be affected. Phenolic compounds have industrial and medicinal and industrial importance due to their antioxidant and anti-inflammatory properties. It was determined that the total phenolic content changed in the initial heating up period, and no significant change was found during the retention period [21]. Similarly, by [89], it was found that the ohmic or traditional heating methods produced an increase in TPC, i.e., the relative total phenolic compound content of the heated samples is more than fresh juice and was greater than 1.0. OH did not cause any decrease in total phenolic content as compared to the conventional methods.

10.6.3.6 Colour Properties

Browning is of critical importance to the juice and purees industries during processing and storage [96]. It is possible to link changes in food color with the experience of heat treatment. To determine the change in fruit juice color during heating, 'Hue angle ratio' can be used. For comparison of raw juices to treated samples, the value of hue angle is taken as 1 for raw juices. The lower values of the hue angle ratio represent a lower browning level. In this respect, ohmically heated pomegranate juice's hue angle ratio values were observed to be higher than that of conventionally heated pomegranate juice [21]. During the heating of fruits and vegetables, numerous reactions such as non-enzymatic browning and pigment destruction reactions may occur and thus, influence their color. In determining the degree of degradation due to thermal processing, color may be used as a quality indicator [97]. Non-enzymatic browning causes off-flavor, color and nutritional loss, and the appearance of brown pigments [89]. During any treatment, changes in the optical color characteristics can be used as necessary measures to determine the total color changes in the samples. The oxidation of ascorbic acid in citrus juices is a significant chemical reaction responsible for browning. The increase in absorbance (at 420 nm) suggests the browning reaction initiation during thermal treatments due to high-temperature exposure. Similar browning-index values were calculated by [89] for both traditional and ohmic pasteurization of orange juice at an equivalent F-value for thermal treatment. The decline in lightness was associated with the rise in browning levels of ohmic-heated juices.

10.6.3.7 Change in Chlorophyll Content

Chlorophyll is more vulnerable to oxidation (degradation) by wet heating techniques than electrical approaches [98]. The highest chlorophyll a, b, and total chlorophyll is found in ohmically heated spinach puree samples at 90 °C for 10 min. The potential enhancing effect of ohmic heating on the formation of chlorophyll derivatives may be the source of this rise. By decarboxylation, pheophytins or phrochlorophylls produce pyropheophytins during steaming or microwave cooking by eliminating the Mg ion.

10.6.3.8 Textural Properties

The thermal processing and temperature significantly affect the texture of vegetables and fruits [99]. Low-temperature, long-term pre-treatment may increase the hardness of the processed vegetables (e.g., asparagus, cauliflower). Several researchers have shown that ohmic pre-treatment, compared to traditional pre-treatment, inhibits the loss of firmness in potato cubes and cauliflower florets [100–102].

10.6.3.9 Sensory Properties

To measure improvements in flavor, texture, taste, and overall acceptance of foods during processing, sensory assessments for liquid food products are carried out. To build a basis for comparison, traditional and ohmic heating treatments are applied to the same thermal background. Nevertheless, there are minimal experiments performed on the sensory assessment of ohmic samples treated. For fresh, conventionally, and ohmically pasteurized orange juices, [89] conducted a sensory assessment. Due to similar taste profiles, panelists were unable to differentiate between fresh and ohmic treated juice. The superior sensory property was found in ohmic-treated juice compared to pasteurized juice.

10.6.4 Dairy Products

High microbial inactivation rates were found during ohmic heating in treated foods including dairy products (Table 10.2). High-fat content allows electrical conductivity to decrease because the fat globules act like an electrical insulator in the milk matrix; hence, lower heating rates around the globule regions are generated, which results in lower microorganism declines [104]. High-fat foods can be heated non-uniformly, which adversely affects the food products' microbiological stability and safety [2]. In this context, the target microorganism for pasteurization of milk is *Coxiella burnetti*, while for sterilization, *Geobacillus stearothermophilus* (8D reduction) or *Bacillus subtilis* (10-12D reduction) are the target microorganisms [105].

To assess the strength of the ohmic heating process, the same thermal process indicators are typically used in traditional processing methods [2]. Several indicators, namely furosine, hydroxymethylfurfural (HMF), and lactulose, FAST index (soluble tryptophan and fluorescence of advanced Maillard products), glycoxidation products, carboxymethyl lysine, protein denaturation and aggregation (β-lactoglobulin), and free fatty acids, are documented in the literature [106]. Several researchers have found the practical application of ohmic heating in dairy processing such as a minimal effect on the product quality such as free fatty acid content, no additional thermal effect on protein denaturation, lower processing time, increase in the protein solubility, and higher retention of vitamin C and color than that of the conventional heating process [106].

Table 10.2 Effect of ohmic heating on viablity of microorganisms in different food materials. (Adapted from Reference [103]).

Food	Microorganism	Processing parameters	Findings
Goat milk	*Escherichia coli* ATCC 25922	Electrical field – 20-54 V/cm Electrode gap- 2 cm Frequency- 50 Hz	D-value for OH sample less than conventional heating.
Milk	*Streptococcus thermophilus 2646*	Voltage- 70-12 V Current- 7.3-2 A Electrode gap- 3 cm	D_{70} = 6.59 min D_{75} =3.09 min D_{80} =0.16 min
	Viable aerobes		D_{57} = 8.64 min D_{60} =6.18 min D_{72} =0.38 min
Buffalo milk	Total visible colonies, yeasts and molds, coliforms, *E.coli*, *Salmonella*	Voltage varied to match the thermal conditions of conventional heating. Temperature- 72°C Electrode gap-2.9 cm	Higher reductions of total visible colonies, yeasts, and molds, *E. coli*, and coliforms (6.04, 6.52, 5.78, and 6.26 log CFU/mL, respectively) than conventional heating; Salmonella- N.D.

(Continued)

Table 10.2 Effect of ohmic heating on viablity of microorganisms in different food materials. (Adapted from Reference [103]). (*Continued*)

Food	Microorganism	Processing parameters	Findings
Orange juice	Bacteria, yeasts and molds	Frequency- 50 Hz Temperature- 90, 120, and 150 °C Time- 1.13, 0.85, and 0.68 s	Complete inactivation of bacteria, yeasts, and molds; The flavor remained intact; Pectin esterase activity and Vitamin C reduced by 98 and15%, respectively & Sensory qualities are conserved.
Orange juice	*E.coli* O157:H7, *Salmonella Typhimurium*, *Listeria monocytogenes*	Electric field- 10-20V/cm Frequency- 60Hz Time – 120-540s	*E. coli* O157:H7 reduced by >5 log CFU/ml after three treatments; Similar results for *Salmonella Typhimurium*, *Listeria monocytogenes*
Tomato juice		Electric field- 10-20V/cm Frequency- 60Hz Time – 90-480 s	
Apple juice	*E.coli* O157:H7, *Salmonella Typhimurium*, *Listeria monocytogenes*	Electric field- 30, 60 V/cm TSS- 36, 48 °Brix Time- 20, 60 s	5 log reduction of all three pathogens.
Gochujang	*Bacillus* vegetative cells	Electric field- 30 V/cm Frequency- 60 Hz Temperature- 100 °C Time- 150 s	*Bacillus* cells were reduced by 99.7% after ohmic heating. Taste, flavor, color, and texture of OH samples were similar to those of conventionally treated samples.

(*Continued*)

Table 10.2 Effect of ohmic heating on viablity of microorganisms in different food materials. (Adapted from Reference [103]). (*Continued*)

Food	Microorganism	Processing parameters	Findings
Beef meatballs	Total mesophilic aerobic bacteria, molds, yeasts, *Staphylococcus aureus*, *Salmonella* spp., *Listeria monocytogenes*	Electric field- 15.26 V/cm Frequency- 50 Hz Temperature- 75 °C Holding time- 0 s	Total mesophilic aerobic bacteria - 2.47 log reduction, Molds & yeasts- N.D., *Staphylococcus aureus* – N.D., *Salmonella* spp., *Listeria monocytogenes* cells not inactivated efficiently.
Meat	*L. innocua*	LTLT: Electric field: 8.33 V/cm Frequency- 50 Hz Temperature- 95 °C Time- 7 min HTST: Electric field: 8.33 V/cm Frequency- 50 Hz Temperature- 72 °C Time- 15 min	Bacteria reduced to N.D. LTLT samples have less cook loss (29.3%) than the conventional method; the HTST sample had better color
Apple juice	*Alicyclobacillus acidoterrestris*	Electric field= 26.7 V/cm Frequency- 25 kHz Temperature- 100 °C Time- 30 s	Total inactivation of cells. No significant change in TSS and color
Cloudberry jam	*Bacillus licheniformis*	Voltage varied to simulate thermal conditions.	D-value for OH sample less than conventional heating.

D_T: Decimal reduction time at T °C temperature; N.D.: Not detected; LTLT: low-temperature long time; HTST: High-temperature short-time; TSS: Total soluble solids.

10.6.5 Seafoods

Surimi is a stabilized fish protein with myofibrillar and a key component in seafood products, namely scallop, surimi crab, lobster, and shrimp. The microbiological content of these kinds of seafood is excellent. For the Food and Drug Administration (FDA) to take regulatory action, the maximum *Escherichia coli* in raw seafood should not be more than $13x10^3$/g without any consumer processing [107]. Similarly, $13x10^4$/g is the maximum limit for *Staphylococcus aureus* according to the guideline, or the food product should be tested toxin positive. The FDA has defined zero tolerance for *Salmonella* and *Listeria monocytogenes* in RTE and refrigerated seafood. For fish mince, ohmic heating tests were performed by [108]. Stabilized mince and pacific whiting surimi paste were tested in the temperature and frequency range of 20-70 °C and 55 Hz to 200 Hz, respectively. With frequency, sample impedance decreased significantly. With temperature and salt concentration, the total dielectric loss and electrical conductivity of Pacific whiting surimi paste increased and stabilized mince increased with the temperature only.

10.7 Advantages of Ohmic Heating

Some of the major advantages of ohmic heating are:

1. Continuous development without surfaces for heat-transfer.
2. Rapid and consistent treatment of liquid and solid phases with minimum damage to heat and loss of nutrients, unlike microwave heating.
3. Ideal method due to low flow velocity for shear-sensitive materials.
4. Optimization of capital investment and product protection due to the loading of high solids.
5. Reduced fouling relative to traditional heating processes.
6. Effective and faster production management with decreased maintenance costs.
7. Environment-friendly system.
8. Maintenance of food color and nutrient content.
9. Less cleaning necessities.
10. Heating of liquid-particle mixtures and particulate foods.
11. Less chance of product damage caused by burning.
12. High-energy conversion efficiency.

10.8 Disadvantages of Ohmic Heating

Some of the disadvantages of ohmic heating are listed below:

1. Lack of generalized information.
2. High installation and operational cost than conventional methods of processing.
3. Ineffective heating of foods containing fat globules due to the non-conductive behavior resulting from lack of water and salt.
4. As the system's temperature increases, the probability of 'runaway' heating; electrical conductivity often increases because of faster electron movements.
5. Chances of fouling during ohmic food heating focus on the importance of the deposit formation on electrodes to the electrodes' corrosion issue.
6. The change demanded depends on the conductivity of the milk product.
7. Band of Narrow Frequency.
8. It is difficult to control and monitor.
9. Complex coupling between temperature and electric field distribution.

10.9 Conclusions

Ohmic heating has tremendous potential for offering rapid and uniform heating throughout the food material. The applicability of OH in a food system depends on various factors, such as electrical conductivity of the heating substance, electric field strength, applied frequency, residence time, and the rate of heat generation. OH is an excellent thermal process for food preservation and stabilization and is applied as a pre-treatment to prepare plant tissues prior to any mass transfer operation (e.g., diffusion, dehydration, or extraction). OH has emerged as a smart technology with numerous actual and future applications. A vast amount of research to thoroughly investigate the post-treatment effects is still going on. Further studies that take into account the changes in flow parameters, electrical behavior, and properties of food, like, electrical conductivity, rheology, and particle size are needed. Economic studies should be undertaken that will play a crucial role in realizing the overall cost and viability of industrial applications. Deep insights relating the influence of electric field on mass transfer, process modeling, and design to establish commercial applications, characterization of electroporation and permeabilization phenomenon, modeling, prediction, and identification of cold spots and the heating pattern of complex foods are required. Ohmic heating has a huge potential for becoming one of the essential food processing technologies in the near future.

References

1. Sakr, M. and Liu, S., A comprehensive review on applications of ohmic heating (OH). *Renewable and Sustainable Energy Reviews*, 39, 262-269, 2014.
2. Sastry, S. K. and Barach, J. T., Ohmic and inductive heating. *Journal of Food Science*, 65, 42-46, 2000.
3. Knirsch, M. C., Dos Santos, C. A., de Oliveira Soares, A. A. M., Penna, T. C. V., Ohmic heating–a review. *Trends in Food Science & Technology*, 21, 9, 436-441, 2010.
4. Goullieux, A., and Pain, J. P., Ohmic heating. In *Emerging technologies for food processing* (pp. 399-426). Academic Press, 2014.
5. Jaeger, H., Roth, A., Toepfl, S., Holzhauser, T., Engel, K. H., Knorr, D., ..., Steinberg, P., Opinion on the use of ohmic heating for the treatment of foods. *Trends in Food Science & Technology*, 55, 84-97, 2016.
6. Halden, K., De Alwis, A. A. P., Fryer, P. J., Changes in the electrical conductivity of foods during ohmic heating. *International Journal of Food Science & Technology*, 25, 1, 9-25, 1990.
7. Barbosa-Cánovas, G. V., Pothakamury, U. R., Gongora-Nieto, M. M., Swanson, B. G., *Preservation of foods with pulsed electric fields*. Elsevier, 1999.
8. Wang, W. C. and Sastry, S. K., Effects of moderate electrothermal treatments on juice yield from cellular tissue. *Innovative Food Science & Emerging Technologies*, 3, 4, 371-377, 2002.
9. Kulshrestha, S. A. and Sastry, S. K., Low-frequency dielectric changes in cellular food material from ohmic heating: effect of end point temperature. *Innovative Food Science & Emerging Technologies*, 7, 4, 257-262, 2006.
10. Zhong, T. and Lima, M., The effect of ohmic heating on vacuum drying rate of sweet potato tissue. *Bioresource Technology*, 87, 3, 215-220, 2003.
11. Praporscic, I., Lebovka, N. I., Ghnimi, S., Vorobiev, E., Ohmically heated, enhanced expression of juice from apple and potato tissues. *Biosystems Engineering*, 93, 2, 199-204, 2006.
12. Kolbe, E. R., Park, J. W., Wells, J. H., Flugstad, B. A., Zhao, Y., *U.S. Patent No. 6,303,166*. Washington, DC: U.S. Patent and Trademark Office, 2001.

13. McKenna, B. M., Lyng, J., Brunton, N., Shirsat, N., Advances in radio frequency and ohmic heating of meats. *Journal of Food Engineering*, 77, 2, 215-229, 2006.
14. Palaniappan, S. and Sastry, S. K., Electrical conductivities of selected solid foods during ohmic heating 1. *Journal of Food Process Engineering*, 14, 3, 221-236, 1991.
15. Castro, I., Teixeira, J. A., Salengke, S., Sastry, S. K., Vicente, A. A., The influence of field strength, sugar and solid content on electrical conductivity of strawberry products. *Journal of Food Process Engineering*, 26, 1, 17-29, 2003.
16. Zareifard, M. R., Ramaswamy, H. S., Trigui, M., Marcotte, M., Ohmic heating behaviour and electrical conductivity of two-phase food systems. *Innovative Food Science & Emerging Technologies*, 4, 1, 45-55, 2003.
17. Bean, R. C., Rasor, J. P., Porter, G. G., Changes in electrical characteristics of avocados during ripening. *California Avocado Society*, 44, 75-78, 1960.
18. Zell, M., Lyng, J. G., Morgan, D. J., Cronin, D. A., Minimising heat losses during batch ohmic heating of solid food. *Food and Bioproducts Processing*, 89, 2, 128-134, 2011.
19. Goullieux, A., Pain, J. P., Baudez, P., A new pilot-scale batch ohmic heater. In *Seventh International Congress on Engineering and Food, Part I*. Sheffield Academic Press, Sheffield, UK, 1997.
20. Lima, M., Heskett, B.F., Sastry, S.K., The effect of frequency and waveform on the electrical conductivity–temperature profiles of turnip tissue. *Journal of Food Process Engineering*, 22, 41–54, 1999.
21. Yildiz, H., Bozkurt, H., Icier, F. İ. L. İ. Z., Ohmic and conventional heating of pomegranate juice: effects on rheology, color, and total phenolics. *Food Science and Technology International*, 15, 5, 503-512, 2009.
22. Icier, F. İ. L. İ. Z. and Ilicali, C., Electrical conductivity of apple and sourcherry juice concentrates during ohmic heating. *Journal of Food Process Engineering*, 27, 3, 159-180, 2004.
23. Icier, F. and Ilicali, C., The effects of concentration on electrical conductivity of orange juice concentrates during ohmic heating. *European Food Research and Technology*, 220, 3-4, 406-414, 2005.
24. Pataro, G., Donsì, G., Ferrari, G., Aseptic processing of apricots in syrup by means of a continuous pilot scale ohmic unit. *LWT-Food Science and Technology*, 44, 6, 1546-1554, 2011.
25. Mercali, G. D., Jaeschke, D. P., Tessaro, I. C., Marczak, L. D. F., Study of vitamin C degradation in acerola pulp during ohmic and conventional heat treatment. *LWT-Food Science and Technology*, 47, 1, 91-95, 2012.
26. Mercali, G. D., Jaeschke, D. P., Tessaro, I. C., Marczak, L. D. F., Degradation kinetics of anthocyanins in acerola pulp: Comparison between ohmic and conventional heat treatment. *Food Chemistry*, 136, 2, 853-857, 2013.
27. Sarkis, J. R., Jaeschke, D. P., Tessaro, I. C., Marczak, L. D., Effects of ohmic and conventional heating on anthocyanin degradation during the processing of blueberry pulp. *LWT-Food Science and Technology*, 51, 1, 79-85, 2013.

28. Eliot-Godéreaux, S. C., Zuber, F., Goullieux, A., Processing and stabilisation of cauliflower by ohmic heating technology. *Innovative Food Science & Emerging Technologies*, 2, 4, 279-287, 2001.
29. Icier, F., Yildiz, H., Baysal, T., Peroxidase inactivation and colour changes during ohmic blanching of pea puree. *Journal of Food Engineering*, 74, 3, 424-429, 2006.
30. Lemmens, L., Tibäck, E., Svelander, C., Smout, C., Ahrné, L., Langton, M., ... Hendrickx, M., Thermal pretreatments of carrot pieces using different heating techniques: Effect on quality related aspects. *Innovative Food Science & Emerging Technologies*, 10, 4, 522-529, 2009.
31. Shirsat, N., Lyng, J. G., Brunton, N. P., McKenna, B., Ohmic processing: Electrical conductivities of pork cuts. *Meat Science*, 67, 3, 507-514, 2004.
32. Tadpitchayangkoon, P., Park, J. W., Yongsawatdigul, J., Gelation characteristics of tropical surimi under water bath and ohmic heating. *LWT-Food Science and Technology*, 46, 1, 97-103, 2012.
33. Icier, F., Izzetoglu, G. T., Bozkurt, H., Ober, A., Effects of ohmic thawing on histological and textural properties of beef cuts. *Journal of Food Engineering*, 99, 3, 360-365, 2010.
34. Shirsat, N., Brunton, N. P., Lyng, J. G., McKenna, B., Scannell, A., Texture, colour and sensory evaluation of a conventionally and ohmically cooked meat emulsion batter. *Journal of the Science of Food and Agriculture*, 84, 14, 1861-1870, 2004.
35. Bansal, B. and Chen, X. D., Effect of temperature and power frequency on milk fouling in an ohmic heater. *Food and Bioproducts Processing*, 84, 4, 286-291, 2006.
36. Lakkakula, N., Lima, M., Walker, T., Rice bran stabilization and rice bran oil extraction using ohmic heating. *Journal of Bioresource Technology*, 92, 157–161, 2004.
37. Icier, F., Influence of ohmic heating on rheological and electrical properties of reconstituted whey solutions. *Food and Bioproducts Processing*, 87, 4, 308-316, 2009.
38. Bozkurt, H. and Icier, F., The change of apparent viscosity of liquid whole egg during ohmic and conventional heating. *Journal of Food Process Engineering*, 35, 1, 120-133, 2012.
39. Vigerstrom, K. B., *U.S. Patent No. 3,997,678*, Washington, DC: U.S. Patent and Trademark Office, 1976.
40. Sensoy, I. and Sastry, S. K., Ohmic blanching of mushrooms. *Journal of Food Process Engineering*, 27, 1, 1-15, 2004.
41. Butz, P. and Tauscher, B., Emerging technologies: chemical aspects. *Food Research International*, 35, 2-3, 279-284, 2002.
42. Demirdöven, A. and Baysal, T., Effects of electrical pre-treatment and alternative heat treatment applications on orange juice production and storage. *Food and Bioproducts Processing*, 94, 443-452, 2015.

43. Stirling, R., Ohmic heating- a new process for the food industry. *Power Engineering Journal*, 6, 365-371, 1987.
44. Quarini, G., Thermalhydraulic aspects of the ohmic heating process. *Journal of Food Engineering*, 24, 4, 561-574, 1995.
45. Amaya, D. B., *A guide to carotenoid analysis in foods*, Washington D.C.: OMNI Research, 2001.
46. Achir, N., Hadjal, T., Madani, K., Dornier, M., Dhuique-Mayer, C., Carotene reactivity in pink grapefruit juice elucidated from model systems and multi-response modeling. *Journal of Agricultural and Food Chemistry*, 63, 15, 3970-3979, 2015.
47. Leizerson, S. and Shimoni, E., Stability and sensory shelf life of orange juice pasteurized by continuous ohmic heating. *Journal of Agricultural and Food Chemistry*, 53, 10, 4012-4018, 2005.
48. Boussetta, N., Lanoisellé, J.-L., Bedel-Cloutour, C., Vorobiev, E., Extraction of soluble matter from grape pomace by high voltage electrical discharges for polyphenol recovery: effect of sulphur dioxide and thermal treatments. *Journal of Food Engineering*, 95, 192–198, 2009.
49. Loypimai, P., Moongngarm, A., Chottanom, P., Moontree, T., Ohmic heating-assisted extraction of anthocyanins from black rice bran to prepare a natural food colourant. *Innovative Food Science & Emerging Technologies*, 27, 102-110, 2015.
50. Castro, I., Teixeira, J.A., Salengke, S., Sastry, S.K., Vicente, A.A., Ohmic heating of strawberry products: Electrical conductivity measurements and ascorbic acid degradation kinetics. *Innovative Food Science and Emerging Technologies*, 5, 27–36, 2004.
51. Gavahian, M., Chu, Y. H., Mousavi Khaneghah, A., Recent advances in orange oil extraction: an opportunity for the valorisation of orange peel waste a review. *International Journal of Food Science & Technology*, 54, 4, 925-932, 2019.
52. Saberian, H., Hamidi-Esfahani, Z., Gavlighi, H. A., Barzegar, M., Optimization of pectin extraction from orange juice waste assisted by ohmic heating. *Chemical Engineering and Processing: Process Intensification*, 117, 154-161, 2017.
53. Nair, G. R., Divya, V. R., Prasannan, L., Habeeba, V., Prince, M. V., Raghavan, G. V., Ohmic heating as a pre-treatment in solvent extraction of rice bran. *Journal of Food Science and Technology*, 51, 10, 2692-2698, 2014.
54. Praporscic, I., Influence of the combined treatment by pulsed electric field and moderate heating on the physical properties and on the pressing behavior of plant products (Doctoral dissertation, Compiègne), 2005.
55. El Darra, N., Grimi, N., Vorobiev, E., Louka, N., Maroun, R., Extraction of polyphenols from red grape pomace assisted by pulsed ohmic heating. *Food and Bioprocess Technology*, 6, 5, 1281-1289, 2013.

56. Katrokha, I., Matvienko, A., Vorona, L., Kupchik, M., Zaets, V., Intensification of sugar extraction from sweet sugar beet cossettes in an electric field. *Sakharnaya Promyshlennost*, 7, 28–31, 1984.
57. Loypimai, P., Moonggarm, A., Chottanom, P., Effects of ohmic heating on lipase activity, bioactive compounds and antioxidant activity of rice bran. *Australian Journal of Basic and Applied Sciences*, 3, 4, 3642–3652, 2009.
58. Schreier, P., Reid, D., Fryer, P., Enhanced diffusion during the electrical heating of foods. *Journal of Food Science and Technology*, 28, 249–260, 1993.
59. Lima, M. and Sastry, S.M., The effects of ohmic heating frequency on hot-air drying rate and juice yield. *Journal of Food Sciences*, 41, 115–119, 1999.
60. Salengke, S. and Sastry, S.K., Effect of ohmic pretreatment on the drying rate of grapes and adsorption isotherm of raisins. *Drying Technology*, 23, 551-564, 2005.
61. Lebovka, N.I., Shynkaryk, M.V., Vorobiev, E., Drying of potato tissue pretreated by ohmic heating. *Drying Technology*, 24, 601-608, 2006.
62. Wang, W.C. and Sastry, S.K., Effects of thermal and electrothermal pretreatment on hot air drying rate of vegetable tissue. *Journal of Food Process Engineering*, 23, 299-319, 2000.
63. Allali, H., Marchal, L., Vorobiev, E., Effect of blanching by ohmic heating on the osmotic dehydration behavior of apples cubes. *Drying Technology*, 27, 739-746, 2009.
64. Allali, H., Marchal, L., Vorobiev, E., Blanching of strawberries on ohmic heating: effects on the kinetics of mass transfer during osmotic dehydration. *Food Bioprocess Technology*, 3, 406-414, 2010.
65. Allali, H., Marchal, L., Vorobiev, E., Effects of vacuum impregnation and ohmic heating with citric acid on the behavior of osmotic dehydration and structural changes in apple fruit. *Biosystems Engineering*, 106, 6-13, 2010.
66. Moreno, J., Gonzales, M., Zúniga, P., Petzold, G., Mella, K., Munoz, O., Ohmic heating and pulsed vacuum effect on dehydration processes and polyphenol component retention of osmodehydrated blueberries (cv. Tifblue). *Innovative Food Science & Emerging Technologies*, 36, 112-119, 2016.
67. Moreno, J., Simpson, R., Estrada, D., Lorenzen, S., Moraga, D., Almonacid, S., Effect of pulsed-vacuum and ohmic heating on the osmodehydration kinetics, physical properties and microstructure of apples (cv. Granny Smith). *Innovative Food Science & Emerging Technologies*, 12, 4, 562–568, 2011.
68. Moreno, J., Simpson, R., Sayas, M., Segura, I., Aldana, O., Almonacid, S., Influence of ohmic heating and vacuum impregnation on the osmotic dehydration kinetics and microstructure of pears (cv. Packham's Triumph). *Journal of Food Engineering*, 104, 621–627, 2011.
69. Moreno, J., Simpson, R., Baeza, A., Morales, J., Muñoz, C., Sastry, S., Effect of ohmic heating and vacuum impregnation on the osmodehydration kinetics and microstructure of strawberries (cv. Camarosa). *LWT - Food Science and Technology*, 45, 2, 148–154, 2012.

70. Gavahian, M. and Tiwari, B. K., Moderate electric fields and ohmic heating as promising fermentation tools. *Innovative Food Science & Emerging Technologies*, 102422, 2020.
71. Cho, H. Y., Yousef, A. E., Sastry, S. K., Growth kinetics of *Lactobacillus acidophilus* under ohmic heating. *Biotechnology and Bioengineering*, 49, 3, 334–340, 1996.
72. Loghavi, L., Sastry, S. K., Yousef, A. E., Effect of moderate electric field frequency and growth stage on the cell membrane permeability of *Lactobacillus acidophilus*. *Biotechnology Progress*, 25, 1, 85–94, 2009.
73. Gally, T., Rouaud, O., Jury, V., Havet, M., Oge, A., Le-Bail, A., Proofing of bread dough assisted by ohmic heating. *Innovative Food Science and Emerging Technologies*, 39, 55–62, 2017.
74. Masure, H. G., Wouters, A. G. B., Fierens, E., Delcour, J. A., Electrical resistance oven baking as a tool to study crumb structure formation in gluten-free bread. *Food Research International*, 116, 925–931, 2019.
75. Reta, Mursalim, Salengke, Junaedi, M., Mariati, Sopade, P., Reducing the acidity of Arabica coffee beans by ohmic fermentation technology. *Food Research*, 1, 5, 157–160, 2017.
76. Luzuriaga, D. A. and Balaban, M. O., Electrical conductivity of frozen shrimp and flounder at different temperatures and voltage levels. *Journal of Aquatic Food Product Technology*, 5, 3, 41-63, 1996.
77. Sriburi, P., Hill, S. E., Barclay, F., Depolymerisation of cassava starch. *Carbohydrate Polymers*, 38, 3, 211-218, 1999.
78. Colonna, P., Buleon, A., Mercier, C., Physically modified starches. *Critical Reports on Applied Chemistry*, 13, 79-114, 1987.
79. Fernando, M. B., Magnolia, L. S., Jose, Z. M. Eduardo, M. S., Preparation and properties of pre-gelatinized cassava (*Manihot esculenta. crantz*) and jıcama (*Pachyrhizus erosus*) starches using ohmic heating. *Cereal Chem.*, 39, 3, 275–283, 2000.
80. An, H. J. and King, J. M., Thermal characteristics of ohmically heated rice starch and rice flours. *Journal of Food Science*, 72, 1, C084-C088, 2007.
81. Wang, W. C. and Sastry, S. K., Changes in electrical conductivity of selected vegetables during multiple thermal treatments. *J. Food Process Eng.*, 20, 499–516, 1997.
82. Wang, W. C. and Sastry, S. K., Starch gelatinization in ohmic heating. *J. Food Eng.*, 34, 225–242, 1997.
83. Piette, G., Buteau, M. L., De Halleux, D., Chiu, L., Raymond, Y., Ramaswamy, H. S., Dostie, M., Ohmic cooking of processed meats and its effects on product quality. *Journal of Food Science*, 69, 2, fep71-fep78, 2004.
84. Shirsat, N., Brunton, N. P., Lyng, J. G., McKenna, B., Scannell, A., Texture, colour and sensory evaluation of a conventionally and ohmically cooked meat emulsion batter. *Journal of the Science of Food and Agriculture*, 84, 14, 1861-1870, 2004.

85. Brunton, N. P., Lyng, J. G., Zhang, L., Jacquier, J. C., The use of dielectric properties and other physical analyses for assessing protein denaturation in beef biceps femoris muscle during cooking from 5 to 85 C. *Meat Science*, 72, 2, 236-244, 2006.
86. Lyng, J. G., Zhang, L., Brunton, N. P., A survey of the dielectric properties of meats and ingredients used in meat product manufacture. *Meat Science*, 69, 4, 589-602, 2005.
87. Zell, M., Lyng, J. G., Cronin, D. A., Morgan, D. J., Ohmic heating of meats: Electrical conductivities of whole meats and processed meat ingredients. *Meat Science*, 83, 3, 563-570, 2009.
88. Lima, M., Heskitt, B. F., Burianek, L. L., Nokes, S. E., Sastry, S. K., Ascorbic acid degradation kinetics during conventional and ohmic heating. *Journal of Food Processing and Preservation*, 23, 5, 421-443, 1999.
89. Leizerson, S., and Shimoni, E., Stability and sensory shelf life of orange juice pasteurized by continuous ohmic heating. *Journal of Agricultural and Food Chemistry*, 53, 10, 4012-4018, 2005.
90. Vikram, V. B., Ramesh, M. N., Prapulla, S. G., Thermal degradation kinetics of nutrients in orange juice heated by electromagnetic and conventional methods. *Journal of Food Engineering*, 69, 1, 31-40, 2005.
91. Assiry, A. M., Sastry, S. K., Samaranayake, C. P., Influence of temperature, electrical conductivity, power and pH on ascorbic acid degradation kinetics during ohmic heating using stainless steel electfrodes. *Bioelectrochemistry*, 68, 1, 7-13, 2006.
92. Assiry, A., Sastry, S. K., Samaranayake, C., Degradation kinetics of ascorbic acid during ohmic heating with stainless steel electrodes. *Journal of Applied Electrochemistry*, 33, 2, 187-196, 2003.
93. Icier, F., Ohmic blanching effects on drying of vegetable byproduct. *Journal of Food Process Engineering*, 33, 4, 661-683, 2010.
94. Yildiz, H., Icier, F., Baysal, T., Changes in β-carotene, chlorophyll and color of spinach puree during ohmic heating. *Journal of Food Process Engineering*, 33, 4, 763-779, 2010.
95. Leizerson, S., and Shimoni, E., Effect of ultrahigh-temperature continuous ohmic heating treatment on fresh orange juice. *Journal of Agricultural and Food Chemistry*, 53, 9, 3519-3524, 2005.
96. Garza, S., Ibarz, A., Pagan, J., Giner, J., Non-enzymatic browning in peach puree during heating. *Food Research International*, 32, 5, 335-343, 1999.
97. Avila, I. M. L. B., and Silva, C. L. M., Modelling kinetics of thermal degradation of colour in peach puree. *Journal of Food Engineering*, 39, 2, 161-166, 1999.
98. Teng, S. S., and Chen, B. H., Formation of pyrochlorophylls and their derivatives in spinach leaves during heating. *Food Chemistry*, 65, 3, 367-373, 1999.
99. Bartolome, L. G., and Hoff, J. E., Firming of potatoes. Biochemical effects of preheating. *Journal of Agricultural and Food Chemistry*, 20, 2, 266-270, 1972.

100. Eliot, S. C., Goullieux, A., Pain, J. P., Combined effects of blanching pre-treatments and ohmic heating on the texture of potato cubes. *Sciences des Aliments (France)*, 1999.
101. Eliot, S. C., Goullieux, A., Pain, J. P., Processing of cauliflower by ohmic heating: influence of precooking on firmness. *Journal of the Science of Food and Agriculture*, 79, 11, 1406-1412, 1999.
102. Eliot, S.C., and Goullieux, A., Application of the firming effect of low-temperature long-time pre-cooking to ohmic heating of potatoes. *Food Sciences (France)*, 2000.
103. Misra, S., and Kumar, S., Ohmic Heating as an Alternative to Conventional Heating for Shelf Life Enhancement of Fruit Juices. *Int. J. Curr. Microbiol. App. Sci*, 9, 3, 01-07, 2020.
104. Kim, S. S., and Kang, D. H., Effect of milk fat content on the performance of ohmic heating for inactivation of *Escherichia coli* O157: H7, *Salmonella enterica Serovar Typhimurium* and *Listeria monocytogenes*. *Journal of Applied Microbiology*, 119, 2, 475-486, 2015.
105. Bylund, G., *Dairy processing handbook*, Tetra Pak Processing Systems AB, 2003.
106. Cappato, L. P., Ferreira, M. V., Guimaraes, J. T., Portela, J. B., Costa, A. L., Freitas, M. Q., ..., Cruz, A. G., Ohmic heating in dairy processing: Relevant aspects for safety and quality. *Trends in Food Science & Technology*, 62, 104-112, 2017.
107. Ward, D. R., and Price, R. J., Food microbiology: exact use of an inexact science. *The NFI green book*, 3, 34-42, 1992.
108. Wu, H., Kolbe, E., Flugstad, B., Park, J. W., Yongsawatdigul, J., Electrical properties of fish mince during multi-frequency ohmic heating, *Journal of Food Science*, 63, 6, 1028-1032, 1998.

11

Microwave Food Processing: Principles and Applications

Jean-Claude Laguerre* and Mohamad Mazen Hamoud-Agha

Institut Polytechnique UniLaSalle, Université d'Artois, Beauvais, France

Abstract
Thermal processes are the most applied methods to make food meet the safety and quality requirements of consumers. However, food heating by conventional methods is associated with several undesirable effects on nutritional and sensorial properties. Designing an innovative thermal process that focuses on minimal, rapid, uniform, and economic heating is of great importance. Microwave heating has numerous advantages over conventional methods. Thanks to its volumetric and rapid heating, microwave technology has several applications in food processing. Nevertheless, non-uniform heating remains the main drawback of microwave heating. Combination of microwave heating with conventional and/or other innovative heating methods have been widely investigated in the literature to overcome this problem. In this chapter, the theorical principles and the governing equations of microwave interactions with food materials are presented. In addition, the applications of microwave heating for food processing and the efficiency of recent improvements are reviewed.

***Keywords*:** Microwave heating, baking, blanching, thawing, drying, sterilization, food safety, food quality

11.1 Introduction

Thermal processing of foods is the most widely applied treatment to produce some physical and/or chemical modifications to make the food edible by improving its texture, flavor, safety, digestibility, and many other

**Corresponding author*: jean-claude.laguerre@unilasalle.fr

Nitin Kumar, Anil Panghal and M. K. Garg (eds.) Thermal Food Engineering Operations, (301–348)

desirable requirements. Conventional heating methods include heat transfer from the heating medium around the product surface to the inner cold regions by convection and conduction. This process is generally slow, as foods are generally of low heat conductivity, which leads to important nutritional losses and quality degradations. To overcome the problems associated with conventional heating methods, several innovative thermal and non-thermal technologies have been studied in the literature [1, 2]. Microwave (MW) heating has several advantages over conventional heating in terms of less processing time, better quality retention and improved energy consumption efficiency thanks to its volumetric and quick heating nature. Microwave technology is successfully applied in many applications of food processing such as baking, blanching, tempering, and thawing, drying, microbial decontamination, extraction, and many other interesting applications [3]. However, Non-uniform MW heating is the major drawback of this technology due to complex interactions between microwaves and food materials and the multiple parameters that affect the heating process. The thermal heterogeneity of MW heating may result in serious safety and quality issues. In recent years, important efforts have been done to solve this problem. The combination of MW with other conventional or innovative heating methods is the principal improvement approach.

This chapter is organized into two parts. The first presents the physical principles of the propagation of a plane electromagnetic wave in free space and then in a dielectric medium. Maxwell's equations governing this propagation are presented and applied in the case of a dielectric material of simple form. The second part of this chapter presents some applications of microwave heating in the field of food processes such as baking, blanching, tempering, drying, pasteurization, and sterilization. The effects of MW heating on food quality and safety are also elaborated along with some innovative solutions to improve the efficiency of this technology.

11.2 Principles of Microwave Heating

11.2.1 Nature of Microwaves

11.2.1.1 Propagation of EM Waves in Free Space

Microwaves (MWs) are part of electromagnetic (EM) radiation, namely the association of an electric and a magnetic field, synchronized perpendicularly and propagating in free space at the speed of light (c). Like any wave, they can be characterized by their related wavelength (λ) and frequency (f) as shown in equation (11.1)

$$f = \frac{c}{\lambda} \tag{11.1}$$

The propagation of the electromagnetic (EM) waves is governed by Maxwell's equations. The EM field is described by the electric (***E***) and magnetic (***H***) field vectors in space. To these, it is necessary to add two other quantities, the electric (***D***) and magnetic (***B***) inductions allowing to take into account the interaction of the EM wave with the material (bold character is used to indicate vector quantity). These four quantities are linked by Maxwell's equations as following in equations (11.2) to (11.9). These equations correspond to the local writing of different fundamental laws of electromagnetism.

Equation (11.2) corresponds to the Maxwell-Gauss (MG) equation. It describes the relationship between the electric flux through any closed Gaussian surface with the electric charge contained in the volume delimited by this surface:

- MG equation

$$\nabla \cdot \boldsymbol{E} = \frac{\rho}{\varepsilon_0} \tag{11.2}$$

where ρ represents the charge giving rise to the electric field (EF) ***E*** and $\varepsilon_0 = 8.8542 \times 10^{-12}\ F/m$ is the permittivity of free space.

This equation can also be written as follows:

$$\nabla \cdot \boldsymbol{D} = \rho \tag{11.3}$$

where ***D*** is the electric flux density:

$$\boldsymbol{D} = \varepsilon_0 \boldsymbol{E} \tag{11.4}$$

Equation (11.5) is the Maxwell-flux or Maxwell-Thompson (MT) equation. It postulates the non-existence of a magnetic monopole. Indeed, the magnetic field is generated by a dipole that has no magnetic charge.

- MT equation

$$\nabla \cdot \boldsymbol{B} = 0 \tag{11.5}$$

Equation (11.6) is the Maxwell-Faraday (MF) equation. It expresses Faraday's law of electromagnetic induction, so it indicates how the variation of a magnetic field induces an electric field.

- MF equation

$$\nabla \times \boldsymbol{E} = -\frac{\partial \boldsymbol{B}}{\partial t} \tag{11.6}$$

Finally, the Maxwell-Ampere (MA) equation (11.7) reflects the fact that a magnetic field can be generated either by a flow of electric current (Ampere's theorem) or by the variation of an electric field

- MA equation

$$\nabla \times \boldsymbol{H} = \boldsymbol{j} + \frac{\partial \boldsymbol{D}}{\partial t} \tag{11.7}$$

where $\boldsymbol{j}$ is the displacement current

Given that:

$$\boldsymbol{B} = \mu_0 \boldsymbol{H} \tag{11.8}$$

Equation (11.7) can still be expressed by:

$$\nabla \times \boldsymbol{B} = \mu_0 \boldsymbol{j} + \mu_0 \frac{\partial \boldsymbol{D}}{\partial t} = \mu_0 \boldsymbol{j} + \mu_0 \varepsilon_0 \frac{\partial \boldsymbol{E}}{\partial t} \tag{11.9}$$

where $\mu_0 = 4\pi \times 10^{-7}\ N/m^2$ is the permeability of free space

If we place ourselves in a region of space devoid of charges ($\rho = 0$) and currents ($\boldsymbol{j} = 0$), that is to say in free space, (considered as a linear, homogeneous, and isotropic medium), two of the four Maxwell equations, the MG and MA equations, will be modified as shown in equations (11.10) and (11.11) respectively. On the other hand, the MT and MF equations remain unchanged because they do not depend on the charges or currents giving rise to the fields **E** and **B**. These equations are said to be structural of the EM field.

- MG equation in the absence of charges

$$\nabla \cdot \boldsymbol{E} = 0 \tag{11.10}$$

- MA equation in the absence of current

$$\nabla \times \boldsymbol{B} = \mu_0 \varepsilon_0 \frac{\partial \boldsymbol{E}}{\partial t} \tag{11.11}$$

The absence of currents reveals a coupling between the fields **E** and **B** through the equations MF (11.6) and modified MA (11.11). Thus, the associated **E** and **B** fields can propagate synchronously in space without being influenced by the charges or currents that gave rise to them. If we consider a very small region of space far from the sources we can consider that the wave is plane, that is to say, that the fields **E** and **H** are perpendicular and located in the plane (x, y) perpendicular to the z-axis of propagation.

The propagation equation for the **E** and **H** fields can be found from Maxwell's equations by performing some transformations. Let us start from the Maxwell-Faraday equation (11.6) and take its curl one while taking into account the equations (11.4), (11.7) and (11.8). We then obtain equation (11.12).

$$\nabla \times (\nabla \times \boldsymbol{E}) = \nabla \times \left(-\frac{\partial \boldsymbol{B}}{\partial t} \right) = \nabla \times \left(-\mu_0 \frac{\partial \boldsymbol{H}}{\partial t} \right) = -\mu_0 \frac{\partial}{\partial t} \left(j + \varepsilon_0 \frac{\partial E}{\partial t} \right) \tag{11.12}$$

We also know that:

$$\nabla \times (\nabla \times \boldsymbol{E}) = \nabla(\nabla \cdot \boldsymbol{E}) - \nabla^2 \boldsymbol{E} \tag{11.13}$$

By equaling (11.12) and (11.13) and using the Maxwell-Gauss equation (11.2) we obtain after transformation and rearrangement the equation (11.14) which is the propagation equation of the electric field **E**. Likewise, using the same procedure as previously on the Maxwell-Ampere equation (11.7) we obtain the equation (11.15) corresponding to the propagation equation of the magnetic field **H**.

$$\nabla^2 \boldsymbol{E} - \mu_0 \varepsilon_0 \frac{\partial^2 \boldsymbol{E}}{\partial t^2} = \nabla\left(\frac{\rho}{\varepsilon_0}\right) + \mu_0 \frac{\partial \boldsymbol{j}}{\partial t} \tag{11.14}$$

$$\nabla^2 \boldsymbol{H} - \mu_0 \varepsilon_0 \frac{\partial^2 \boldsymbol{H}}{\partial t^2} = -\nabla \times (\mu_0 \boldsymbol{j}) \tag{11.15}$$

In the case where the EM wave propagates in free space far from the charges and currents that gave rise to it, we can consider that $\rho = 0$ and $\boldsymbol{j} = 0$. Under these conditions, equation (11.14) and (11.15) are simplified in equations (11.16) and (11.17), respectively:

$$\nabla^2 \boldsymbol{E} - \frac{1}{c^2} \frac{\partial^2 \boldsymbol{E}}{\partial t^2} = 0 \tag{11.16}$$

$$\nabla^2 \boldsymbol{H} - \frac{1}{c^2} \frac{\partial^2 \boldsymbol{H}}{\partial t^2} = 0 \tag{11.17}$$

With:

$$c = \frac{1}{\sqrt{\mu_0 \varepsilon_0}} \tag{11.18}$$

where c is the wave propagation speed in the free space.

11.2.1.2 Propagation of EM Waves in Matter

The propagation of a plane EM wave in a lossy material has been depicted in Figure 11.1. If we consider that the electric and magnetic fields have a time-harmonic form with a pulsation ω, then they can be expressed by equations (11.19) and (11.20) respectively:

$$\boldsymbol{E} = \boldsymbol{E}(x, y)\, e^{j\omega t} \tag{11.19}$$

$$\boldsymbol{H} = \boldsymbol{H}(x, y)\, e^{j\omega t} \tag{11.20}$$

If we start again from the propagation equations (11.14) and (11.15), that we express them for a medium without charges or currents (e.g., a dielectric material) and that we replace **E** and **H** using the expressions of equations (11.19) and (11.20) respectively we get equations (11.21) and (11.22) also called Helmholtz equations [4]:

$$\nabla^2 \boldsymbol{E} = -\mu\varepsilon\omega^2 \boldsymbol{E} \tag{11.21}$$

$$\nabla^2 \boldsymbol{H} = -\mu\varepsilon\omega^2 \boldsymbol{H} \tag{11.22}$$

With $\varepsilon = \varepsilon_0\varepsilon_r$ and $\mu = \mu_0\mu_r$, where ε and ε_r are the absolute and relative permittivity respectively, and μ and μ_r the absolute and relative magnetic permeability of the medium. In a non-magnetic dielectric medium $\mu = \mu_0$ and *e* is a complex number that is expressed by

$$\varepsilon = \varepsilon_0\varepsilon_r = \varepsilon_0(\varepsilon' - j\varepsilon_e'') \tag{11.23}$$

where ε' is the relative dielectric constant of the medium and ε_e'' relative effective loss factor which considers all energy losses due to dielectric relaxation and ionic conduction. We will come back to this point in the following paragraph.

Assuming that the wave propagates in the z-direction and approximating a unidirectional electric field **E** depending only on x, we can rewrite Equation (11.21) as follows

$$\frac{d^2 \boldsymbol{E}_x}{dz^2} = -\omega^2\mu\varepsilon \boldsymbol{E}_x \tag{11.24}$$

the general solution of this equation is

$$E_x(z) = Ae^{+\gamma z} + Be^{-\gamma z} \tag{11.25}$$

In this equation, the constants A and B correspond to the magnitude of waves propagation in the *-z* and + *z* directions, respectively, while $\gamma = \omega\sqrt{\mu\varepsilon}$ is the wave propagation constant. It is a complex number that is still expressed in the form:

$$\gamma = \alpha + j\beta \tag{11.26}$$

where α is the attenuation coefficient and b the phase constant. These quantities are related to the material dielectric properties as follows:

$$\alpha = \omega\sqrt{\frac{\mu\varepsilon_0\varepsilon'}{2}\left[\sqrt{1+\left(\frac{\varepsilon_e''}{\varepsilon'}\right)^2}-1\right]} \text{ Np/m} \tag{11.27}$$

$$\beta = \omega\sqrt{\frac{\mu\varepsilon_0\varepsilon'}{2}\left[\sqrt{1+\left(\frac{\varepsilon_e''}{\varepsilon'}\right)^2}+1\right]} \text{ rad/m} \tag{11.28}$$

If we apply equation (11.25) to a semi-infinite dielectric slab in the direction of z, so at $z = o$, $A = 0$ and $B = E_0$ (magnitude of the electric field at $z = 0$), then equation (11.25) can be rewritten as follows:

$$E_x(z) = E_0 e^{-\gamma z} \tag{11.29}$$

If we remember that our electric field is sinusoidal and we use the expression for γ given in equation (11.26) then we can write:

$$E_x(z) = \mathbf{Re}\,[E_0 e^{-\gamma z} e^{j\omega t}] = E_0 e^{-\alpha z}\cos(\omega t - \beta z) \tag{11.30}$$

where **Re** allows to consider only the real part of the expression.

The amplitude of E_x along the z-axis is therefore:

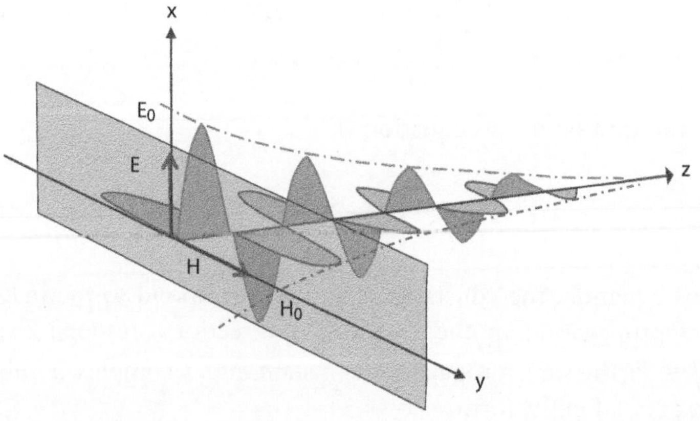

Figure 11.1 Propagation of a plane EM wave in a lossy material.

$$|E_z| = E_0 e^{-\alpha z} \quad (11.31)$$

It appears from this equation that the amplitude of the field decreases exponentially while passing through the dielectric material, hence the expression "lossy material" often used for dielectric media.

11.2.2 Mechanism of Microwave Heating

11.2.2.1 Dielectric Characteristic of a Material

When an EM wave passes through a dielectric material, it interacts with it through its permittivity ε and its magnetic permeability μ. If it is a non-magnetic material, its permeability will be identical to that of free space ($\mu = \mu_0$) and consequently, the magnetic field will have no interaction with it. On the other hand, the electric field will interact with the material because its absolute permittivity will be different from the permittivity of free space ($\varepsilon \neq \varepsilon_0$). We can recall here the expression of this permittivity given previously in equation (11.32).

$$\varepsilon = \varepsilon_0 \varepsilon_r = \varepsilon_0 (\varepsilon' - j\varepsilon_e'') \quad (11.32)$$

In this equation, the relative permittivity ε_r is a complex whose real (ε') and imaginary (ε_e'') parts, dielectric constant, and effective loss factor, respectively, constitute the dielectric properties of the material. The effective loss factor combines the phenomena of energy dissipation of the wave by the Joule effect (ionic conduction) and by dipole rotation (dielectric relaxation). It is expressed by the following relation:

$$\varepsilon_e'' = \frac{\sigma_d}{\omega \varepsilon_0} + \varepsilon'' \quad (11.33)$$

Where the first term to the right of the equal sign is attributable to conduction losses (σ_d representing dielectric conductivity) and the second to losses related to dielectric relaxation. If the medium does not have free charges, we will get $\varepsilon_e'' = \varepsilon''$ and the relative permittivity of the material often noted ε^* will be expressed by:

$$\varepsilon^* = \varepsilon' - j\varepsilon'' \quad (11.34)$$

The dielectric properties of most dielectric materials vary with several factors such as the frequency of the electric field, the water content of the material, its temperature, its bulk density, its composition, etc. [5].

- Variation with the frequency

The relative permittivity of a dielectric material depends on the frequency of the applied electric field as shown by the formula established by Debye for a liquid medium (equation (11.35)) [4]

$$\varepsilon^* = \varepsilon' - j\varepsilon'' = \varepsilon_\infty + \frac{\varepsilon_s - \varepsilon_\infty}{1 + j\omega\tau} \tag{11.35}$$

where ε_s and ε_∞ are dielectric constants at d.c. and at high frequencies respectively and τ is the relaxation time, i.e., the time taken for dipoles initially oriented with respect to the applied field to return to a random orientation after removal of the applied field. The separation of equation (11.35) into its real and imaginary part leads to relations (11.36) and (11.37) respectively.

$$\varepsilon' = \varepsilon_\infty + \frac{\varepsilon_s - \varepsilon_\infty}{1 + \omega^2\tau^2} \tag{11.36}$$

$$\varepsilon'' = \frac{(\varepsilon_s - \varepsilon_\infty)\omega\tau}{1 + \omega^2\tau^2} \tag{11.37}$$

In fact, microwave heating is essentially a problem of phase shift between the frequency of the electric field and the rotation frequency of the polar molecules of the medium trying to align with the field. Debye's equation (11.35) makes it possible to understand the mechanism of microwave heating of a dielectric material.

To better understand this interaction between the EF and the molecule, consider the two situations presented in Figure 11.2. In this figure, the dielectric response of an isolated polar molecule (water vapor) (a) and of the same polar molecule in a condensed medium (liquid water) (b) is analyzed in function of an alternating EF whose frequency varies from low frequency (LF) to infrared (IR) frequency. We note that in the first case, the molecule is perfectly able to rotate at the field frequency (because unhindered by intermolecular interactions) as long as the frequency of the field E remains at appropriate levels. The dielectric constant of the molecule (ε') remains constant throughout this phase.

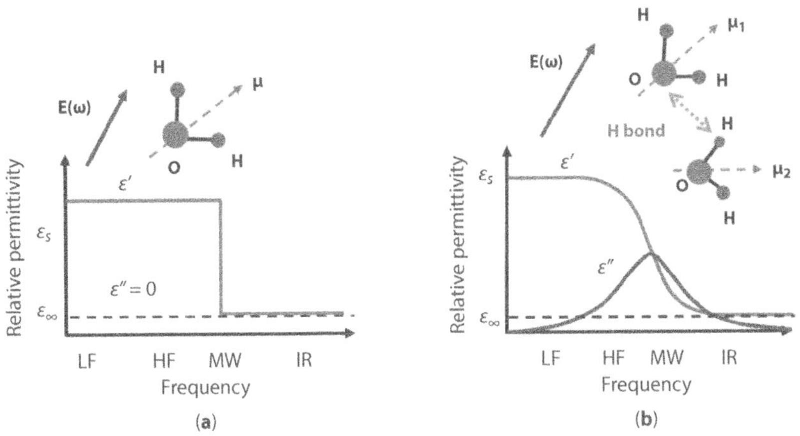

Figure 11.2 (a) Variation of dielectric permittivity of an isolated water molecule when subjected to an EF frequency variation, (b) variation of dielectric permittivity of a water molecule in a condensed medium when subjected to an EF frequency variation.

On the other hand, as soon as one reaches too high frequencies (close from those of MW) the molecule of water can no longer be aligned with the field E and therefore stops turning, the dielectric constant ε' falls instantly to a zero, without the emission of heat, while the absorption coefficient (ε") remains zero throughout the operation. This behavior explains why it is impossible to heat steam with microwaves.

If now one carries out the same operation in condensed medium, we notice that the dielectric response is not the same anymore. As before, the water molecule tries to align with the electric field, and achieves this for low frequencies, ε' remains high but decreases slightly as the frequency of the field increases. When the frequency reaches very high values, the molecule cannot align itself on the field because of its too many interactions with the medium (H bonds). Unlike before, the dipolar rotation does not stop instantly, it appears a phase shift between the speed of rotation of the water molecule and the frequency of the field E. This phase shift is measured by d the loss factor angle with $\tan\delta = \left(\varepsilon''/_{\varepsilon'}\right)$. Thus, the kinetic energy contained in the molecule gradually dissipates into heat, hence the gradual increase of the absorption coefficient ε" up to a maximum value for MW frequencies.

- Variation with the water content

As the water molecule is one of the most absorbent polar molecules found in food products, the dielectric properties of these are greatly dependent

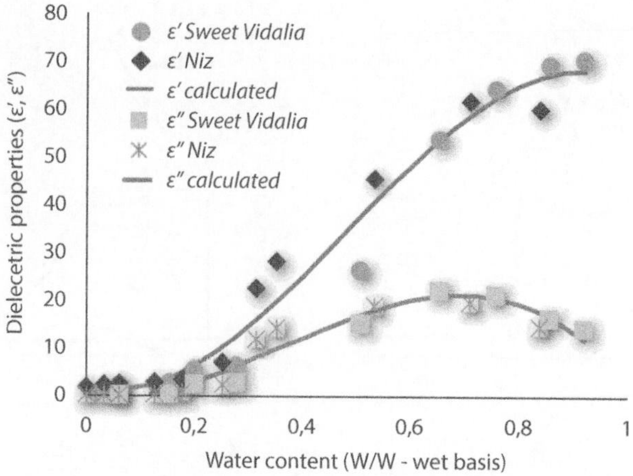

Figure 11.3 Variation of the dielectric properties of two varieties of onions as a function of their moisture content (——calculated values using equation (11.35) and (11.39)) [6].

on their water content. Figure 11.3 illustrates this fact for two varieties of onions, *Niz* and *Sweet Vidalia* [6]. Two polynomial equations (11.38) and (11.39) allowed a good fit of the experimental data regardless of the variety.

$$\varepsilon' = -217.151X^3 + 316.439X^2 - 35.562X - 2.434 \quad (R^2 = 0.973) \quad (11.38)$$

Where X is the moisture content in wet basis of the onions

$$\varepsilon'' = -153.291X^3 + 168.390X^2 - 12.116X - 0.186 \quad (R^2 = 0.941) \quad (11.39)$$

11.2.2.2 Waves-Product Interactions

- *According to the chemical nature of the matter*

For MW energy to be absorbed and dissipated as heat in a medium, it is first necessary that the wave can be transmitted into it. The transmission of the wave depends on the nature of the medium. There are three types of medium:

1. Reflecting media: this type of medium reflects almost all of the incident wave; only a very small part of the wave is transmitted and dissipated into heat on a very small thickness of the material. Metals correspond to this type of medium and

this explains why they are used to transport the waves in a device called a waveguide in microwave equipment.

2. Transparent media: most of the wave is transmitted in the medium but it is almost not absorbed. It passes through the medium and comes out without being altered. Examples of transparent materials are plastic polymers, teflon, glass, ice, ... all these materials are characterized by their very low loss factor ε".
3. Absorbent media: also called dielectric media, a large part of the incident wave is transmitted to them and the totality of this is absorbed if the material is thick enough. The representatives of this type of media are for example liquid water, alcohols, carbohydrates, food, etc. They are characterized by their high loss factor value.

- *According to the orientation of the electrical field and the product shape - Polarization effect*

When an electric field is applied to a material containing polar molecules, these molecules orient themselves with respect to the field, creating an overall polarization of the material. The electric flux density is related to the polarization vector by the equation (11.40)

$$\boldsymbol{D} = \varepsilon \boldsymbol{E} = \varepsilon_0 \boldsymbol{E} + \boldsymbol{P} \tag{11.40}$$

and

$$\boldsymbol{P} = \varepsilon_0 \chi_e \boldsymbol{E} \tag{11.41}$$

where χ_e which is a complex that corresponds to the electrical susceptibility of the material, a quantity characterizing the polarization induced by the electric field in the material. It comes from equations (11.40) and (11.41) that the permittivity of the material is expressed by

$$\varepsilon = \varepsilon_0(1 + \chi e) \tag{11.42}$$

and its relative permittivity by:

$$\varepsilon_r = \frac{\varepsilon}{\varepsilon_0} = 1 + \chi_e \tag{11.43}$$

from where:

$$\chi_e = \varepsilon_r - 1 \tag{11.44}$$

The polarization of the material caused by the applied electric field will be associated an electric field oriented in the opposite direction to the external field. Thus, the material reacts to the external field applied to it by developing an opposite-field tending to bring it back to its initial state. Therefore, the internal electric field (E_{int}) will be a resultant of the applied electric field (E_{app}) and a depolarizing field (E_{dep}) as shown in equation (11.45).

$$E_{int} = E_{app} - E_{dep} \tag{11.45}$$

It is this internal field that will condition the heat dissipation of the energy inside the material due to the friction generated by the dielectric relaxation of the dipolar molecules and the oscillation of the free charges of the medium. The depolarizing field (E_{dep}) depends on the polarization P as well as on the shape and dimensions of the material. For an isotropic medium, It is expressed by the equation (11.46)

$$E_{dep} = \frac{b}{\varepsilon_0} P \tag{11.46}$$

where b represents a form factor depending on the shape of the material.

P can be expressed from equations (11.41) and (11.44), it then comes:

$$P = \varepsilon_0(\varepsilon_r - 1) \tag{11.47}$$

By substituting equation (11.47) in equation (11.46) and equation (11.46) in equation (11.45), we obtain the expression of the electric field as a function of the form factor b of the material:

$$E_{int} = \frac{E_{app}}{1 + b(\varepsilon_r - 1)} \tag{11.48}$$

Table 11.1 gives the value of the form factor b and the expression of the internal electric field Ei_{nt} for some simple shapes.

Table 11.1 Form factor b and internal electric field for some simple geometrical shapes.

Shape	b	E_{int}	Comment
Sphere 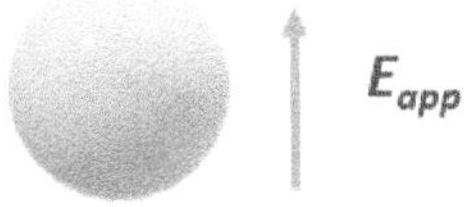	$b = 1/3$	$E_{int} = \frac{3E_{aPP}}{2+\varepsilon_r}$	-
Cylinder 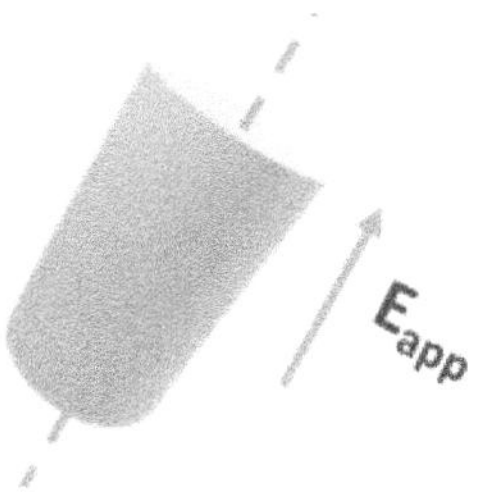	$b = 0$	$E_{int} = E_{app}$	The action of the applied electric field is maximal
Thick disk 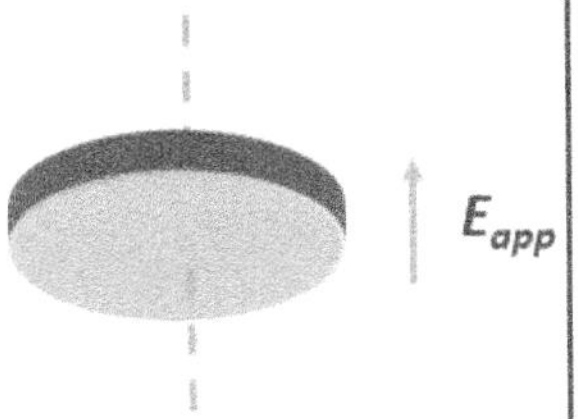	$b = 1$	$E_{int} = \frac{E_{aPP}}{\varepsilon_r}$	The action of the applied electric field is minimal

- According to the composition of the product

It is not enough that a material has a high loss factor for it to be easily heated by microwaves. It is indeed necessary that a significant part of the wave can be transmitted. This transmission depends essentially on the dielectric constant of the material. An example will help to illustrate this. Let's consider a water puddle that receives microwave radiation. We know that at the air/water interface that part of the incident wave will be transmitted while the other part will be reflected. Let's try to estimate the part of transmitted radiation. Knowing that the dielectric constant of water at 2450 MHz is $\varepsilon'_w \approx 80$, the reflection coefficient of the wave at the water surface will be given by:

$$R = \left| \frac{1-\sqrt{\varepsilon'_w}}{1+\sqrt{\varepsilon'_w}} \right| = \frac{\sqrt{\varepsilon'_w}-1}{\sqrt{\varepsilon'_w}+1} \approx 0.8 \tag{11.49}$$

On the other hand

$$\frac{P_r}{P_i} = R^2 \approx 0.65 \tag{11.50}$$

where P_r and P_i are the reflected and incident powers, respectively. From this, it appears that

$$\begin{cases} P_r = 65\%\ P_i \\ P_t = 35\%\ P_i \end{cases} \tag{11.51}$$

where P_t, is the transmitted power.

This calculation shows us that the water, although having a high loss factor ($\varepsilon'_w \approx 12$ at 2450 MHz), will not heat up strongly at the first impact of the wave because its surface behaves like a real mirror to the wave by reflecting 65% of the incident energy. However, in practice, a glass of water exposed to microwave radiation in an oven will eventually heat up because at each impact of the wave 25% of the incident energy will be transmitted and dissipated as heat.

Now if we consider a wet powder with a dielectric constant of 10, using the same calculation as before we find R = 0.5 and R^2 = 0.25. So, the reflected and transmitted powers will be 25% and 75%, respectively. Thus, we obtain a better coupling between the waves and the wet powder compared to water. It will be easier to heat this wet powder than pure water.

11.2.3 Transmission and Absorption of a Wave in a Material

11.2.3.1 Expression of Transmitted Power

When an electromagnetic wave is directed towards a material, a part of the wave is reflected on the surface thereof while the other part penetrates it to be absorbed. The absorption of the wave during its crossing results in a decrease in the amplitude of the internal EF and so of its power. It is generally accepted that this attenuation follows Lambert's law, namely an exponential decay of the EF inside the material as seen previously in equation (11.31):

$$|E_z| = E_0 e^{-\alpha z} \tag{11.31}$$

It should be noted, however, that this attenuation does not follow an exponential law in very thin materials. Thus, the calculation of the internal electric field in the case of microwave heating of a thick plate of beef using either Maxwell's equations or equation (11.31) result in completely different profiles for both methods, for thicknesses ranging from 0.02 to 4.59 cm. Indeed, the field profile obtained at 0.02 cm using Maxwell's equations is sinusoidal and not exponential. On the other hand, the two profiles converge completely from a thickness of 3.68 cm, indicating that Lambert's law becomes valid [7].

As we will see in section 11.2.3.3 the transmitted power is proportional to $|E|^2$. Thus, assuming that the thickness of the material is sufficient, the transmitted power (P_t) can be expressed by:

$$P_t = P_i e^{-2\alpha z} \tag{11.52}$$

where P_i Represents the incident power of the wave.

Thus, like the electric field, the transmitted power undergoes an exponential decay depending on the attenuation coefficient *a*.

11.2.3.2 Penetration Depths

- *Skin penetration depth*

The penetration depth is defined as the penetration distance in the material for which the magnetic of the electric field is divided by e. From this definition we can rewrite equation (11.31) by introducing the penetration depth δ:

$$|E_z| = E_0 e^{-\frac{z}{\delta}} \tag{11.53}$$

where:

$$\delta = \frac{1}{\alpha} \tag{11.54}$$

δ corresponds to the penetration depth of the electric field but is also called skin penetration depth. In conclusion, at a distance δ in the material, the amplitude of the electric field will be reduced by 63% of its initial value.

- *Power penetration depth*

Let us substitute the expression of the attenuation coefficient *a* from equation (11.54) in equation (11.52), the transmitted power is then expressed by:

$$P_t = P_i e^{-2\alpha z} = P_i e^{-2z/\delta} \tag{11.55}$$

At skin penetration depth *δ*, this power becomes:

$$P_t = \frac{P_i}{e^2} = 0.14\, P_i \tag{11.56}$$

indicating that 86% of the incident power has been dissipated.

The power penetration depth is defined as the distance D_p at which the power falls to 1/e of its value at the surface. Thus, the transmitted power is expressed by equation (11.57):

$$P_t = P_i e^{-z/D_P} \tag{11.57}$$

where

$$D_P = \frac{1}{2\alpha} = \frac{\delta}{2} \tag{11.58}$$

The power penetration depth depends on the dielectric constant of the material as well as the frequency of the wave or its wavelength. It can be calculated for a food product with the following equation:

$$D_P = \frac{\lambda_0\sqrt{2}}{2\pi}\left[\varepsilon'\left(\sqrt{1+\left(\frac{\varepsilon''}{\varepsilon'}\right)^2}-1\right)\right]^{0.5} \tag{11.59}$$

where λ_0 is the free space wavelength (12.2 cm for a frequency of 2450 MHz).

If ε" is small, equation (11.59) can be simplified as follows:

$$D_P = \frac{\lambda_0 \sqrt{\varepsilon'}}{2\pi\varepsilon''} \tag{11.60}$$

This equation remains applicable to most food products although many have high ε"-values. It is noted that D_p is inversely proportional to ε", in other words, the wave penetrates less deeply into high lossy materials. Equation (11.60) can be rearranged to show the frequency f (Hz) of the wave ($\lambda_0 = c/f$):

$$D_P = \frac{4.78\,10^7}{f\sqrt{\varepsilon'}\tan\delta} \tag{11.61}$$

where $\tan\delta = \frac{\varepsilon''}{\varepsilon'}$ is called the loss factor, and d, the loss angle is the phase shift.

This last expression shows that the depth of penetration is inversely proportional to the frequency, which makes it possible to direct the choice of the frequency (915 or 2450 MHz) according to the application.

11.2.3.3 Power Dissipation

We have just seen that when a material is subjected to microwave heating, the transmitted power decreases as it passes through it. This decrease is related to the absorption of the wave energy by the material and its dissipation as heat. This dissipation depends on the dielectric characteristics of the material, its dimensions, the frequency of the wave, and the electrical field. Equation (11.62) gives the relationship between the power dissipated P_d and the various factors mentioned above.

$$P_d = 2\pi f \varepsilon_0 \varepsilon_e'' E_{int}^2 \tag{11.62}$$

Where f is the wave frequency, E_{int} the internal electric field and ε_e'' the effective loss factor taking into account the losses by dielectric relaxation and by ionic conduction (Joule effect). It is expressed as follows:

$$\varepsilon_e'' = \varepsilon'' + \frac{\sigma}{\omega\varepsilon_0} \tag{11.63}$$

where σ represents the electrical conductivity of the medium and w the pulsation of the wave. In most MW food heating applications, the heat dissipation by Joule effect is neglected (thus $\varepsilon''_{eff} = \varepsilon''$). Under these conditions, equation (11.62) can be simplified as follows:

$$P_d = 5.563\,10^{-11} f \varepsilon'' E_{int}^2 \; (W/m^3) \tag{11.64}$$

This equation indicates that the higher the loss factor, the greater the power dissipated.

11.3 Applications

11.3.1 Microwave Baking

Baking is a complex process that involves several and simultaneous physicochemical and biochemical transformations such as starch gelatinization, protein denaturation, water evaporation, volume development by carbon dioxide liberation, crust formation, and non-enzymatic browning reactions.

The baking method and process conditions influence the quality of the baked products. In conventional baking, the heat transfer takes place from the oven cavity to the dough surface by convection, then, gradually into the core by conduction [8]. This mechanism may result in moisture and cooking gradients within the product and excessive drying and hardening of the crust which becomes a barrier to further heat penetration. Microwave baking is characterized by volumetric and rapid heating by, principally, frictional energy of rotating dipoles under the alternating electromagnetic field of microwaves. The considerable popularity of microwave oven uses, on industrial and domestic levels, motivated the researchers to study and improve the MW baking process. As well as other MW heating applications, saving time and energy, and avoiding overheating problems are the major advantages of MW baking. For example, MW baking was successfully applied in the case of biscuit baking. MW baking enhances the quality preservation in terms of textural changes compared to conventional baking [9, 10]. Besides, a more uniform moisture distribution was reported in the case of MW baked biscuits compared to conventional biscuits [9, 10]. However, the MW baked products are generally of inferior quality compared to conventional baked ones and are not fully accepted by consumers [11]. The rapid MW heating does not allow enough time for starch enzymatic conversion and gelatinization, gluten structure development, dough expansion, browning reactions, and crust formation [12]. For these reasons, the MW baked products are characterized by fast staling, firm tough and dry texture, non-uniform color, lack of flavor, and desirable crust formation [13, 14]. Several studies reported also that the MW baked products had lower volume compared to conventional baked ones [15].

The poor texture of MW baking products may be due to lower starch granules swelling [16] or a lower degree of gelatinization [17] as compared to conventional baking. Furthermore, several studies showed that MW baking products are of lower moisture contents compared to conventional ones [18–20]. MW heating produces internal pressure gradients toward the product boundaries, which, thereby, increases the moisture removal. The evaporated water condenses at the product surface because of cold air in the microwave oven around the product which prevents Maillard and browning reactions and crisping [21]. The authors of [22] observed insufficient browning in MW baking breads. For this reason, also, typical flavors of baked products generated as a result of browning reactions do not exist in MW baking products [8].

Improving the quality of MW baking products requires adapted ingredient formulations. MW heating depends on the dielectric properties of food products. Studies have shown that adding certain components such as fat, dough emulsifiers, and modified starches may procure good dielectric properties and enhance the quality of baked products. The authors of [17] showed that the presence of fat influenced the starch gelatinization and the quality of MW baked cake. Other researchers reported that dough emulsifiers, such as lecithin, gums, soy protein, and some enzymes, like α-amylase, xylanase, and lipase, improve the dough volume and reduce the firmness of MW baking products [23–25]. Several additives have been also tested to improve the moisture retention of MW baking products, like cellulose powder, alginates, and milk protein gums [24, 26]. Furthermore, specialty modified starches, sugar powders, and browning agents with using of susceptors may also improve the browning and crisping of MW baking products [19, 27].

On the other hand, several studies focused on combining MW heating with other heating methods, such as infrared (IR) heating, hot air, or steam [28, 29] to overcome the disadvantages of MW baking technology and to enhance the desirable baked goods qualities. The MW heating combined with IR heating results in enhancement of crust formation and surface browning [29, 30]. For example, the authors of [29] studied the effects of the Microwave-halogen lamp baking combination on the quality of cookies. The halogen lamp and microwave powers and baking time have a significant effect on the cookie's color, hardness, moisture contents, and spread ratio. At optimum baking conditions (20% microwave power and 70% halogen lamp levels for 5.5 min), the microwave-halogen lamp combination method produces similar cookies to conventionally baked ones while reducing baking time to 50%. Furthermore, the authors of [31] reported that the MW-IR combination method produced breads with comparable staling degrees as conventionally baked ones. The authors of [15] compared the quality of cakes baked in microwave, IR, and microwave-IR (MW-IR) combination ovens. Cakes baked in microwave oven had the lowest quality in terms of color, firmness, weight loss, and specific volume. In contrast, MW-IR combining method produced cakes with similar quality to conventional baked products with reducing baking time by about 75%. The authors of [32] studied the effects of different cake formulations (fat-free, 25% fat, and 25% Simplesse™ containing cake) and three baking ovens (microwave, MW-IR combination, and conventional) on starch gelatinization. It was found that fat improved the starch gelatinization in MW and MW-IR baking, whereas it reduced the degree of starch gelatinization in conventional baking. This result can be explained by the higher dielectric properties of the samples with higher fat content since fat-containing samples had lower porosity compared to fat-free cakes [32]. In the case of conventional baking, fats may delay starch gelatinization by delaying heat and water transport into the starch granules due to the formation of complexes between the lipid and amylose during baking [33]. Adding of Simplesse™ as a fat replacer decreased the starch gelatinization in all baking methods. In accordance with the previous published studies, rapid MW baking does not allow starch gelatinization. The degree of gelatinization ranged from 55% to 78% depending on formulation for MW baking compared to 85% to 93% in conventionally baked cakes. MW-IR technology significantly improves the degree of starch gelatinization to 70–90%.

Two cycles of microwave baking were also studied in the literature to eliminate the quality problems inherent to this technology [12, 14, 15]. The first cycle, with low power long time processing conditions, ensures the internal baking, and the second cycle, with high power and short time; contributes to the overall finishing baking. The authors of [14] studied innovative two cycles microwave-toaster oven. This technology combines time-saving of MW baking and browning and crisping advantages of the electric toaster. At optimum operating conditions in this study (204 W/120 s for the first cycle, 937 W/70 s for the second cycle, and final toasting step of five at 200°C), two-cycle microwave-toaster oven produced a pound cake with better volume and luminosity. The cake baked by conventional and two-cycle microwave-toaster methods was equally appreciated by consumers [14].

To conclude, baked products by microwave, when applied alone, are of lower quality compared to conventional ones. Improvement additives and combined microwave heating with other heating methods, particularly infra-red heating, are interesting improvement trends.

11.3.2 Microwave Blanching

Blanching is a basic operation in fruits and vegetables processing. It is defined as a mild thermal pre-treatment applied usually before other food processes such as canning, frying, drying, or freezing. This operation mainly aims to inactivate the quality deterioration enzymes, decrease microbial load, remove pesticides and other chemical contaminants residue, eliminate dissolved oxygen, improve dehydration rate and facilitate peeling and packing [34–37]. Blanching, hence, leads to enhance color, flavor, and sensory characteristics of the product and preserve its quality upon long-term storage [38, 39]. This process involves rapid exposure of the product to a heat source (generally hot water or steam). The time-temperature combination is an essential element to preserve the production quality. The normal blanching temperature is between 50-70 °C for 1 to less than 10 min [40] to inactivate polyphenol oxidase (PPO), peroxidase (POD), lipoxygenase, and pectin enzymes which causes the textural softening, enzymatic browning, off-flavors, and degradation of nutritional quality [34]. Due to its thermal resistance, POD inactivation is considered an indicator to assess the blanching efficiency [41, 42]. Nevertheless, total POD inactivation implies over-processing with consequent quality losses [42]. Ascorbic acid contents have been also used as an indicator of nutritional quality to evaluate the nutrient losses after the blanching process [34].

Hot water blanching is the most traditionally adopted blanching method in the food industry. Besides its effectiveness in enzymatic inactivation, this method is characterized by the high consumption of water and energy and important loss of nutrients such as bioactive compounds, minerals, carbohydrates, and proteins [43–45]. For example, the reduction of ascorbic acid during blanching of potato, broccoli, cauliflower, and carrot slices [46, 47] was reported principally due to leaching into water treatment medium. Furthermore, blanching wastewater, as it is highly polluted with soluble components, can cause environmental problems and needs to be treated [48].

To overcome the limitations of water-based (or wet) blanching process, innovative blanching technologies with less quality degradation, nutrient loss, and higher energy efficiency have recently been developed [34, 39, 49], such as microwave blanching. Recent research reported the advantages of using the microwave for blanching fruits and vegetables [37]. It is considered a dry blanching treatment since water is not required [50]. In MW blanching, the heat is generated inside the product thanks to direct interaction between the electromagnetic field and food. Comparing to traditional blanching methods, MW blanching is of great interest as it allows faster heating rate, volumetric heating, greater nutrients retention, and preservation of sensorial quality [51, 52]. Therefore, because of the above reasons, MW blanching has been successfully applied as an alternative to conventional treatment. Several pieces of research have been realized on the MW blanching of fruits and vegetables like peas [53], bell pepper [54], green beans [55], mushroom [56], carrots [57], asparagus [58], mango [50] and fruit peelings [59].

Particularly, many studies reported higher retention of vitamin C under MW blanching compared to conventional hot water blanching in various fruits and vegetables [60]. For example, the authors of [55] found that MW blanching of green beans required shorter processing time and resulted in higher ascorbic acid retention. Similarly, [54] reported that MW blanching reduced the loss of vitamin C by 18, 8.5, and 33.5% for spinach, bell peppers, and carrots, respectively, in comparison to water blanching. Recently, the authors of [50] compared the effect of MW and conventional water blanching on vitamin C content in mango (Mangifera indica) under two blanching scenarios, High-Temperature Short Time (HTST, at 90°C for 5 min) and Low-Temperature Long Time (LTLT, at 70°C for 12 min). MW blanching led to higher retention of total vitamin C in both HTST and LTLT treatments. A higher inactivation rate of ascorbic acid oxidase by MW blanching was also reported in this study. Furthermore, MW blanching may enhance the antioxidant activity of pepper [61] and asparagus [58], and bell peppers [52] through the formation of derivatives of phenolic compounds during blanching. Recently, the authors of [59] studied in detail the antioxidant activity of mango, apple, orange, and banana peelings under MW blanching at 720 W for different times (1 to 5 min). MW blanching resulted in a significant increase of total phenolics content and antioxidant activities and reduction of browning rate. In accordance with these results, complete inactivation of browning PPO enzyme in mushroom through microwave heating step combined to water bath treatment was successfully reported [56]. Numerous studies showed also that MW heating has a higher effect on POD inactivation than conventional thermal treatments in different food products such as broccoli, green beans, asparagus, carrots, red beets, potatoes, savoy cabbage, and white cabbage [51, 62–64]. The authors of [52] confirmed these results. The authors demonstrated minimum weight loss and maximum retention of ascorbic acid, red pigments, total antioxidant activity in red bell peppers blanched by MW compared to hot water blanching. The inactivation of PPO and POD enzymes depends on microwave power, temperature, and treatment duration.

However, despite MW blanching advantages, this technology has some serious drawbacks that limit its commercial application. As described earlier, MW heating efficiency depends on several parameters such as MW frequency and power, geometry, mass, position, and/or mobility of the product inside the MW oven and the chemical composition of the product and its dielectric and physical properties [51]. All these factors should be considered during MW blanching. The blanching temperature is difficult to be controlled precisely because of the natural heterogeneity of MW heating that may lead to non-uniform enzymatic inactivation [36]. The authors of [51] studied the effect of geometric shape (cylindrical, cubed, parallelepiped, and slabbed) on MW heating inactivation of POD in several vegetables (potatoes, savoy cabbage, and white cabbage). In all the samples, the enzymatic inactivation was initially attained at the core of the sample. Despite the application of a prolonged and intensive treatment, the total inactivation of POD was not achieved. The authors reported that MW blanching is not suitable for reaching the required blanching temperature on the surface. The limited penetration depth of microwave radiations into the product is another important limitation of the MW blanching process. For example, it has been measured that the microwave penetration depth for sweet potato, red bell pepper, and broccoli is about 1.5 - 3.5 cm [65]. Furthermore, it was reported that moisture might evaporate during MW blanching causing water loss. In addition, high intensity of microwave power and high internal vapor pressure, which produced when cellular water converted into steam during MW heating, can lead to collapse or folding of cell walls as reported by [66] who showed that MW blanching of carrots did not improve the texture of carrot compared to water blanching treatment. The negative impact of MW blanching on carrot texture was also confirmed by [57]. In this study, mild (60°C, 40 min) and strong (90°C, 1 min) microwave blanching were applied. The microstructure of samples treated under mild conditions was relatively like the fresh ones. Strong MW blanching conditions caused the dispersion of some cell walls.

Microwave blanching provides rapid heating and reduces the thermal degradation and the loss of soluble nutritious components. However, more fundamental researches are needed for a better understanding of this technology and for designing suitable industrial equipment.

11.3.3 Microwave Tempering and Thawing

Tempering and thawing are common food processes to handle the frozen raw materials, particularly meats and sea products, before its further processing [67]. Tempering is a thermal process to heat the frozen foods from the storage temperature (generally -25 to -18°C) to below the freezing point (-5 to -2°C). Under these conditions, the product is still relatively firm but readily suitable for mechanical chopping and processing [68, 69]. In the thawing process, the temperature of the frozen product is homogeneously raised to 0°C and no free ice is still present.

The magnitude of quality changes in tempered and/or thawed products is influenced by many factors including applied method and processing time and temperature [70]. There are several conventional methods employed for tempering and thawing based on heat transfer in water or in still and moving air. In these methods, heat transfers slowly from the product surface to the center by conduction. According to operating conditions, such as surrounding air velocity and humidity, heating medium temperature, product shape, and size, tempering and thawing processes may take few hours to several days, due to low thermal conductivity of frozen food, which causes serious quality and safety problems [71–73]. Reducing tempering time will minimize the chemical changes, microbial growth, and water loss caused by dehydration or dripping [74].

Microwave thawing and tempering is one of the most popular applications of this technology in the food industry. Compared to conventional methods, the energy is delivered directly to the inner regions of the product during microwave tempering which leads to, effectively, shorten tempering time, reduced energy consumption, improve product quality and prevent the growth of microbial pathogens [75, 76].

Microwave tempering and thawing may be carried out in batch or continuous systems. This technology has been successfully applied in the meat, poultry, and fish industries [77]. For example, the authors of [78] studied the microwave and air thawing of bread. The result revealed that microwave thawed bread had generally better quality. The authors of [79] evaluated the effects of different thawing methods on color, polyphenol, and ascorbic acid retention in strawberries. The results showed that ascorbic acid and anthocyanins were preserved under microwave thawing (10 min). Conventional thawing at 4°C (for 24 hr) induced the most pigment and ascorbic acid losses. The authors of [80] compared the effect of different thawing methods (air, immersion, microwave, and high-pressure) on the quality of hami melon. The results showed that less texture damage, drip loss, and discoloration were observed for the samples thawed by microwave compared to other methods. Samples thawed by microwave at 500 W had the best preserved ascorbic acid (about 50%) and the highest average overall sensory score than other samples.

In fact, despite the advantages and the widespread use of microwave thawing and tempering systems, this technology is constrained by thermal heterogeneity. MW heating is a sensible and complex process. Numerous intrinsic factors may affect the thawing and tempering process. The dielectric properties variation between the frozen and the relatively warmer or defrosted zones within the product is the major cause of non-uniform and runaway heating problems if not carefully controlled [81]. The dielectric loss factor of frozen water sharply increases after thawed [82]. Considering the limited penetration depth of microwaves in frozen foods, accompanied by the focusing heating effect at corners and edges [83], the product surface will be heated before its core. The thawed water at the outer layers will absorb much more energy than do the frozen water at the inner layers which will further limit the penetration depth and accelerate more and more, as a consequence, surface heating. This phenomenon is defined as thermal runaway [84]. The authors of [85] studied the tempering of a frozen block of shrimp by microwave system at 915 MHz and 500W and 1 kW. At both power setting, local surface overheating was observed. However, microwave tempering at 500 W yielded relatively more uniform internal temperature.

To overcome the runaway and heterogeneous heating problems in microwave tempering and thawing applications, some solutions were proposed in the literature. The lower frequency (915 MHz) is recommended over the higher frequency (2450 MHz) for heating the thick products as the penetration depth is deeper. Furthermore, lowering power density, multiple feed points, and mode stirrers may also be useful to improve heating uniformity [81]. Cooling the product surface by different means is also an interesting method. A combined microwave-vacuum thawing system is studied [67]. Thanks to vacuum, evaporated water cools the surface. Furthermore, the authors of [86] described a microwave-cold air combined system for the same purpose. The air temperature can be easily controlled. The microwave power cycle can also be monitored (On/off) to respect predetermined temperature-imposed gradients between hot and cold points and to allow thermal conduction between warm and frozen regions to avoid runaway heating.

Microwave tempering and thawing is an interesting application thanks to several proven advantages over conventional systems. However, its industrial use is still limited because of runaway heating problems inherent in this technology when heating frozen foods. Despite considerable interest in this technology, there are relatively limited scientific studies to propose reliable microwave tempering and thawing apparatus.

11.3.4 Microwave Drying

Drying is one of the most applied preservation processes in food industries by removing moisture and consequently reducing the water activity of agricultural products to safe limits to prevent microbiological, chemical, and biochemical degradation during storage. Drying may be applied also for economic interests, consummation aspects, and/or for desired technological modifications of raw materials [87].

Hot air (HA) drying is widely used to produce dried fruits, vegetables, herbs, and other food products at an industrial scale. This method involves simultaneous energy and mass transfer by exposing the product to contentious hot airflow under controlled conditions. During conventional drying, internal heat and mass (water and volatile compounds) transfer is a limiting factor of the drying rate which sharply decreases with time. Therefore, HA drying requires the application of severe conditions at the end of the operation, which result in a huge amount of energy consumption, overheating and/or over-drying of surface layers, and undesirable degradation of dried products including, lipid oxidation, browning reactions, and other sensory, nutritional and functional properties deterioration [88, 89]. Therefore, alternative innovative and efficient drying methods have been proposed [90, 91].

Microwave drying was successfully applied in the food industry [92]. Several studies about this subject were summarized in the literature [93]. During MW drying, thanks to the volumetric and rapid heating of microwaves, the vapor is generated within the product which leads to an increase of internal pressure that forces the moisture towards the surface [94]. This forced water outflow increases the drying rate and reduces the drying time [95]. For example, the authors of [96] reported a decrease in drying time by about 56% by increasing the microwave power density from 10 to 20 w/g of lotus seeds. Furthermore, the internal pressure prevents food shrinkage, promotes greater porosity, and enhances the rehydration ability. MW drying results also in better thermal efficiency and improved overall product quality compared to conventional HA drying [97, 98]. Samples of coriander leaves dried by microwave showed similar color properties to fresh ones [99]. Similarly, the authors of [100] reported better preservation chlorophylls and carotenoids of coriander samples dried in microwave oven compared to conventional oven-dried ones. Furthermore, in MW heating, the energy absorption is defined by wet regions. Heating selectivity is another advantage of this technology to avoid the overheating problem of dried parts in conventional hot air drying thereby enriching the product's quality [89]. However, due to moisture pumping mechanism in MW drying accompanied by uncontrolled and/or low temperature of surrounding air in the MW oven, moisture accumulation at the surfaces may lead to surface excessive heating, product carbonization and off-flavors development. This phenomenon was reported, for example, in the case of MW drying of onion [101]. For this reason, MW drying is usually combined with other conventional or innovative drying methods to improve the water transfer at the surface which enhances the drying efficiency and the quality of dried products [102–104]. Applications of combined MW drying, principally, include microwave-assisted hot air drying (MWHD), microwave assisted vacuum drying (MWVD) and microwave-assisted freeze drying (MWFD).

11.3.4.1 Microwave-Assisted Hot Air Drying

In this method, the accumulated moisture at the product surface due to microwave internal heating can be removed continuously by convective hot air flow [103, 105] which increases the drying rate. Three types of microwave heating and hot air flow combination are possible [106]. MW heating can be applied alone at the start of the drying process to heat the interior of the product and pump the vapor toward the surface layers to be removed later by airflow. MW heating can be applied at the constant drying rate period under hot airflow. At this stage, the product surface is relatively dry whereas the inner core is still moist. The application of MW heating can promote moisture movement to the surfaces. MW heating can be applied, finally, at the end of hot air drying during the second falling rate period to remove bound water and to avoid product shrinkage at this stage. In addition to the cited combination methods, simultaneous microwave and hot air drying have been also proposed [103]. Numerous studies about MWHD have been published in the literature. Saving drying time and improved quality have been reported, using MWHD, apple, and mushroom [107], cranberries [108], macadamia nuts [109], rice [110], soybeans [111], and bananas [112]. The authors of [113] reported that MWHD of garlic reduced the drying time by up to 80% compared to hot air drying alone and produced a product of better quality. For Kiwifruit, MWHD also reduced the drying time by up to 89% and produced products with less shrinkage and better rehydration properties compared to convective hot air drying alone [114].

11.3.4.2 Microwave-Assisted Vacuum Drying

Microwave drying can be associated with the vacuum to reduce the boiling point of water and to prevent oxidation reactions and, thus, to maintain the quality of dried products [115]. MWVD was successfully carried out for several products [106], such as apple [116], cranberries [117], tomatoes [118], potato cubes and slices [119,120], green tea [121] among several other agricultural products [106]. MWVD reduced the drying time up to 90% and improved the rehydration capacity of mint leaves compared to hot air drying [122]. Similarly, in the case of MWVD of bananas, dried samples were comparable to freeze-dried ones in terms of color and flavor. The authors reported also that microwave vacuum dried samples absorbed twice the water quantity compared to conventionally hot air-dried samples [123]. In MWVD, microwave power is the most important factor to control the process. High microwave power may lead to considerable loss of ascorbic acid and phenolics contents [124].

11.3.4.3 Microwave-Assisted Freeze-Drying

In a freeze-drying process, also known as lyophilization, the frozen product is dried by sublimation under low pressure and temperature conditions. It is used usually for heat-sensitive materials particularly in food and pharmaceutical industries. This method preserves the structure and the porosity of the food product enabling good rehydration capacity compared to hot air-drying treatment [87, 125]. Furthermore, the lack of oxygen in the processing environment prevents the oxidation reaction and preserves the valuable sensitive compounds. However, along with high installation costs, freeze-drying is a slow process and demands high energy to the child the product and to maintain it under vacuum for several days in some cases to achieve the desired moisture level. The authors of [126] reported that the freeze drying process has a 200-500% higher cost than the traditional hot air drying process. Microwave heating combined with freeze-drying has been studied as a potential solution to reduce the drying time required. MW heating can provide the heat necessary for the sublimation process during drying and increase, thus, its rate [127]. For example, MWFD reduced the drying time and energy consumption by 40% and 35% respectively in the case of banana drying [128]. Similarly, the authors of [129] reported that MWFD reduced the drying time by up to 50%

compared to freeze-drying with a comparable quality of dried products. MWFD was applied to onion slices [130], potato [131], beef meat [132] and to dried vegetable soup mixes [127, 133]. These studies confirmed a comparable quality of microwave freeze-dried and lyophilized products. However, because of complex interactions between microwave, ice, and thawed water, thermal runaway may take place which leads to uneven heating and poor product quality [102]. Furthermore, Corona discharge and arcing may be occurred because of high vacuum levels in this process [134]. MWFD has no industrial-scale application due to its high costs [68].

Microwave drying is also associated with other drying processes such as fluidized bed, osmotic dehydration, and infrared heating [93]. Microwave-assisted drying systems have many advantages and may become the trends in drying technologies in the near future.

11.3.5 Microwave Pasteurization and Sterilization

The conventional thermal decontamination process is the oldest and the most applied method to preserve most foods. Pasteurization is widely used to inactivate vegetative pathogens, spoilage microorganisms, and alteration enzymes. Pasteurized products need to be stored under refrigerated conditions as this method does not destroy bacterial spores. Sterilization, which can be defined as severe pasteurization, destroys bacterial spores (12 D reduction of *Clostridium botulinum* spores) and extends the shelf life of foods for prolonged periods at room temperature [95, 135]. For thermal preservation treatments, time and temperature combinations are determined according to the product's nature (solid, liquid, ...) and physical properties (thermal conductivity, specific heat, porosity, ...), and target microorganisms or enzymes to retain a good quality of treated products. In conventional heating, due to slow heat transfer from the heating medium to the inner regions, defined as the cold point of the product, severe conditions (higher temperatures and/or longer heating times) need to be applied which lead to important degradation of sensorial and nutritional qualities [136, 137]. Innovative microwave-assisted pasteurization and sterilization process has been intensively studied to overcome the drawbacks of conventional thermal treatments.

Due to direct and volumetric interactions between microwaves and food, MW heating offers the possibility to reduce the deleterious overheating effects of conventional thermal processes and to produce, thus, safe, high quality, and minimally treated products [138]. Several studies approved the efficiency of MW heating to kill microorganisms and their spores and to inactivate alteration enzymes [139]. The microwave decontamination process may be particularly advantageous for viscus and solid products [138]. MW heating is generally faster than conventional heating which reduces considerably the processing time and leads to better retention of nutrients and sensitive valuable components [140]. The authors of [141] reported that microwave pasteurization of pickled asparagus at 915 MHz was two times faster than conventional heating in a water bath, which significantly reduced the thermal degradation. Furthermore, decontamination of packaged products is possibly by MW heating for different packaging materials [142]. Other studies reported also better energy performance for microwave pasteurization compared to conventional thermal treatment (UHT) while delivering comparable microbiological lethality in orange juice for example [143]. Microwave pasteurization and sterilization have been well reviewed in the literature [139, 144, 145]. For example, a comprehensive and detailed review about the microwave pasteurization and sterilization of dairy products was recently published [137]. MW heating has been considered as an efficient decontamination method for destroying several foodporn pathogens and spores [146] such as *Listeria monocytogenes* [144], *Salmonella typhimurium* [147], *E. coli*, *Enterococcus spp*, and spores of *Clostridium sporogeneses* [148, 149]. Many papers demonstrated also the effectiveness of MW heating for the pasteurization and sterilization of different foods such as fruit juices [145], fruit and vegetables [147, 150, 151], meat products [148,152]. The effects of microwave decontamination on food quality attributes were also investigated [153]. For example, the authors of [154] studied the pasteurization of orange juice in a pilot-scale microwave oven. Carotenoid content decreased by 13% after microwave pasteurization. Ascorbic acid retention was evaluated from 97 to 96% depending on treatment temperature. In another study, the authors of [155] reported better preservation of polyphenols, anthocyanins, and vitamin C contents of strawberry puree heated by microwave compared to the conventional method. The authors reported, in addition, the inactivation of polyphenol oxidase and peroxidase by 98% and 100% respectively. Similarly, the authors of [156] showed that microwave treatment of grape tomatoes resulted in a 1.45 log reduction of *Salmonella enterica* without affecting the color, pH value, and nutritional quality after heating. The authors of [157] reported also better color retention of pre-packaged carrots in brine treated by microwave-assisted pasteurization system and conventional hot water processes resulting in equivalent microbial safety. No significant differences of texture, carotenoids and PME activities were found for samples processed by microwave and hot water.

Microbial killing under MW heating has been explained by different mechanisms such as selective and fast heating of microorganism cells, and electroporation or cell membrane rupture due to the potential difference across it. For example, the study of [158] reported an increase of nucleotides and protein released from *E. coli* and *B. subtilis* under microwave heating. This leak is correlated to microwave applied power. Likewise, damage of cellular membrane and increased permeability was also reported by [159].

The nature of microwaves' effect on microorganisms (thermal or non-thermal) is a permanent subject of debate. Several studies reported that microwaves can inactivate microorganisms at sublethal temperatures or that the microwave bactericidal effects are enhanced compared to conventional heating under comparable conditions [160, 161]. On the other hand, controversial results have been reported to demonstrate the absence of a non-thermal effect [162]. The authors of [139] reviewed recently, in detail, this subject. The accurate distinguishing between thermal and non-thermal microwaves on the microbial inactivation is a complicated task. The same heating profile (time-temperature history) needs to be applied under microwave and water bath heating conditions, which is not practically possible due to different heating principles between these technologies. Furthermore, the heating homogeneity needs to be perfect in all points of the sample either with conventional or MW heating. Following these considerations, the existence of the non-thermal effect of microwaves on microorganisms has not been confirmed; even if this effect exists, it has no significant consequence on the microwave pasteurization and sterilization efficiency [139, 163].

Microwave pasteurization and sterilization processes have been successfully adopted at an industrial success in recent years for heating ready-to-eat meals, fresh pasta, and milk, for example [137, 138]. However, non-uniform heating is still an engineering challenge. The presence of cold spots, mainly in solid products, is a serious safety issue [164, 165] and the main obstacle to the wide application of microwave decontamination process in food industries. Several studies reported the survival of pathogens such as *Salmonella spp*, *Staphylococcus aureus*, *L. monocytogenes* in food heated in microwave ovens [166]. Also, non-uniform heating of microwaves may lead to quality degradation. The authors of [167] studied the microwave decontamination of *E. coli* and *Campylobacter jejuni* inoculated on chicken meat. The authors reported a temperature variation up to 20°C between the different parts of the sample accompanied by partial cooking of hot parts and the survival of microorganisms.

Several solutions have been proposed to reduce the temperature non-uniformity and to design reliable microwave pasteurization and sterilization process. The combination of microwave heating with other food processing techniques has a lot of interest. Combination of microwave heating with a water bath, steam, infrared, air-jet, and ultraviolet radiation heating [28] decreases the non-uniformity, ensures required decontamination, and enhances the quality of treated products. Other methods and systems have been also studied such as Microwave-Circulated Water Combination (MCWC) heating system [168], sample rotation device to rotate the product during treatment [150], and pulsed microwave heating technique [169]. Microwave-assisted thermal sterilization (MATS) technology is an industrial system, developed originally by Washington State University for the sterilization of pre-packaged homogeneous and nonhomogeneous foods. This system combines the energy from microwaves of long-wavelength (915 MHz) along with hot water immersion. This technology has been approved by FDA. MATS system has several advantages including relatively short treatment time, elimination of the edge-heating issues, flexibility in the packaging used, reducing the effect on sensory properties, and improving nutrient retention [170, 171].

Although MW heating has been efficiently used for food decontamination, it is necessary to strengthen further studies to solve the temperature heterogeneity and to ensure the safety of treated products.

References

1. Sun, D.-W. *Thermal food processing: new technologies and quality issues.*; 2016; ISBN 978-1-138-19963-7.
2. Chauhan, O.P. *Non-thermal processing of foods*; CRC Press: Boca Raton, 2019; ISBN 978-1-138-03584-3.
3. *The microwave processing of foods*; Regier, M., Knoerzer, K., Schubert, H., Eds.; Woodhead Publishing Series in Food Science, Technology and Nutrition; Second edition.; Woodhead Publishing: Duxford, United Kingdom, 2017; ISBN 978-0-08-100528-6.
4. Metaxas, A.C. *Foundations of electroheat: a unified approach*; Wiley: Chichester, 1996; ISBN 978-0-471-95644-0.
5. Nelson, S.O.; Datta, A.K. Dielectric properties of food materials and electric field interactions. In *The Handbook of Microwave Technology for Food Applications*; Datta, A.K., Anantheswaran, R.C., Eds.; New York, 2001; pp. 69–114.
6. Abhayawick, L.; Laguerre, J.C.; Tauzin, V.; Duquenoy, A. Physical properties of three onion varieties as affected by the moisture content. *Journal of Food Engineering* **2002**, *55*, 253–262, doi:10.1016/S0260-8774(02)00099-7.
7. Ayappa, K.G.; Davis, H.T.; Crapiste, G.; Davis, E.A.; Gordon, J. Microwave heating: an evaluation of power formulations. *Chemical Engineering Science* **1991**, *46*, 1005–1016, doi:10.1016/0009-2509(91)85093-D.

8. Yolacaner, E.T.; Sumnu, G.; Sahin, S. Microwave assisted baking. In *Microwave processing of foods*; Regier, M., Knoerzer, K., Schubert, Helmar, Eds.; Woodhead Publishing Series in Food Science, Technology and Nutrition; Woodhead Publishing, 2017; pp. 117–141.
9. Ahmad, S.S.; Morgan, M.T.; Okos, M.R. Effects of microwave on the drying, checking and mechanical strength of baked biscuits. *Journal of Food Engineering* **2001**, *50*, 63–75, doi:10.1016/S0260-8774(00)00186-2.
10. Arepally, D.; Reddy, R.S.; Goswami, T.K.; Datta, A.K. Biscuit baking: A review. *LWT* **2020**, *131*, 109726, doi:10.1016/j.lwt.2020.109726.
11. Garg, A.; Malafronte, L.; Windhab, E.J. Baking kinetics of laminated dough using convective and microwave heating. *Food and Bioproducts Processing* **2019**, *115*, 59–67, doi:10.1016/j.fbp.2019.02.007.
12. Sumnu, G. A review on microwave baking of foods. *International Journal of Food Science & Technology* **2001**, *36*, 117–127, doi:10.1046/j.1365-2621.2001.00479.x.
13. Sánchez-Pardo, M.E.; Ortiz-Moreno, A.; Mora-Escobedo, R.; Chanona-Pérez, J.J.; Necoechea-Mondragón, H. Comparison of crumb microstructure from pound cakes baked in a microwave or conventional oven. *LWT - Food Science and Technology* **2008**, *41*, 620–627, doi:10.1016/j.lwt.2007.05.003.
14. Sánchez-Pardo, M.E.; Ortiz-Moreno, A.; García-Zaragoza, F.J.; Necoechea-Mondragón, H.; Chanona-Pérez, J.J. Comparison of pound cake baked in a two cycle microwave-toaster oven and in conventional oven. *LWT - Food Science and Technology* **2012**, *46*, 356–362, doi:10.1016/j.lwt.2011.08.013.
15. Sumnu, G.; Sahin, S.; Sevimli, M. Microwave, infrared and infrared-microwave combination baking of cakes. *Journal of Food Engineering* **2005**, *71*, 150–155, doi:10.1016/j.jfoodeng.2004.10.027.
16. Palav, T.; Seetharaman, K. Mechanism of starch gelatinization and polymer leaching during microwave heating. *Carbohydrate Polymers* **2006**, *65*, 364–370, doi:10.1016/j.carbpol.2006.01.024.
17. Sakiyan, O.; Sumnu, G.; Sahin, S.; Meda, V.; Koksel, H.; Chang, P. A study on degree of starch gelatinization in cakesbaked in three different ovens. *Food Bioprocess Technol* **2011**, *4*, 1237–1244, doi:http://dx.doi.org/10.1007/s11947-009-0210-2.
18. Patel, B.K.; Waniska, R.D.; Seetharaman, K. Impact of different baking processes on bread firmness and starch properties in breadcrumb. *Journal of Cereal Science* **2005**, *42*, 173–184, doi:10.1016/j.jcs.2005.04.007.
19. Seyhun, N.; Sumnu, G.; Sahin, S. Effects of different starch types on retardation of staling of microwave-baked Cakes. *Food and Bioproducts Processing* **2005**, *83*, 1–5, doi:10.1205/fbp.04041.
20. Chavan, R.S.; Chavan, S.R. Microwave baking in food industry: A review. *International J. of Dairy Science* **2010**, *5*, 113–127, doi:10.3923/ijds.2010.113.127.
21. Schiffmann, R.F. Critical factors in microwave genetrated aromas. In *Thermally generated flavors: Maillard, microwave and extrusion processes*;

McGorrin, J., Parlimant, T.H., Morello, M.J., Eds.; ACS symposium series; American Chemical Society: Washington D.C., USA, 1994; pp. 386–394.

22. İçöz, D.; Sumnu, G.; Sahin, S. Color and texture development during microwave and conventional baking of breads. *International Journal of Food Properties* **2004**, *7*, 201–213, doi:10.1081/JFP-120025396.
23. Keskin, S.O.; Sumnu, G.; Sahin, S. Usage of enzymes in a novel baking process. *Nahrung* **2004**, *48*, 156–160, doi:10.1002/food.200300412.
24. Ozmutlu, O.; Sumnu, G.; Sahin, S. Effects of different formulations on the quality of microwave baked breads. *European Food Research and Technology* **2001**, *213*, 38–42.
25. Clarke, C.I.; Farrell, G.M. The effects of recipe formulation on the textural characteristics of microwave-reheated pizza bases. *Journal of the Science of Food and Agriculture* **2000**, *80*, 1237–1244.
26. Mast, R.L. Application of microwave technology to bakery products.; 2000.
27. Torrealba-Meléndez, R.; Sosa-Morales, M.E.; Olvera-Cervantes, J.L.; Corona-Chávez, A. Dielectric properties of cereals at frequencies useful for processes with microwave heating. *J Food Sci Technol* **2015**, *52*, 8403–8409, doi:10.1007/s13197-015-1948-3.
28. Datta, A.K.; Rakesh, V. Principles of microwave combination heating: Principles of microwave combination heating. *Comprehensive Reviews in Food Science and Food Safety* **2013**, *12*, 24–39, doi:10.1111/j.1541-4337.2012.00211.x.
29. Ozge Keskin, S.; Sumnu, G.; Sahin, S. Bread baking in halogen lamp-microwave combination oven. *Food Research International* **2004**, *37*, 489–495, doi:10.1016/j.foodres.2003.10.001.
30. Ohlsson, T.; Bengtsson, N. Minimal processing of foods with nonthermal methods. In *inimal Processing Technologies in the Food Industry*; Ohlsson, T., Bengtsson, N., Eds.; Woodhead Publishing: Cambridge, 2002; pp. 34–60.
31. Ozge Ozkoc, S.; Sumnu, G.; Sahin, S. The effects of gums on macro and micro-structure of breads baked in different ovens. *Food Hydrocolloids* **2009**, *23*, 2182–2189, doi:10.1016/j.foodhyd.2009.04.003.
32. Sakiyan, O.; Sumnu, G.; Sahin, S.; Meda, V. The effect of different formulations on physical properties of cakes baked with microwave and near infrared-microwave combinations. *J Microw Power Electromagn Energy* **2007**, *41*, 20–26, doi:10.1080/08327823.2006.11688551.
33. Elliasson, A.C. Starch gelatinization in the presence of emulsifiers: A morphological study of wheat starch. *Starch* **1985**, *37*, 411–415.
34. Xiao, H.-W.; Pan, Z.; Deng, L.-Z.; El-Mashad, H.M.; Yang, X.-H.; Mujumdar, A.S.; Gao, Z.-J.; Zhang, Q. Recent developments and trends in thermal blanching – A comprehensive review. *Information Processing in Agriculture* **2017**, *4*, 101–127, doi:10.1016/j.inpa.2017.02.001.
35. Munyaka, A.W.; Makule, E.E.; Oey, I.; Van Loey, A.; Hendrickx, M. Thermal Stability of L-Ascorbic Acid and Ascorbic Acid Oxidase in Broccoli (Brassica oleracea var. italica). *Journal of Food Science* **2010**, *75*, 336–340.

36. Xanthakis, E.; Valdramidis, V.P. Impact of heating operations on the microbial ecology of foods. In *Modeling the microbial ecology of foods: Quantitative microbiology in food processing*; Sant'Ana, A., Ed.; John Wiley & Sons, Ltd: New Jersey, 2017; pp. 117–142.
37. Xin, Y.; Zhang, M.; Xu, B.; Adhikari, B.; Sun, J. Research trends in selected blanching pretreatments and quick freezing technologies as applied in fruits and vegetables: A review. *International Journal of Refrigeration* **2015**, *57*, 11–25, doi:10.1016/j.ijrefrig.2015.04.015.
38. Bahceci, K.S.; Serpen, A.; Gokmen, V.; Acar, J. Study of lipoxygenase and peroxidase as indicator enzymes in green beans: change of enzyme activity, ascorbic acid and chlorophylls during frozen storage. *Journal of Food Engineering* **2005**, *66*, 187–192.
39. Ranjan, S.; Dasgupta, N.; Walia, N.; Thara Chand, C.; Ramalingam, C. Microwave blanching: An emerging trend in food engineering and its effects on Capsicum annuum L. *Journal of Food Process Engineering* **2017**, *40*, e12411, doi:10.1111/jfpe.12411.
40. Rahman, M.S. *Handbook of Food Preservation*; 2nd ed.; CRC Press: New York, 2007.
41. Nurhuda, H.H.; Maskat, M.Y.; Mamot, S.; Afiq, J.; Aminah, A. Effect of blanching on enzyme and antioxidant activities of rambutan (Nephelium lappaceum) peel. *International Food Research Journal* **2013**, *20*, 1725–1730.
42. Severini, C.; Giuliani, R.; De Filippis, A.; Derossi, A.; De Pilli, T. Influence of different blanching methods on colour, ascorbic acid and phenolics content of broccoli. *J Food Sci Technol* **2016**, *53*, 501–510, doi:10.1007/s13197-015-1878-0.
43. Jaworska, G.; Bernaś, E.; Biernacka, A.; Maciejaszek, I. Comparison of the texture of fresh and preserved Agaricus bisporus and Boletus edulis mushrooms: Texture of fresh and preserved mushrooms. *International Journal of Food Science & Technology* **2010**, *45*, 1659–1665, doi:10.1111/j.1365-2621.2010.02319.x.
44. Gonçalves, E.M.; Cruz, R.M.S.; Abreu, M.; Brandão, T.R.S.; Silva, C.L.M. Biochemical and colour changes of watercress (Nasturtium officinale R. Br.) during freezing and frozen storage. *Journal of Food Engineering* **2009**, *93*, 32–39, doi:10.1016/j.jfoodeng.2008.12.027.
45. Volden, J.; Borge, G.I.A.; Hansen, M.; Wicklund, T.; Bengtsson, G.B. Processing (blanching, boiling, steaming) effects on the content of glucosinolates and antioxidant-related parameters in cauliflower (Brassica oleracea L. ssp. botrytis). *LWT - Food Science and Technology* **2009**, *42*, 63–73, doi:10.1016/j.lwt.2008.05.018.
46. Haase, N.U.; Weber, L. Ascorbic acid losses during processing of French fries and potato chips. *Journal of Food Engineering* **2003**, *56*, 207–209, doi:10.1016/S0260-8774(02)00252-2.

47. Lisiewska, Z.; Kmiecik, W. Effects of level of nitrogen fertilizer, processing conditions and period of storage of frozen broccoli and cauliflower on vitamin C retention. *Food Chemistry* **1996**, *57*, 267–270.
48. Kinalski, T.; Noreña, C.P.Z. Effect of blanching treatments on antioxidant activity and thiosulfinate degradation of garlic (Allium sativum L.). *Food Bioprocess Technol* **2014**, *7*, 2152–2157, doi:10.1007/s11947-014-1282-1.
49. Dorantes-Alvarez, L.; Ortiz-Moreno, A.; Guzmán-Gerónimo, R.; Parada-Dorantes, L. Microwave-assisted blanching. In *The Microwave Processing of Foods*; Regier, M., Purnell, G., Schubert, H., Eds.; Elsevier, 2017; pp. 179–199 ISBN 978-0-08-100528-6.
50. Xanthakis, E.; Gogou, E.; Taoukis, P.; Ahrné, L. Effect of microwave assisted blanching on the ascorbic acid oxidase inactivation and vitamin C degradation in frozen mangoes. *Innovative Food Science & Emerging Technologies* **2018**, *48*, 248–257, doi:10.1016/j.ifset.2018.06.012.
51. Liburdi, K.; Benucci, I.; Esti, M. Effect of microwave power and blanching time in relation to different geometric shapes of vegetables. *LWT* **2019**, *99*, 497–504, doi:10.1016/j.lwt.2018.10.029.
52. Wang, J.; Yang, X.-H.; Mujumdar, A.S.; Wang, D.; Zhao, J.-H.; Fang, X.-M.; Zhang, Q.; Xie, L.; Gao, Z.-J.; Xiao, H.-W. Effects of various blanching methods on weight loss, enzymes inactivation, phytochemical contents, antioxidant capacity, ultrastructure and drying kinetics of red bell pepper (Capsicum annuum L.). *LWT* **2017**, *77*, 337–347, doi:10.1016/j.lwt.2016.11.070.
53. Lin, S.; Brewer, M.S. Effects of blanching method on the quality characteristics of frozen peas. *Journal of Food Quality* **2005**, *28*, 350–360.
54. Ramesh, M.N.; Wolf, W.; Tevini, D.; Bognar, A. Microwave blanching of vegetables. *J Food Science* **2002**, *67*, 390–398, doi:10.1111/j.1365-2621.2002.tb11416.x.
55. Ruiz-Ojeda, L.M.; Peñas, F.J. Comparison study of conventional hot-water and microwave blanching on quality of green beans. *Innovative Food Science & Emerging Technologies* **2013**, *20*, 191–197, doi:10.1016/j.ifset.2013.09.009.
56. Devece, C.; Rodríguez-López, J.N.; Fenoll, L.G.; Tudela, J.; Catalá, J.M.; de Los Reyes, E.; García-Cánovas, F. Enzyme inactivation analysis for industrial blanching applications: comparison of microwave, conventional, and combination heat treatments on mushroom polyphenoloxidase activity. *J Agric Food Chem* **1999**, *47*, 4506–4511, doi:10.1021/jf981398+.
57. Lemmens, L.; Tibäck, E.; Svelander, C.; Smout, C.; Ahrné, L.; Langton, M.; Alminger, M.; Van Loey, A.; Hendrickx, M. Thermal pretreatments of carrot pieces using different heating techniques: Effect on quality related aspects. *Innovative Food Science & Emerging Technologies* **2009**, *10*, 522–529, doi:10.1016/j.ifset.2009.05.004.
58. Sun, T.; Tang, J.; Powers, J.R. Antioxidant activity and quality of asparagus affected by microwave-circulated water combination and conventional sterilization. *Food Chemistry* **2007**, *100*, 813–819, doi:10.1016/j.foodchem.2005.10.047.

59. Feumba Dibanda, R.; Panyoo Akdowa, E.; Rani P., A.; Metsatedem Tongwa, Q.; Mbofung F., C.M. Effect of microwave blanching on antioxidant activity, phenolic compounds and browning behaviour of some fruit peelings. *Food Chemistry* **2020**, *302*, 125308, doi:10.1016/j.foodchem.2019.125308.
60. Jeevitha, G.C.; Umesh Hebbar, H.; Raghavarao, K.S.M.S. Electromagnetic radiation-based dry blanching of red bell peppers: A comparative study. *Journal of Food Process Engineering* **2013**, *36*, 663–674.
61. Dorantes-Alvarez, L.; Jaramillo-Flores, E.; González, K.; Martinez, R.; Parada, L. Blanching peppers using microwaves. *Procedia Food Science* **2011**, *1*, 178–183, doi:10.1016/j.profoo.2011.09.028.
62. Başkaya Sezer, D.; Demirdöven, A. The effects of microwave blanching conditions on carrot slices: optimization and comparison: Effects of microwave blanching on carrot slices. *Journal of Food Processing and Preservation* **2015**, *39*, 2188–2196, doi:10.1111/jfpp.12463.
63. Benlloch-Tinoco, M.; Igual, M.; Rodrigo, D.; Martínez-Navarrete, N. Superiority of microwaves over conventional heating to preserve shelf-life and quality of kiwifruit puree. *Food Control* **2015**, *50*, 620–629, doi:10.1016/j.foodcont.2014.10.006.
64. Latorre, M.E.; de Escalada Plá, M.F.; Rojas, A.M.; Gerschenson, L.N. Blanching of red beet (Beta vulgaris L. var. conditiva) root. Effect of hot water or microwave radiation on cell wall characteristics. *LWT - Food Science and Technology* **2013**, *50*, 193–203, doi:10.1016/j.lwt.2012.06.004.
65. Koskiniemi, C.B.; Truong, V.-D.; Simunovic, J.; McFeeters, R.F. Improvement of heating uniformity in packaged acidified vegetables pasteurized with a 915MHz continuous microwave system. *Journal of Food Engineering* **2011**, *105*, 149–160, doi:10.1016/j.jfoodeng.2011.02.019.
66. Kidmose, U.; Martens, H.J. Changes in texture, microstructure and nutritional quality of carrot slices during blanching and freezing. *Journal of the Science of Food and Agriculture* **1999**, *79*, 1747–1753.
67. James, S.J.; James, C.; Purnell, G. Microwave-assisted thawing and tempering. In *The Microwave Processing of Foods*; Elsevier, 2017; pp. 252–272 ISBN 978-0-08-100528-6.
68. Verma, D.K.; Mahanti, N.K.; Thakur, M.; Chakraborty, S.K.; Srivastav, P.P. Microwave Heating: Alternative Thermal Process Technology for Food Application. In *Emerging Thermal and Nonthermal Technologies in Food Processing*; Srivastav, P.P., Verma, D.K., Patel, A.R., Al-Hilphy, A.R., Eds.; Apple Academic Press: Includes bibliographical references and index., 2020; pp. 25–67 ISBN 978-0-429-29733-5.
69. Chen, Y.; He, J.; Li, F.; Tang, J.; Jiao, Y. Model food development for tuna (Thunnus Obesus) in radio frequency and microwave tempering using grass carp mince. *Journal of Food Engineering* **2021**, *292*, 110267, doi:10.1016/j.jfoodeng.2020.110267.
70. Wang, B.; Du, X.; Kong, B.; Liu, Q.; Li, F.; Pan, N.; Xia, X.; Zhang, D. Effect of ultrasound thawing, vacuum thawing, and microwave thawing on

gelling properties of protein from porcine longissimus dorsi. *Ultrasonics Sonochemistry* **2020**, *64*, 104860, doi:10.1016/j.ultsonch.2019.104860.
71. Zhao, Y.; Gao, R.; Zhuang, W.; Xiao, J.; Zheng, B.; Tian, Y. Combined single-stage tempering and microwave vacuum drying of the edible mushroom Agrocybe chaxingu: Effects on drying characteristics and physical-chemical qualities. *LWT* **2020**, *128*, 109372, doi:10.1016/j.lwt.2020.109372.
72. Oliveira, M.; Gubert, G.; Roman, S.; Kempka, A.; Prestes, R. Meat quality of chicken breast subjected to different thawing methods. *Rev. Bras. Cienc. Avic.* **2015**, *17*, 165–171, doi:10.1590/1516-635x1702165-172.
73. Choi, E.J.; Park, H.W.; Chung, Y.B.; Park, S.H.; Kim, J.S.; Chun, H.H. Effect of tempering methods on quality changes of pork loin frozen by cryogenic immersion. *Meat Science* **2017**, *124*, 69–76, doi:10.1016/j.meatsci.2016.11.003.
74. Seyhun, N.; Ramaswamy, H.; Sumnu, G.; Sahin, S.; Ahmed, J. Comparison and modeling of microwave tempering and infrared assisted microwave tempering of frozen potato puree. *Journal of Food Engineering* **2009**, *92*, 339–344, doi:10.1016/j.jfoodeng.2008.12.003.
75. Li, B.; Sun, D.-W. Novel methods for rapid freezing and thawing of foods – a review. *Journal of Food Engineering* **2002**, *54*, 175–182, doi:10.1016/S0260-8774(01)00209-6.
76. Di Rosa, A.R.; Bressan, F.; Leone, F.; Falqui, L.; Chiofalo, V. Radio frequency heating on food of animal origin: a review. *Eur Food Res Technol* **2019**, *245*, 1787–1797, doi:10.1007/s00217-019-03319-8.
77. Jasim, A.; Ramaswamy, H.S. Microwave pasteurization and sterilization of foods. In *Handbook of food preservation*; Rahman, M.S., Ed.; CRC Press: Boca Raton, FL, USA, 2007; pp. 691–711.
78. Fik, M.; Macura Quality changes during frozen storage and thawing of mixed bread. *Food / Nahrung* **2001**, *45*, 138–142.
79. Holzwarth, M.; Korhummel, S.; Carle, R.; Kammerer, D.R. Evaluation of the effects of different freezing and thawing methods on color, polyphenol and ascorbic acid retention in strawberries (Fragaria×ananassa Duch.). *Food Research International* **2012**, *48*, 241–248, doi:10.1016/j.foodres.2012.04.004.
80. Xin Wen; Rui Hu; Jin-Hong Zhao; Yu Peng; Yuan-Ying Ni Evaluation of the effects of different thawing methods on texture, colour and ascorbic acid retention of frozen hami melon (Cucumis melo var. saccharinus). *Int J Food Sci Technol* **2015**, *50*, 1116–1122, doi:10.1111/ijfs.12755.
81. Zhang, R.; Wang, Y.; Wang, X.; Luan, D. Study of heating characteristics for a continuous 915 MHz pilot scale microwave thawing system. *Food Control* **2019**, *104*, 105–114, doi:10.1016/j.foodcont.2019.04.030.
82. Sosa-Morales, M.E.; Valerio-Junco, L.; López-Malo, A.; García, H.S. Dielectric properties of foods: Reported data in the 21st Century and their potential applications. *LWT - Food Science and Technology* **2010**, *43*, 1169–1179, doi:10.1016/j.lwt.2010.03.017.

83. Miran, W.; Palazoğlu, T.K. Development and experimental validation of a multiphysics model for 915 MHz microwave tempering of frozen food rotating on a turntable. *Biosystems Engineering* **2019**, *180*, 191–203, doi:10.1016/j.biosystemseng.2019.02.008.
84. Curet, S.; Rouaud, O.; Boillereaux, L. Microwave tempering and heating in a single-mode cavity: Numerical and experimental investigations. *Chemical Engineering and Processing: Process Intensification* **2008**, *47*, 1656–1665, doi:10.1016/j.cep.2007.09.011.
85. Koray Palazoğlu, T.; Miran, W. Experimental comparison of microwave and radio frequency tempering of frozen block of shrimp. *Innovative Food Science & Emerging Technologies* **2017**, *41*, 292–300, doi:10.1016/j.ifset.2017.04.005.
86. Virtanen, A.J.; Goedeken, D.L.; Tong, C.H. Microwave assisted thawing of model frozen foods using feed-back temperature control and surface cooling. *J Food Science* **1997**, *62*, 150–154, doi:10.1111/j.1365-2621.1997.tb04388.x.
87. Nijhuis, H.H.; Torringa, H.M.; Muresan, S.; Yuksel, D.; Kloek, W. Approaches to improving the quality of dried fruit and vegetables. *Trends in Food Science and Technology* **1998**, *9*, 13–20.
88. Zhang, M.; Tang, J.; Mujumdar, A.S.; Wang, S. Trends in microwave-related drying of fruits and vegetables. *Trends in Food Science & Technology* **2006**, *17*, 524–534, doi:10.1016/j.tifs.2006.04.011.
89. Sumnu, G.; Sahin, S. Recent developments in microwave heating. In *Emerging Technologies for Food Processing*; Sun, D.-W., Ed.; Academic Press: London, 2005; pp. 419–444 ISBN 978-0-12-676757-5.
90. Witrowa-Rajchert, D.; Wiktor, A.; Sledz, M.; Nowacka, M. Selected emerging technologies to enhance the drying process: A review. *Drying Technology* **2014**, *32*, 1386–1396, doi:10.1080/07373937.2014.903412.
91. Cullen, P.J.; Tiwari, B.K.; Valdramidis, V.P. Status and trends of novel thermal and non-thermal technologies for fluid foods. In *Novel thermal and non-thermal technologies for fluid foods*; Cullen, P.J., Tiwari, B.K., Valdramidis, V.P., Eds.; Academic Press: San Diego, 2012; pp. 1–6 ISBN 978-0-12-381470-8.
92. Gaukel, V.; Siebert, T.; Erle, U. Microwave-assisted drying. In *The Microwave Processing of Foods*; Regier, M., Knoerzer, K., Schubert, H., Eds.; Elsevier, 2017; pp. 152–178 ISBN 978-0-08-100528-6.
93. Zielinska, M.; Ropelewska, E.; Xiao, H.-W.; Mujumdar, A.S.; Law, C.L. Review of recent applications and research progress in hybrid and combined microwave-assisted drying of food products: Quality properties. *Critical Reviews in Food Science and Nutrition* **2019**, 1–53, doi:10.1080/10408398.2019.1632788.
94. Li, Y.; Lei, Y.; Zhang, L.; Peng, J.; Li, C. Microwave drying characteristics and kinetics of ilmenite. *Transactions of Nonferrous Metals Society of China* **2011**, *21*, 202–207, doi:10.1016/S1003-6326(11)60700-0.
95. Laguerre, J.-C.; Mazen Hamoud-Agha, M. Microwave heating for food preservation. In *Food Preservation and Waste Exploitation*; A. Socaci, S.,

C. Fărcaş, A., Aussenac, T., Laguerre, J.-C., Eds.; IntechOpen, 2020 ISBN 978-1-78985-425-1.

96. Zhao, G.; Zhang, R.; Liu, L.; Deng, Y.; Wei, Z.; Zhang, Y.; Ma, Y.; Zhang, M. Different thermal drying methods affect the phenolic profiles, their bioaccessibility and antioxidant activity in Rhodomyrtus tomentosa (Ait.) Hassk berries. *LWT - Food Science and Technology* **2017**, *79*, 260–266, doi:10.1016/j.lwt.2017.01.039.
97. Aydogdu, A.; Sumnu, G.; Sahin, S. Effects of microwave-infrared combination drying on quality of eggplants. *Food Bioprocess Technol* **2015**, *8*, 1198–1210, doi:10.1007/s11947-015-1484-1.
98. Horuz, E.; Bozkurt, H.; Karataş, H.; Maskan, M. Effects of hybrid (microwave-convectional) and convectional drying on drying kinetics, total phenolics, antioxidant capacity, vitamin C, color and rehydration capacity of sour cherries. *Food Chemistry* **2017**, *230*, 295–305, doi:10.1016/j.foodchem.2017.03.046.
99. Sarimeseli, A. Microwave drying characteristics of coriander (Coriandrum sativum L.) leaves. *Energy Conversion and Management* **2011**, *52*, 1449–1453, doi:10.1016/j.enconman.2010.10.007.
100. Divya, P.; Puthusseri, B.; Neelwarne, B. Carotenoid content, its stability during drying and the antioxidant activity of commercial coriander (Coriandrum sativum L.) varieties. *Food Research International* **2012**, *45*, 342–350, doi:10.1016/j.foodres.2011.09.021.
101. Abhayawick, L.; Laguerre, J.-C.; Duquenoy, A. Runaway heating of onions during microwave drying. In *IDS 2000—12th International Drying Symposium*; Kerkhof, P.J.A.M., Coumans, W.J., Mooiwer, G.D., Eds.; Noordwijkerhout, 2000.
102. Guo, Q.; Sun, D.-W.; Cheng, J.-H.; Han, Z. Microwave processing techniques and their recent applications in the food industry. *Trends in Food Science & Technology* **2017**, *67*, 236–247, doi:10.1016/j.tifs.2017.07.007.
103. Onwude, D.I.; Hashim, N.; Chen, G. Recent advances of novel thermal combined hot air drying of agricultural crops. *Trends in Food Science & Technology* **2016**, *57*, 132–145, doi:10.1016/j.tifs.2016.09.012.
104. Moses, J.A.; Norton, T.; Alagusundaram, K.; Tiwari, B.K. Novel drying techniques for the food industry. *Food Engineering Reviews* **2014**, *6*, 43–55.
105. Amiri Chayjan, R.; Kaveh, M.; Khayati, S. Modeling drying characteristics of hawthorn fruit under microwave-convective conditions. *Journal of Food Processing and Preservation* **2015**, *39*, 239–253, doi:10.1111/jfpp.12226.
106. Wray, D.; Ramaswamy, H.S. Novel concepts in microwave drying of foods. *Drying Technology* **2015**, *33*, 769–783, doi:10.1080/07373937.2014.985793.
107. Funebo, T.; Ohlsson, T. Microwave-assisted air dehydration of apple and mushroom. *Journal of Food Engineering* **1998**, *38*, 353–367, doi:10.1016/S0260-8774(98)00131-9.
108. Beaudry, C.; Raghavan, G.S.V.; Rennie, T.J. Microwave finish drying of osmotically dehydrated cranberries. *Drying Technology* **2003**, *21*, 1797–1810.

109. Silva, F.A.; Marsaioli, A.; Maximo, G.J.; Silva, M.A.A.P.; Gonçalves, L.A.G. Microwave assisted drying of macadamia nuts. *Journal of Food Engineering* **2006**, *77*, 550–558, doi:10.1016/j.jfoodeng.2005.06.068.
110. Jiao, A.; Xu, X.; Jin, Z. Modelling of dehydration–rehydration of instant rice in combined microwave-hot air drying. *Food and Bioproducts Processing* **2014**, *92*, 259–265, doi:10.1016/j.fbp.2013.08.002.
111. Gowen, A.A.; Abu-Ghannam, N.; Frias, J.; Oliveira, J. Modeling dehydration and rehydration of cooked soybeans subjected to combined microwave–hot-air drying. *Innovative Food Science & Emerging Technologies* **2008**, *9*, 129–137, doi:10.1016/j.ifset.2007.06.009.
112. Maskan, M. Microwave/air and microwave finish drying of banana. *Journal of Food Engineering* **2000**, *44*, 71–78.
113. Sharma, G.P.; Prasad, S. Optimization of process parameters for microwave drying of garlic cloves. *Journal of Food Engineering* **2006**, *75*, 441–446, doi:10.1016/j.jfoodeng.2005.04.029.
114. Maskan, M. Kinetics of colour change of kiwifruits during hot air and microwave drying. *Journal of Food Engineering* **2001**, *48*, 169–175, doi:10.1016/S0260-8774(00)00154-0.
115. Chandrasekaran, S.; Ramanathan, S.; Basak, T. Microwave food processing—A review. *Food Research International* **2013**, *52*, 243–261, doi:10.1016/j.foodres.2013.02.033.
116. Schulze, B.; Hubbermann, E.M.; Schwarz, K. Stability of quercetin derivatives in vacuum impregnated apple slices after drying (microwave vacuum drying, air drying, freeze drying) and storage. *LWT - Food Science and Technology* **2014**, *57*, 426–433, doi:10.1016/j.lwt.2013.11.021.
117. Zielinska, M.; Ropelewska, E.; Markowski, M. Thermophysical properties of raw, hot-air and microwave-vacuum dried cranberry fruits (Vaccinium macrocarpon). *LWT - Food Science and Technology* **2017**, *85*, 204–211, doi:10.1016/j.lwt.2017.07.016.
118. Monteiro, R.L.; Link, J.V.; Tribuzi, G.; Carciofi, B.A.M.; Laurindo, J.B. Microwave vacuum drying and multi-flash drying of pumpkin slices. *Journal of Food Engineering* **2018**, *232*, 1–10, doi:10.1016/j.jfoodeng.2018.03.015.
119. Bondaruk, J.; Markowski, M.; Błaszczak, W. Effect of drying conditions on the quality of vacuum-microwave dried potato cubes. *Journal of Food Engineering* **2007**, *81*, 306–312, doi:10.1016/j.jfoodeng.2006.10.028.
120. Song, X.; Zhang, M.; Mujumdar, A.S.; Fan, L. Drying characteristics and kinetics of vacuum microwave–dried potato slices. *Drying Technology* **2009**, *27*, 969–974, doi:10.1080/07373930902902099.
121. Hirun, S.; Utama-ang, N.; Vuong, Q.V.; Scarlett, C.J. Investigating the commercial microwave vacuum drying conditions on physicochemical properties and radical scavenging ability of Thai green Tea. *Drying Technology* **2014**, *32*, 47–54, doi:10.1080/07373937.2013.811249.

122. Therdthai, N.; Zhou, W. Characterization of microwave vacuum drying and hot air drying of mint leaves (Mentha cordifolia Opiz ex Fresen). *Journal of Food Engineering* **2009**, *91*, 482–489, doi:10.1016/j.jfoodeng.2008.09.031.
123. Drouzas, A.E.; Schubert, H. Microwave application in vacuum drying of fruits. *Journal of Food Engineering* **1996**, *28*, 203–209, doi:10.1016/0260-8774(95)00040-2.
124. Böhm, V.; Kühnert, S.; Rohm, H.; Scholze, G. Improving the nutritional quality of microwave-vacuum dried strawberries: A preliminary study. *Food sci. technol. int.* **2006**, *12*, 67–75, doi:10.1177/1082013206062136.
125. Azarpazhooh, E.; Ramaswamy, H.S. Microwave-osmotic dehydration of apples under continuous flow medium spray conditions: Comparison with other methods. *Drying Technology* **2009**, *28*, 49–56, doi:10.1080/07373930903430611.
126. Zhang, M.; Jiang, H.; Lim, R.-X. Recent developments in microwave-assisted drying of vegetables, fruits, and aquatic products—Drying kinetics and quality considerations. *Drying Technology* **2010**, *28*, 1307–1316, doi:10.1080/07373937.2010.524591.
127. Wang, R.; Zhang, M.; Mujumdar, A.S.; Sun, J.-C. Microwave freeze–drying characteristics and sensory quality of instant vegetable soup. *Drying Technology* **2009**, *27*, 962–968, doi:10.1080/07373930902902040.
128. Jiang, H.; Zhang, M.; Liu, Y.; Mujumdar, A.S.; Liu, H. The energy consumption and color analysis of freeze/microwave freeze banana chips. *Food and Bioproducts Processing* **2013**, *91*, 464–472, doi:10.1016/j.fbp.2013.04.004.
129. Duan, X.; Zhang, M.; Mujumdar, A.S. Studies on the microwave freeze drying technique and sterilization characteristics of cabbage. *Drying Technology* **2007**, *25*, 1725–1731, doi:10.1080/07373930701591044.
130. Abbasi, S.; Azari, S. Novel microwave–freeze drying of onion slices. *International Journal of Food Science and Technology* **2009**, *44*, 974–979.
131. Wang, R.; Zhang, M.; Mujumdar, A.S. Effects of vacuum and microwave freeze drying on microstructure and quality of potato slices. *Journal of Food Engineering* **2010**, *101*, 131–139, doi:10.1016/j.jfoodeng.2010.05.021.
132. Wang, Z.H.; Shi, M.H. Microwave freeze drying characteristics of beef. *Drying Technology* **1999**, *17*, 434–447, doi:10.1080/07373939908917544.
133. Wang, R.; Zhang, M.; Mujumdar, A.S. Effect of food ingredient on microwave freeze drying of instant vegetable soup. *LWT - Food Science and Technology* **2010**, *43*, 1144–1150, doi:10.1016/j.lwt.2010.03.007.
134. Lombraña, J.I.; Rodríguez, R.; Ruiz, U. Microwave-drying of sliced mushroom. Analysis of temperature control and pressure. *Innovative Food Science & Emerging Technologies* **2010**, *11*, 652–660, doi:10.1016/j.ifset.2010.06.007.
135. Barbosa-Cánovas, G.V.; Medina-Meza, I.; Candoğan, K.; Bermúdez-Aguirre, D. Advanced retorting, microwave assisted thermal sterilization (MATS), and pressure assisted thermal sterilization (PATS) to process meat products. *Meat Science* **2014**, *98*, 420–434, doi:10.1016/j.meatsci.2014.06.027.

136. Bhushand, D.M.; Vyawarea, A.N.; Wasnik, P.G.; Agrawal, A.K.; Sandey, K.K. Microwave processing of milk: A rewiew. In *Processing technologies for milk and milk products: Methods, applications, and energy usage*; Agrawal, A.K., Goyal, M.R., Eds.; CRC Press: Boca Raton, FL, USA, 2017; pp. 219–251.
137. Martins, C.P.C.; Cavalcanti, R.N.; Couto, S.M.; Moraes, J.; Esmerino, E.A.; Silva, M.C.; Raices, R.S.L.; Gut, J.A.W.; Ramaswamy, H.S.; Tadini, C.C.; *et al.* Microwave processing: Current background and effects on the physicochemical and microbiological aspects of dairy products. *Comprehensive Reviews in Food Science and Food Safety* **2019**, *18*, 67–83, doi:10.1111/1541-4337.12409.
138. Tang, J.; Hong, Y.-K.; Inanoglu, S.; Liu, F. Microwave pasteurization for ready-to-eat meals. *Current Opinion in Food Science* **2018**, *23*, 133–141, doi:10.1016/j.cofs.2018.10.004.
139. Kubo, M.T.; Siguemoto, É.S.; Funcia, E.S.; Augusto, P.E.; Curet, S.; Boillereaux, L.; Sastry, S.K.; Gut, J.A. Non-thermal effects of microwave and ohmic processing on microbial and enzyme inactivation: a critical review. *Current Opinion in Food Science* **2020**, *35*, 36–48, doi:10.1016/j.cofs.2020.01.004.
140. Vadivambal, R.; Jayas, D.S. Changes in quality of microwave-treated agricultural products—a review. *Biosystems Engineering* **2007**, *98*, 1–16, doi:10.1016/j.biosystemseng.2007.06.006.
141. Lau, M.H.; Tang, J. Pasteurization of pickled asparagus using 915 MHz microwaves. *Journal of Food Engineering* **2002**, *51*, 283–290, doi:10.1016/S0260-8774(01)00069-3.
142. Tang, Z.; Mikhaylenko, G.; Liu, F.; Mah, J.-H.; Pandit, R.; Younce, F.; Tang, J. Microwave sterilization of sliced beef in gravy in 7-oz trays. *Journal of Food Engineering* **2008**, *89*, 375–383, doi:10.1016/j.jfoodeng.2008.04.025.
143. Atuonwu, J.C.; Tassou, S.A. Quality assurance in microwave food processing and the enabling potentials of solid-state power generators: A review. *Journal of Food Engineering* **2018**, *234*, 1–15, doi:10.1016/j.jfoodeng.2018.04.009.
144. Bahrami, A.; Moaddabdoost Baboli, Z.; Schimmel, K.; Jafari, S.M.; Williams, L. Efficiency of novel processing technologies for the control of *Listeria monocytogenes* in food products. *Trends in Food Science & Technology* **2020**, *96*, 61–78, doi:10.1016/j.tifs.2019.12.009.
145. Salazar-González, C.; San Martín-González, M.F.; López-Malo, A.; Sosa-Morales, M.E. Recent studies related to microwave processing of fluid foods. *Food Bioprocess Technol* **2012**, *5*, 31–46, doi:10.1007/s11947-011-0639-y.
146. Gedikli, S.; Tabak, Ö.; Tomsuk, Ö.; Çabuk, A. Effect of microwaves on some gram negative and gram positive bacteria. *Journal of Applied Biological Sciences* **2008**, *2*, 67–71.
147. Valero, A.; Cejudo, M.; García-Gimeno, R.M. Inactivation kinetics for *Salmonella enteritidis* in potato omelet using microwave heating treatments. *Food Control* **2014**, *43*, 175–182, doi:10.1016/j.foodcont.2014.03.009.
148. Jamshidi, A.A.; Seifi, H.A.; Kooshan, M. The effect of short-time microwave exposures on *Escherichia coli* O157: H7 inoculated onto beef slices. *African Journal of Microbiology Research* **2010**, *4*, 2371–2374.

149. Bauza-Kaszewska, J.; Skowron, K.; Paluszak, Z.; Dobrzański, Z.; Śrutek, M. Effect of microwave radiation on microorganisms in fish meals. *Annals of Animal Science* **2014**, *14*, 623–636, doi:10.2478/aoas-2014-0020.
150. Koskiniemi, C.B.; Truong, V.-D.; Simunovic, J.; McFeeters, R.F. Improvement of heating uniformity in packaged acidified vegetables pasteurized with a 915MHz continuous microwave system. *Journal of Food Engineering* **2011**, *105*, 149–160, doi:10.1016/j.jfoodeng.2011.02.019.
151. Benlloch-Tinoco, M.; Pina-Pérez, M.C.; Martínez-Navarrete, N.; Rodrigo, D. *Listeria monocytogenes* inactivation kinetics under microwave and conventional thermal processing in a kiwifruit puree. *Innovative Food Science & Emerging Technologies* **2014**, *22*, 131–136, doi:10.1016/j.ifset.2014.01.005.
152. Shaheen, M.S.; El-Massry, K.F.; El-Ghorab, A.H.; Anjum, F.M. Microwave Applications in Thermal Food Processing. In *The Development and Application of Microwave Heating*; Wenbin Cao, Ed.; InTech, 2012 ISBN 978-953-51-0835-1.
153. Khan, I.; Tango, C.N.; Miskeen, S.; Lee, B.H.; Oh, D.-H. Hurdle technology: A novel approach for enhanced food quality and safety – A review. *Food Control* **2017**, *73*, 1426–1444, doi:10.1016/j.foodcont.2016.11.010.
154. Cinquanta, L.; Albanese, D.; Cuccurullo, G.; Di Matteo, M. Effect on orange juice of batch pasteurization in an improved pilot-scale microwave oven. *Journal of Food Science* **2010**, *75*, E46–E50, doi:10.1111/j.1750-3841.2009.01412.x.
155. Marszałek, K.; Mitek, M.; Skąpska, S. Effect of continuous flow microwave and conventional heating on the bioactive compounds, colour, enzymes activity, microbial and sensory quality of strawberry purée. *Food Bioprocess Technol* **2015**, *8*, 1864–1876, doi:10.1007/s11947-015-1543-7.
156. Lu, Y.; Turley, A.; Dong, X.; Wu, C. Reduction of Salmonella enterica on grape tomatoes using microwave heating. *International Journal of Food Microbiology* **2011**, *145*, 349–352, doi:10.1016/j.ijfoodmicro.2010.12.009.
157. Peng, J.; Tang, J.; Luan, D.; Liu, F.; Tang, Z.; Li, F.; Zhang, W. Microwave pasteurization of pre-packaged carrots. *Journal of Food Engineering* **2017**, *202*, 56–64, doi:10.1016/j.jfoodeng.2017.01.003.
158. Woo, I.-S.; Rhee, I.-K.; Park, H.-D. Differential damage in bacterial cells by microwave radiation on the basis of cell wall structure. *Appl. Environ. Microbiol.* **2000**, *66*, 2243–2247, doi:10.1128/AEM.66.5.2243-2247.2000.
159. Shamis, Y.; Taube, A.; Mitik-Dineva, N.; Croft, R.; Crawford, R.J.; Ivanova, E.P. Specific electromagnetic effects of microwave radiation on *Escherichia coli*. *Appl. Environ. Microbiol.* **2011**, *77*, 3017–3022, doi:10.1128/AEM.01899-10.
160. Barnabas, J.; Siores, E.; Lamb, A. Non-Thermal Microwave Reduction of Pathogenic Cellular Population. *International Journal of Food Engineering* **2010**, *6*, doi:10.2202/1556-3758.1878.
161. Kozempel, M.; Cook, R.D.; Scullen, O.J.; Annous, B.A. Development of a process for detecting nonthermal effects of microwave energy on microorganisms at low temeperature. *J Food Processing Preservation* **2000**, *24*, 287–301, doi:10.1111/j.1745-4549.2000.tb00420.x.

162. Fujikawa, H.; Ushioda, H.; Kudo, Y. Kinetics of *Escherchia coli* destruction by microwave irradiation. *T I M E* **1992**, *58*, 5.
163. Hamoud-Agha, M.M.; Curet, S.; Simonin, H.; Boillereaux, L. Holding time effect on microwave inactivation of *Escherichia coli* K12: Experimental and numerical investigations. *Journal of Food Engineering* **2014**, *143*, 102–113, doi:10.1016/j.jfoodeng.2014.06.043.
164. Kim, J.E.; Choi, H.-S.; Lee, D.-U.; Min, S.C. Effects of processing parameters on the inactivation of *Bacillus cereus* spores on red pepper (*Capsicum annum* L.) flakes by microwave-combined cold plasma treatment. *International Journal of Food Microbiology* **2017**, *263*, 61–66, doi:10.1016/j.ijfoodmicro.2017.09.014.
165. Mishra, V.K.; Ramchandran, L. Novel thermal methods in dairy processing. In *Emerging Dairy Processing Technologies*; Datta, N., Tomasula, P.M., Eds.; John Wiley & Sons, Ltd: Chichester, UK, 2015; pp. 33–70 ISBN 978-1-118-56047-1.
166. Heddleson, R.A.; Doores, S.; Anantheswaran, R.C.; Kuhn, G.D. Viability loss of Salmonella species, *Staphylococcus aureus*, and *Listeria monocytogenes* in complex foods heated by microwave energy. *J Food Prot* **1996**, *59*, 813–818, doi:10.4315/0362-028X-59.8.813.
167. Göksoy, E.O.; James, C.; Corry, J.E.L. The effect of short-time microwave exposures on inoculated pathogens on chicken and the shelf-life of uninoculated chicken meat. *Journal of Food Engineering* **2000**, *45*, 153–160, doi:10.1016/S0260-8774(00)00054-6.
168. Guan, D.; Gray, P.; Kang, D.-H.; Tang, J.; Shafer, B.; Ito, K.; Younce, F.; Yang, T.C.S. Microbiological validation of microwave-circulated water combination heating technology by inoculated pack studies. *J Food Science* **2003**, *68*, 1428–1432, doi:10.1111/j.1365-2621.2003.tb09661.x.
169. Yang, H.W.; Gunasekaran, S. Comparison of temperature distribution in model food cylinders based on Maxwell's equations and Lambert's law during pulsed microwave heating. *Journal of Food Engineering* **2004**, *64*, 445–453, doi:10.1016/j.jfoodeng.2003.08.016.
170. Soni, A.; Smith, J.; Thompson, A.; Brightwell, G. Microwave-induced thermal sterilization- A review on history, technical progress, advantages and challenges as compared to the conventional methods. *Trends in Food Science & Technology* **2020**, *97*, 433–442, doi:10.1016/j.tifs.2020.01.030.
171. Tang, J. Unlocking potentials of microwaves for food safety and quality: microwaves for safety of ready-to-eat meals. *Journal of Food Science* **2015**, *80*, E1776–E1793, doi:10.1111/1750-3841.12959.

12

Infrared Radiation: Principles and Applications in Food Processing

Puneet Kumar[1]*, Subir Kumar Chakraborty[2] and Lalita[2]

[1]*ICAR-Central Institute of Temperate Horticulture, Srinagar, India*
[2]*ICAR-Central Institute of Agricultural Engineering, Bhopal, India*

Abstract

In the electromagnetic spectrum, wavelengths ranging between 0.78–1000 μm are classified as Infrared (IR) radiation. Applications of IR radiation can be far and wide, including as a source of heat, thus IR radiation is becoming an important source of energy for various food processing operations. IR is a component of solar radiation and has always been responsible for uncontrolled heating. Artificial IR sources emit the radiations of the desired wavelength of Near, Mid, or Far IR radiations and offer more control over the operating parameters. Near-IR to Mid-IR radiation can be used for drying operations and Far-IR radiations are used for blanching and sterilization processes. The use of IR heating offers uniform heating, reduced process time, and retention of food quality, and significant energy-saving IR radiations help in the processing of heat-sensitive and perishable food products having a thin peel. This chapter elucidates the principle of IR radiation and its applications as stand-alone and its synergic use with different technologies for various food processing operations. Also detailed are the effects of IR radiations on the different aspects of food quality.

***Keywords*:** Infrared radiations, IR Emitters, IR assisted technologies, IR peeling, IR microbial decontamination

**Corresponding author*: puneetkumar2292@gmail.com

Nitin Kumar, Anil Panghal and M. K. Garg (eds.) Thermal Food Engineering Operations, (349–374)

12.1 Introduction

Infrared (IR) radiations are the region of electromagnetic spectrum sandwiched between visible (0.38 to 0.78 μm) and microwaves (10^3 to 10^6 μm) as shown in Figure 12.1 [1]. Since ancient times IR heating, as in sun drying, has been used for drying food aimed at reducing water activity and extending shelf life. The advantages of IR heating over the conventional heating process makes it a promising alternative for the heat treatment of various food material. The major advantages of the application of IR heating are high heat transfer coefficient, shorter processing time, and lower cost of operation. As the air is transparent to IR, the operation can be done at ambient temperature without heating the surrounding air. Ordinary IR heating has many advantages but has some limitations too, like low penetration depth, insensitivity to reflective properties, loss of nutritional properties, and length of exposure causing discoloration due to non-enzymatic browning. Therefore, there is a need for more efficient IR heating which utilizes the advantages and overcomes the limitations. Nowadays, IR radiations of desired wavelength and intensity can be generated artificially. The equipment used for IR radiation generation is compact and has a higher degree of control over the range and intensity of emitted IR radiations. The wavelength of the desired rays can be obtained by using some specific optical filters and regulating the surface temperature of the emitter [2].

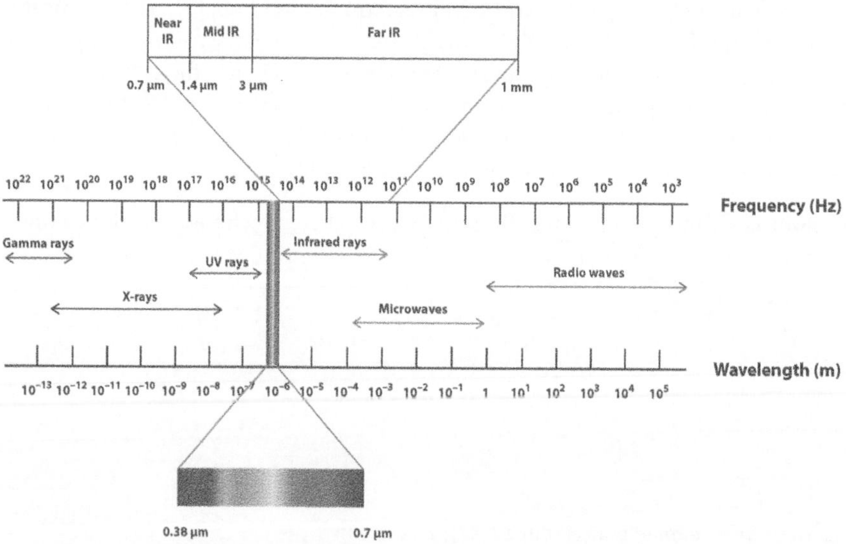

Figure 12.1 Electromagnetic spectrum.

There are many classifications for IR radiations; as per ISO 20473, IR radiations are divided into three major regions, Near-IR, Mid-IR, and Far-IR, having wavelength ranged between 0.78-3 μm, 3-50 μm, and 50-1000 μm, respectively. Another classification is based on the CIE scheme which is majorly used in food processing interventions and researches [3, 4]. According to the CIE division scheme, Near-IR has a range of 0.7 to 1.4 μm, Mid-IR has 1.4 to 3 μm and Far-IR has 3 to 1000 μm. The Near-IR region is easily detected by human eyes, whereas, mid-IR and far-IR are invisible to human eyes.

12.2 Mechanism of Heat Transfer

When electromagnetic radiations are incident on any material, the energy absorbed by material follows any one of the three mechanisms, i.e., change in electronic, vibrational, and rotational state of atoms. The mechanism of energy absorption is directly associated with the wavelength of incoming electromagnetic radiations. The wavelength of electromagnetic radiations ranging between 0.2 - 0.7 μm absorbs energy by a change in electronic state, 2.5 - 100 μm by a change in vibrational state, and above 100 μm by change rotational state [1]. As the Far-IR region is the largest among the Near and Mid IR, therefore IR treated products absorb the energy by a change in the vibrational state of their atoms. Most of the commercial Far-IR lamps emit a wavelength of 2.5 to 100 μm.

12.2.1 Principles of IR Heating

Since the electromagnetic spectrum is composed of various waves based on their wavelength or frequency and these waves have both spectral and directional characteristics, therefore the spectral characteristics of IR radiations are important to consider while choosing for application. The radiation emitted from the IR source comprises different wavelengths and the range of radiations depends on the temperature and emissivity of the emitter. Radiation transfer and absorption are a function of both spectral and directional characteristics of incoming radiations, which makes the radiation phenomenon more complex to understand. In a black body, the maximum radiations emitted with a certain wavelength are directly associated with its temperature. The basic laws which describe this relationship are Planck's law, Stefan–Boltzmann's law, and Wien's displacement law [5].

12.2.1.1 *Planck's Law*

Planck's law describes the spectral density of electromagnetic radiation emitted by a black body at a given temperature *T*, then there is no net flow of matter or energy between the body and its environment.

$$E_{b\lambda}(T,\lambda)=\frac{2\pi hc_0^2}{n^2\lambda^5[e^{hc_0/n\lambda kT}-1]}$$

where,

K is Boltzmann's constant (1.3806×10^{-23} J/K), *n* is the refractive index of the medium (for vacuum, $n = 1$), λ is the wavelength (μm), *T* is the IR source temperature (K), C_o is the speed of the light in a vacuum (km/s), and *h* is Planck's constant (6.626×10^{-34} J·s).

12.2.1.2 *Wien's Displacement Law*

"The Wien's displacement law explains the peak wavelength (λ_{max}) where the spectral distribution of radiation emitted by a blackbody reaches the maximum emissive power". The temperature (*T*) of the source required to achieve the desired spectral distribution can be estimated as,

$$\lambda_{max}=\frac{2898}{T}$$

The emissive power of the IR source of unknown temperature can be measured by a Fourier transformation infrared spectrometer.

12.2.1.3 *Stefan–Boltzmann's Law*

This law states that the total energy radiated per unit surface area of a black body across all wavelengths per unit time is directly proportional to the fourth power of the black body's thermodynamic temperature.

$$Q = \sigma A T^4$$

where,

Q is the rate of heat emission (J s^{-1}), *A* is the surface area of the emitter (m^2), *T* is the absolute temperature (K), and σ is Stefan–Boltzmann constant (5.670×10^{-8} J $s^{-1}m^{-2}K^{-4}$).

Table 12.1 Emissivity of different materials.

Material	Emissivity
Burnt toast	1.00
Ice	0.97
Water	0.95
Dough	0.85
White paper, painted metal or wood	0.9
Unpolished metal	0.7- 0.25
Polished metal	<0.05

Stefan–Boltzmann's law expresses the emissive power of the black body at a specific temperature. In the case of a pure black body, the heat flux estimated by this law should be consistent with Planck's law. As biological materials are not perfect absorbers, and radiation emitters are also not perfect radiators, therefore, the concept of grey bodies comes which says that grey bodies emit and absorb a fraction of theoretical maximum available radiations. Therefore, for grey bodies, the Stefan–Boltzmann's equation is given as,

$$Q = \varepsilon \sigma A T^4$$

where,

ε is the emissivity of the grey body (ranges between 0 to 1). The emissivity also varies with grey body temperature and wavelength of emitted radiation. The emissivity of the different materials used in food processing is given in (Table 12.1) [6].

12.2.2 Source of IR Radiations

Many IR sources are popular in the food processing sector. The emittance is a function of the source of IR radiations, therefore the intensity received on food materials is directly related to the IR source. Some food materials can bear higher intensity but some cannot; it is therefore important to select the right IR source based on the product to be treated. IR source is categorized broadly into two classes,

1. Natural Source
2. Artificial sources

12.2.2.1 Natural Source

The sun is the sole natural source of IR radiation for the earth. Nearly half of the electromagnetic radiations of the sun reaching the earth's surface are infrared radiations; amongst them most of the radiations are near-IR, having a wavelength shorter than 4 μm. The total irradiance received on earth as the sun is at its zenith is just over 1 kW/sq. meter, of which infrared radiations contribute about 527 watts, visible light 445 watts, and ultraviolet radiations 32 watts [6, 7].

12.2.2.2 Artificial Sources

Solar drying is common in the tropical part of the globe. IR is part of solar radiation and is used for heat-treatment of foods because of its free availability throughout the year. The major disadvantage of natural IR radiations is that we do not have full control of the heating process. There is a need to emit controlled intensity and desired range of IR radiations for better control and efficient application of IR radiations (Figure 12.2).

Artificially, IR radiations can be emitted from electric or gas-fired emitters (Figure 2.2) [3]. The operational cost of the electric emitter is more but it is preferred over the gas-fired emitter because of its higher degree of operation control and non-polluting nature.

Gas-fired emitter

These are emitters in which liquified gas or natural gas is used as fuel for generating IR radiation. The gas-fired emitter may be of two types i) the direct flame radiator and ii) burner type radiation. The "direct flame radiators" generate the IR radiations by directly supplying the gas flames to the radiating surface, whereas in the case of "burner type radiators" the combustion of gas takes place inside the perforated ceramics or thin steel wire mesh. The catalytic gas-fired emitter is another type of emitter that produces

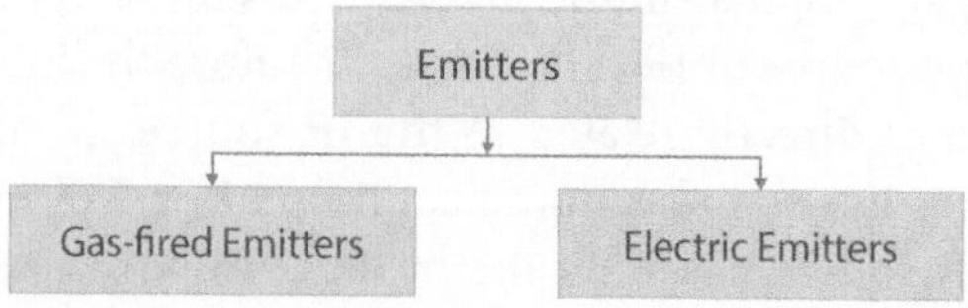

Figure 12.2 Artificial source of IR radiations.

Table 12.2 Performance characteristics of infrared emitters.

Type of emitter	Maximum running temperature (°C)	Maximum intensity (kW/m^2)	Maximum process temperature (°C)	Radiant heat (%)	Convection heat (%)	Heating-cooling time (s)
Short wavelength						
Heat lamp	2200	10	300	75	25	1
IR gun	2300	2	1600	98	2	1
Quartz tube	2233	80	600	80	20	1
Medium wavelength						
Quartz tube	950	60	500	55	45	30
Long-wavelength						
Element	800	40	500	50	50	<120
Ceramic	700	40	400	50	50	<120

invisible flames. This type of emitter has a peak wavelength from Mid-IR to the Far-IR range. The generation of radiations across a broader spectrum of wavelength helps in multiple post-harvest operations. The catalytic gas-fired infrared heater can be efficiently used in drying, peeling, blanching and disinfection [7, 8].

Electric emitter

The electric IR emitting system comprises an IR heater and a reflector to guide the radiations to the targeted material. The shape, size, and location of the reflector mounted on the IR heater affects the application efficacy of the IR radiations. Improper mounting may result in multiple reflections and lead to the radiations being captive between the source and the reflectors. These captive radiations are a major reason behind the excessive local heating and loss of IR yield. Therefore, it is required to attach two parabolic arcs mounted symmetrically around the source of IR radiations, this will make the centroid the location for efficient application without constricting the radiations [9]. This type of emitters includes ceramic, quartz, or halogen tubes fitted with electric filament (Table 12.2) [6].

12.3 Factors Affecting the Absorption of Energy

All the food materials don't absorb the same extent of IR radiations from the total available. Several factors play a role in the absorption of the IR radiations by the food. Broadly these factors can be categorized as shown in Figure 12.3.

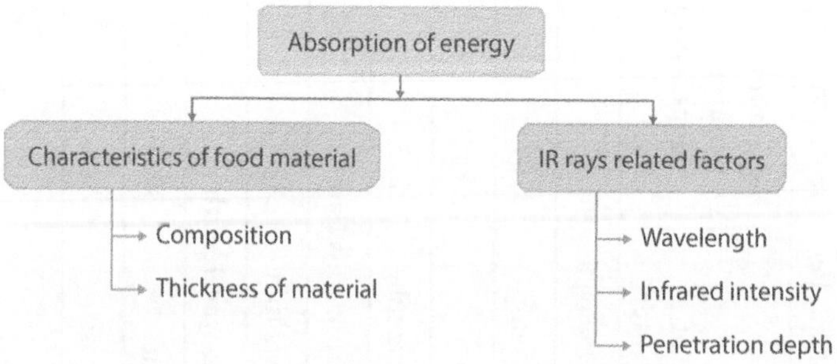

Figure 12.3 Factors associated with the extent of IR energy absorbed from total IR energy available.

12.3.1 Characteristics of Food Materials

The nature of food treated with IR radiations affects the efficiency of the application. Optical properties such as absorptivity, reflectivity, and transmissivity govern how the material will respond to the incident radiations [3]. Some of the major characteristics of food materials are explained below.

12.3.1.1 Composition

The constituents of food material play a significant role during IR processing; this is because water containing organic components and inorganic salts of food materials absorb the radiation at the near-to-mid range of IR [4, 10]. Different biomolecules tend to absorb IR radiations of different wavelengths. The carbohydrates have two major absorption bands of wavelength 3 μm and 7 to 10 μm, amino acids and polypeptides have two strong absorption bands of wavelength 3 to 4 μm and 6 to 9 μm, whereas lipids absorb almost entire IR regions and have three major absorption bands of wavelengths 3 to 4, 6, and 9 to 10 μm [11]. It is a well-established fact that the food with a higher amount of moisture can be rapidly sterilized; the same holds true when the heating source is IR radiations as well.

12.3.1.2 Layer Thickness

The penetration depth of IR radiation is a function of the thickness of the food material and the wavelength of incident radiation. When the thickness of the material increases, the radiant energy takes a longer time to reach the core of the food material; this is because the top surface absorbs more energy by radiation than by conduction. The increase in the thickness of material hinders the penetration of IR inside the food which subsequently increases the drying time and reduces the chances of elimination of microbes present inside the food.

12.3.2 IR Parameters

Apart from the characteristics of food materials, the properties of incident radiation also play a major role in governing the efficacy of IR treatment to attain the desired food processing operation. The major IR parameters which affect the efficacy of IR treatment are explained below.

12.3.2.1 *Wavelength of IR Rays*

Mostly, mid-to-far infrared rays are preferred in food processing as the water and organic compounds such as protein, carbohydrates, and fats, absorb maximum radiations in this range. These wavelengths are preferably used in drying, peeling, blanching, and pasteurization operations. When apple slices are dried with near-infrared rays the energy consumption is reduced by 50% [2], in comparison with conventional hot air oven drying. Near-to-mid infrared drying gives dried products that are of superior quality with better drying efficiency within a short period of time [12]; whereas, whenever there is a need to preserve the antioxidant properties during rapid pasteurization then far infrared gives promising results [13].

12.3.2.2 *IR Intensity*

As the intensity of the infrared rays increases, the energy available per unit area also increases. It is a well-known fact that high-intensity IR radiations help achieve the desired moisture content rapidly. But increase in intensity may also result in the weakening of the microstructure of the product resulting in a complete breakdown of the matrix during further processing operations [14]. It can therefore be concluded that for obtaining superior quality end product with lower energy consumption, the intensity of IR radiation need not be too high.

12.3.2.3 *Depth of Penetration*

During drying, the transmitted energy attenuates with penetration depth. The intensity of incoming IR rays decreases gradually while passing through the absorbing or scattering medium. The energy transmitted to a certain depth inside a food product is indicated by the depth of penetration. Depth of penetration is the measure of the depth radiation can penetrate with intensity decreases to $1/e^{th}$ (about 37%) of its original value at the surface of the material (Table 12.3). This depth is a function of the nature of the material (physical properties, physicochemical state, and chemical composition) and wavelength spectrum of radiations; [15, 16] all this can be measured by the partial least squares-discriminant analysis technique.

Table 12.3 The depth of penetration of near-infrared rays into some food products.

Food product	Spectral peak (μm)	Depth of penetration (μm)
Wheat grain	1.0	2
Biscuits	1.0	4
Raw apples	1.16	4.1
Wheat dough	1.0	4 to 6
Raw potatoes	1.0	6

12.3.3 Advantages of IR Heating Over Conventional Heating Methods

There are numerous advantages of IR heating over conventional heating; some of them are as follows:

1) The rate of heat transfer is fast, so the working temperature can be reached very rapidly, which may reduce the processing time up to 70%.
2) The surrounding air is transparent to IR rays, so heat energy can be supplied to the target material without much loss; this automatically leads to a higher rate of heat transfer to the product.
3) There is an overall reduction in the cost of operation.
4) The equipment required for the application of IR is very simple, and easy to construct and control.
5) The application of IR in food processing is non-hazardous to the environment as it doesn't produce any pollutants.
6) No special skill is required to operate the equipment.

12.4 Applications of IR in Food Processing

Now a days, the food processing sector is booming due to rapid urbanization especially in developing countries like India, Brazil, China, Russia, and South Africa. Increasing demand for food to feed the huge population poses a big challenge for the safety standards during food processing. Thermal processing is one of the methods for ensuring the safety of the processed food. In conventional thermal processes, the heat is generated by the direct combustion of fuel and electric heaters, and the most common conventional thermal processing method is the hot air method. In conventional thermal processing, the food surface is subjected to heat by convection and conduction. At times, prolonged exposure to heat results in the hardening of the outer surface (case-hardening), development of off-flavor, browning, and nutritional losses. These conventional methods have their limitations and they also pose a direct and indirect threat to the environment in terms of pollution. Considering these shortcomings of conventional thermal processing methods there is an immediate need to come up with an environment-friendly and more efficient method of thermal processing. The merits of infrared heating over conventional thermal processing make it a popular choice for thermal processing operations.

12.4.1 Drying

Sunlight, being the source of solar energy, has been one of the most common methods used for drying food products since time immemorial. Drying techniques have improved over time; there are many new methods used for drying foods—hot air oven drying, freeze-drying, microwave drying, and refractive-window drying, to name only a few. With the prevalence of technologies developed to control the wavelength of emitted rays, IR heating is getting a place as an alternative drying method due to its higher energy conversion rate, low cost of operation, and use of simple equipment. The medium-to-far IR rays have been proved to be suitable for drying agricultural produce [17]. Ceramic heaters are most preferred in drying operations because the wavelength of emitted rays is 3 μm. Due to the less penetration depth of IR rays, the moisture content from the top layer is removed by radiation up to a few millimeters from the surface followed by conduction beneath the top layer [18].

The drying unit consists of a closed chamber with trays, an emitter, a vibrating unit, and a data monitoring unit. The depth of grains, IR intensity, initial moisture content, the distance between emitter and grains, and frequency/amplitude of shaking play a major role in the uniform drying of grains [17, 19]. As the distance between grains and emitter increases, there is a distinct decrease in the extent of IR heat that reaches the grains. The drying time increases with an increase in the thickness of food material; this is due to the higher moisture gradient and conductive resistance

[20, 21]. Therefore, in thin-layer drying where the overall distance between emitter and food product is less, drying takes place at a faster rate. While drying thin slices of fruits or vegetables, it has been observed that the central core of the product achieves higher temperature faster. Therefore, in thin-layer drying, the food product is directly exposed to IR rays.

The non-uniform drying is overcome by intermittent shaking of the trays. Heat travels from the grains in the top layer to the colder grains beneath it. Vibrations cause more frequent contact between the hot and cold grains, heat transfers faster resulting in a uniform heat distribution which ultimately leads to uniform moisture removal. In a mixed bed dryer the grains are dried twice as fast as compare with non-mixing types [19]. The IR intensity and initial moisture content of the products play a major role in the rate of moisture removal. At the beginning of the drying operation, the moisture content decreases exponentially; as the drying operation proceeds, the rate of moisture removal decreases and plateaus. Besides, the IR intensity is also a very important factor for drying. The drying rate is directly proportional to the intensity of applied IR radiation. As the intensity increased, it also increases the energy absorbed by the food products resulting in faster removal of moisture.

12.4.2 Peeling

Peeling of tender fruits and vegetables is a primary unit operation. The most common peeling methods are chemical (lye peeling, using sodium hydroxide or potassium hydroxide) and mechanical methods. These methods have many drawbacks, like use of a huge quantity of water, longer peeling time, and also the generation of effluents. Enhancement of the peeling efficiency at the industrial level is carried out by many modern techniques, such as pressurized steam-vacuum peeling, steam-assisted lye peeling techniques, etc. But these techniques render inconsistent quality of peeled products resulting in significant changes in some physical properties (color, texture, and aroma). The mechanical methods have a great potential for peeling while exhibiting a better efficiency, but it damages the cellular structure of the product and contributes to loss of product during removal of the peel. IR-assisted peeling is an emerging technique with a lot of promise that can overcome the problem of mass loss during peel removal by traditional techniques.

Presently, IR assisted peeling is not common at the industrial level but it has the potential for successful adoption. Attempts have been made to develop batch type or continuous type IR treated peeling system which is a non-contact and dry peeling method where the product is treated with the

IR rays for a short period. The IR-assisted peeling is effluent free, thus it is an environment-friendly technology. Agricultural produce (such as peach, tomato, pears, jujube, etc.) which have a very thin outer skin can be successfully peeled by IR treatment because its penetration depth is limited to 1-4 mm; consequently IR does not affect the product beyond its outer skin.

The phenomenon of IR-assisted peeling is very complex, it involves all three modes of heat transfer. The product exposed to an IR heater has its top surface exposed to radiation. Once the top surface receives the heat, any further transfer of heat within the product is by conduction and convection. The main advantage of IR-assisted peeling is that the convective cooling of the product can be avoided as the surrounding air is also heated due to the presence of water molecules (which act as a conductor) in the air. A comprehensive understanding of the heat transfer mechanism requires a thorough knowledge of the thermal properties of the product as well as the air around it. The loosening of the peel by this method is governed by the heating which in turn causes loss of integrity of cells located in the hypodermal layer [22]. IR radiations initiate conductive heating within the food material, as the vaporization temperature is achieved the water molecule from the peel acquires enough energy to leave the material matrix and escape from the outer skin of the product; thus it results in easy removal of the peel. The efficacy of this method is influenced by process parameters (radiation intensity, IR emitter temperature, emitter to product gap, the thickness of the product, and treatment time) as well as properties of the product (moisture content, thermal properties, etc.). The heat flux imparted to the product at high IR intensity results in a more effective process but exposure for a longer time may deteriorate the qualities of the product. When a product is exposed to an IR emitter, only the front-facing side gets maximum IR radiation, and the rest of the sides gets less or no rays. This limitation is overcome by using multiple emitters which makes the radiations available to all parts of the product. During peeling operation, control is also required for the product to emitter as well as the emitter-to-emitter distance [23]. The peeling associated attributes are expressed in terms of peelability, skin remaining, peeling yield, peeling losses, and peeled skin thickness [22, 24].

a) **Peelability:** It is the measure to quantify the degree of peel removed and measured as the ratio of the area of peel removed to the overall surface area of the food product.
b) **Peel yield:** it is calculated as the ratio of product mass change before and after peeling.

c) **Peeled skin thickness:** It is the thickness of the peel removed after applying IR heating.
d) **Peeling losses:** It is the proportion of weight losses to initial weight after peel removal.

The peelability of a product increases with an increase in IR exposure time. Manual scrubbing peelability is comparable to lye peeling method when treated for 180 s especially in the case of soft textured fruits [24]. IR heating increases the elastic properties of skin and only minor disruption of the hypodermal layer takes place [23], thus enabling the separation of the peel layer from the flesh beneath it. This helps in reducing the peeling losses and thickness of the peel. The peeling losses are inversely proportional to exposure time because longer exposure time reduces the loss of flesh with the peel. While adjusting the gap between product and emitter, one needs to take extra precaution because as the gap increases the intensity of IR absorbed on the product also decreases and peelability may also get hampered [22, 24].

The accomplishment of peeling operation is indicated by the firmness and color of the food product; these properties have a direct effect on the product acceptability among the consumers. The color changes more as the intensity of applied IR and its exposure time increases. The L* (lightness) of the product decreases with an increase in intensity due to the enzymatic browning. As the spacing between product and emitter is increased the color change decreases rapidly since the absorbed IR rays only contribute to the evaporation of water but not in enzymatic browning. The firmness of the products increases with an increase in exposure time and intensity of IR radiation. This increase in firmness is due to the gelatinization and subsequent drying of the starch leading to the formation of a hard layer on outer skin [23, 25]. When products are lye treated, they soften the tissue, and discoloration of surface skin takes place, whereas, in the case of the IR treatment the integrity of products is maintained as the hypodermal layer remains intact to a greater extent as compared to the lye method [26]. Thus, to achieve optimal peeling, an appropriate residence time, the intensity of IR, and the number of emitters need to be decided and designed.

12.4.3 Blanching

Many enzymes are present in fruits and vegetables, among them a few cause deterioration in the intrinsic qualities, consequently reducing the shelf life and marketability of the product. Polyphenoloxidase, catalase, and phenolase are enzymes that are responsible for browning and off-flavor development during processing operations. Blanching is a food processing operation that is done before drying to inactivate these enzymes. Exposing the food product to hot water and steam are the most common blanching methods. But these methods are associated with the leaching of essentials compounds, loss of water-soluble compounds, and wastewater. The problems associated with these blanching methods can be checked by dry methods such as IR blanching. This method not only retains essentials compounds but is also a rapid method as heat transfer very fast which results in reducing the blanching time and the associated energy consumption [27]. The efficacy of blanching is determined by the inactivation of most heat-resistant peroxidase enzymes. Time, temperature, and product thickness are the major process parameters which are needed to be considered during IR blanching. In this method, the surface temperature ranges from 140 °C to 240 °C with varying operational times. Increase in operational time and temperature results in reducing the moisture content of the product. Prolong exposure to IR radiation may lead to decolorization of the product. Thick product and more than 90% inactivation of peroxidase requires long exposure time, in such case IR radiation should be applied in pulses. In the case of potato chips made from IR blanched potato, there is a significant reduction in oil uptake, Vitamin C is well preserved in carrot and garlic. IR blanched products possess crispy and appealing textural properties as well. Combining IR blanching with the conventional drying process, can be combined into a single operation.

12.4.4 Microbial Decontamination

The proliferation of mold and other microorganisms is the main foe to the safe storage of food and its value-added products. This section reports about the microbial decontamination of food by applying IR radiations. The IR waves are divided into far, medium, and near radiations with a specific range of wavelengths. Different chemical groups present in food and microorganisms absorb the different wavelengths of IR. For example, water and sugar have an OH group that can absorb 2.7-3.3 μm whereas carbonyl group and N-H bond in protein can absorb 5.92 and 2.83-3.33 μm, respectively. Aliphatic carbon-hydrogen bond in lipids, sugars, proteins can absorb 3.25–3.7 μm. Exposure to IR radiation induces a heating effect in food resulting in thermal denaturation of proteins and nucleic acids. IR heating firstly disturbs the cell wall of the microorganism and shrinks the cytoplasmic membrane causing the mesosome integration which leads to leakage of internal cellular contents. Thus, this results in DNA damage, which is responsible for the inactivation of microorganisms.

The system is well developed for cumin seeds with electrically operated ceramic emitters having maximum power and maximum temperature at 650 W and 553 °C, respectively, for microbial decontamination [28]. The surface temperature of cumin seeds varies between 200 and 300 °C for 1.57-4.4 min. Reduction in microbial populations was evaluated by total plate counts for total mesophilic aerobic bacteria (TMAB). The treatment time of 2.12, 3.40, and 5.35 min at 200 °C resulted in TMAB counts reduced to 104 CFU/g. When combined treatment of IR and UV is applied to cumin seeds it reduces the treatment time and TMAB could be decreased to 104 CFU/g. *B. cereus* spore reduction in paprika at 90 and 100 °C for 6 min holding time at 0.88 a_w was 2.5 log10 CFU/g. However, at 0.84 a_w and same temperature and holding time, the same reduction (1–2.5 log10 CFU/g) could not be achieved [29]. It means that apart from a_w, temperature and pH of the heating medium also affects *B. cereus* spore reduction in paprika powder. The power and distance between product and emitter are also very important. Emitter with 200 W power and food product held for about 1 min at 5 cm distance had a reduction in the magnitude of 2.3 log CFU/g [30]. In meat products, *L. monocytogenes* is one of the most common microorganisms. The inactivation of targeted *L. monocytogenes* microorganisms in turkey frankfurters was achieved at 3.5–4.5 log reduction [31]. This inactivation was achieved at 70–80 °C surface temperature. Several studies have reported the attempts on inactivation of microorganisms in the pasteurization process by the IR system. *L. monocytogenes* was reduced to 1 log CFU/g from the initial population of 7.32 log CFU/g [32]. The treatment was given at temperature of 80 and 85 °C for 3 and 2 min holding time, respectively. The process parameter for pasteurization of dried lime juice was reported to be temperature and power [33]. Maximum reduction was represented in term of total plate count (2.60 log CFU/ml) and was achieved at 62.2 °C and 700 W. IR heating exhibits a lethal effect on vegetative bacterial cells during the pasteurization process, it is due to rapid heat energy impinged by IR radiations [34, 35]. There are pieces of evidence that in raw almonds a maximum of 2 log CFU/g *salmonella* is present (Almond Board of California, 2003). The common methods for inactivation of *salmonella* in almonds are chemical methods, but they leave residue on the product which may pose a threat to human health upon consumption of the same. Log reduction of 1.03 log could be achieved at a radiation density of 5458 W/m^2 for 35 s [7]. When the degree of exposure was increased for a holding time of up to 60 min, it resulted in a rapid increase in temperature of more than 100 °C, consequently, 4.2 log CFU/g reduction of *Salmonella enteritidis* was achieved. The reduction of *Pediococcus* spp. in almonds was carried out at 120-80 °C with varying holding periods. It has been observed that efficient inactivation did not take place at a temperature below 80 °C with a wide range of holding time. The surface temperature of 120 °C and cooling to 90 °C of 5 min holding time resulting in an inactivation of up to 5 log reduction [36]. IR treatment is a successful method for grain disinfestation and is one of the best alternatives to conventional disinfection (i.e., chemical method). In early 2000, IR disinfection of grains gained importance at an industrial scale. The efficacy of IR disinfection of grains is depending upon targeted micro-organisms, their stage of growth, temperature applied at the surface, distance from the emitter, exposure time, moisture content of the grain, and radiation intensity [37]. Studies have shown the relationship between the radiation characteristics of rough rice and adult insects. When infested rice grains were IR treated, comparing with rough rice, insects *Sitophilus zeamais* and *Tribolium castaneum* absorbed high IR radiations (i.e., 1800 to 1308 cm^{-1}). This is attributed to the fact that *Sitophilus zeamais* and *Tribolium castaneum* have more protein and lipid content than rough rice. Complete inactivation was achieved at a heating temperature of 60.2 ± 0.5 °C with an IR intensity of 2780 W/m^2 for 110 s [38].

12.5 IR-Assisted Hybrid Drying Technologies

The penetration depth of IR radiations is less; therefore it is preferable to use IR in combination with some other thermal treatment especially for drying, pasteurization, and baking. Some of the most common combinations where IR acts as a subsidiary technique are IR-assisted freeze-drying, IR-assisted hot air drying, IR-assisted low-pressure superheated steam heating, IR-assisted pulsed vacuum drying, IR-assisted microwave drying, etc. The combination of IR with other thermal techniques makes the operation more efficient and time-saving. The application of IR-assisted thermal processing for dry-roasting of dry fruits is faster and more energy-efficient in comparison with other treatments. Some of the IR-assisted common thermal processing operations are explained as follows:

12.5.1 IR-Freeze-Drying

Freeze-drying (FD) is a common method of drying which is used to protect the microstructure, prevent hard crust formation and volume change during the drying of fruits and vegetables [39]. The freeze-drying operation is expensive and often needs a prolonged drying period [40, 41]. Therefore, many drying techniques are employed to assist freeze-drying to make it relatively less expensive and reduce the drying period. Hot air drying (HAD), microwave drying, and infrared (IR) drying are much sought-after drying techniques to assist freeze-drying. Many researchers have emphasized using assisted drying techniques. In one of the studies, IR heating and conventional HAD drying was used as a pre-treatment followed by freeze-drying for banana slices. It was concluded that IR pre-treated slices had a significant increase in drying rate as compared to conventional HAD pre-treatment. IR pre-treated slices also resisted the volume change during drying and IR heating resulted in large pore formation at the core of the slices. The IR pre-treatment yields larger pores due to the rapidly evaporating moisture from the center and also due to internal vaporization which leads to a hard crust formation at the outer surface of slices [42–44]. The size of pores formed during drying is a function of both IR pre-heat treatment as well as its intensity (kW/m^2) [43]. IR pre-heating results in a rapid rise in the surface temperature leading to higher moisture evaporation compared to cell wall collapsing moisture migration from the interior to the surface of the food product [44]. Thus, IR heating could reduce the drying time to attain the same level of dehydration; at times this reduction can be one-half as compared with freeze-drying alone [45]. One of the important factors considered while drying any food product is the rehydration ratio (RR). The RR is directly associated with the type of drying technique applied, microstructural changes, heat transfer rate, drying rate, and drying period. When any product is dried using FD, HAD followed by FD (HAD-FD) and IR-followed by FD (IR-FD) it affects its RR. The RR is directly associated with the extent of microstructural or cellular changes that occurred while the drying took place. Prolonged drying and a higher temperature damages cellular structure to a greater extent due to which RR reduces. The higher heating rate leads to a greater water vapor diffusion from the interior of food products and retains its porous cellular structure [27]. Among FD, HAD-FD, and IR-FD, HAD-FD shows the least RR that is due to a greater extent of cell wall disruption followed by FD and IR-FD. The IR-FD has a higher RR attributed to rapid heating and higher water diffusion rate [41, 46, 47].

12.5.2 Hot Air-Assisted IR Heating

When IR and hot air convection heating are applied individually to the same product, they take a long time and give a product of inferior quality. In the case of hot air-assisted IR heating, the drying time is reduced considerably and the quality of the food is also preserved [48]. In comparison with conventional heating, hot air-assisted IR heating reduces the drying period of vegetables by around 48% and energy requirement by 63% [49].

Major drying of food products takes place during the falling rate period and the drying rate is directly proportional to the intensity of IR radiations and inversely proportional to the velocity of hot air blast [50, 51]. The drying rate is independent of relative humidity and bed of depth for grains [52]. IR radiation is more preferred for drying after parboiling because it dries slowly and leads to lower breakage of rice during milling. The hot air-assisted IR heating does not denature the protein and starch present in food products. But during heating minor cracks occur on the grain surface resulting in increased volume of the grain, increased water uptake, and leaching losses [53].

The major drawback of using the high-intensity IR and less velocity of hot air is its effect on the color of the food product. The longer exposure of grains to high-intensity IR results in discoloration or yellowness of the surface [51, 54]. This discoloration is caused by the decomposition of and loss of total phenolic compounds during exposure to hot air-assisted IR. The physicochemical properties such as expansion ratio, rehydration ratio, water diffusivity of grains increases with the increase in the intensity of IR and decreases in the velocity of air. The thickness of the product layer also affects the physicochemical properties. Shrinkage and diffusivity in the product increase with thickness, whereas intensity has an inverse effect on shrinkage and a positive effect on diffusivity [50, 55].

The effective application can be ensured by applying vibration while heating to avoid the localized drying and obtaining uniform drying. A frequency of 20-22 Hz with 8-9 mm amplitude is sufficient to obtain uniform drying [56].

12.5.3 Low-Pressure Superheated Steam Drying with IR

It is a novel drying technology for heat-sensitive products like a banana. At low temperatures i.e., 80 °C and 7 kPa, drying time was more as compared to vacuum-assisted IR drying. However, the quality of the product was superior in terms of color and crispiness [57]. Even the pores of produce are larger in size and more in number which helps to achieve a better rehydration ability.

12.6 Conclusion

IR heating is seen as the alternative to various conventional food processing operations like drying, blanching, peeling, sterilization and roasting. Near-IR to Mid-IR can be used for drying operation and Far-IR can be found worthy for blanching and sterilization process. IR heating can be adopted at large in scale food industries for various applications as this method has salient features, such as low operational cost, higher thermal efficiency with a higher rate of energy conversion, a higher degree of control, and easy to operate equipment. As the penetration depth of IR radiation is less, for applications where heat is needed to reach deeper inside the food materials IR assisted technologies such as IR assisted freeze-drying, IR assisted low-pressure superheated steam heating, IR assisted pulsed vacuum drying, IR-assisted microwave heating, etc., are used. Various thermal techniques are coupled with IR heating to make the operation more efficient and time-saving. In comparison with conventional drying operation, hot air-assisted IR drying takes 48% less time and 68% less energy. The organic compounds with different functional groups present in food and microorganisms absorb the different wavelengths of the IR radiations. Like, OH group in water and sugar, N-H bond in protein and aliphatic carbon-hydrogen bond in lipids, sugars, proteins absorb IR radiations ranged between 2.7-3.3 μm, 2.83-3.33 μm, and 3.25–3.7 μm respectively. Therefore, the functional group of organic compounds present in the food matrix forms the basis for the selection of the range of IR radiation for microbial decontamination. The trend of increased use of IR radiation heating for food processing indicates it has a good potential to replace the conventional thermal processes with improved product quality and production economics.

References

1. Decareau, R. V., Microwaves. In *The Food Processing Industry,* Academic Press Orlando, FL, 1985.
2. Nowak, D., and Lewicki, P. P., Infrared drying of apple slices. *Innovative Food Science and Emerging Technologies*, 5, 3, 353–360, 2004.
3. Krishnamurthy, K., Khurana, H. K., Soojin, J., Irudayaraj, J., and Demirci, A., Infrared heating in food processing: an overview. *Comprehensive Reviews in Food Science and Food Safety*, 7, 1, 2-13, 2008.
4. Sakai, N., and Hanzawa, T., Applications and advances in far-infrared heating in Japan. *Trends in Food Science and Technology,* 5, 11, 357–362, 1994.
5. Dagerskog, M., and Osterstrom, L., Infra-red radiation for food processing: 1. A study of the fundamental properties of infra-red radiation. *Lebensmittel-Wissenschaft und Technologie*, 12, 4, 237-242, 1979.
6. Fellows, P. J. (Ed. 4th ed.), *Food Processing Technology: Principles and Practice,* Woodhead Publishing, 2017.

7. Brandl, M. T., Pan, Z., Huynh, S., Zhu, Y., and McHugh, T. H., Reduction of Salmonella enteritidis population sizes on almond kernels with infrared heat. *Journal of Food Protection*, 71, 5, 897–902, 2008.
8. Zhu, Y., and Pan, Z., Processing and quality characteristics of apple slices under simultaneous infrared dry-blanching and dehydration with continuous heating. *Journal of Food Engineering*, 90, 4, 441–452, 2009.
9. Rabl A, Clodic D, and Dehausse R, inventors; ARMINES, assignee, Apparatus emitting an electromagnetic radiation, in particular infrared, comprising a plane source of rays and a reflector, United States patent US 4,922,107. 1990 May 1.
10. Sandu, C., Infrared radiative drying in food engineering: a process analysis. *Biotechnology Progress*, 2, 3, 109-119, 1986.
11. Rosenthal, I., *Electromagnetic radiations in food science* (Vol. 19), Springer Science & Business Media, 2012.
12. Chen, Q., Bi, J., Wu, X., Yi, J., Zhou, L., and Zhou, Y., Drying kinetics and quality attributes of jujube (Zizyphus jujuba Miller) slices dried by hot-air and short-and medium-wave infrared radiation. *LWT - Food Science and Technology*, 64, 2, 759–766, 2015.
13. Wanyo, P., Siriamornpun, S., and Meeso, N., Improvement of quality and antioxidant properties of dried mulberry leaves with combined far-infrared radiation and air convection in Thai tea process. *Food and Bioproducts Processing*, 89, 1, 22–30, 2011.
14. Dondee, S., Meeso, N., Soponronnarit, S., and Siriamornpun, S., Reducing cracking and breakage of soybean grains under combined near-infrared radiation and fluidized-bed drying. *Journal of Food Engineering*, 104, 1, 6–13, 2011.
15. Erdogdu, S. B., Eliasson, L., Erdogdu, F., Isaksson, S., and Ahrné, L., Experimental determination of penetration depths of various spice commodities (black pepper seeds, paprika powder and oregano leaves) under infrared radiation. *Journal of Food Engineering*, 161, 75–81, 2015.
16. Pawar, S. B., and Pratape, V. M., Fundamentals of Infrared Heating and Its Application in Drying of Food Materials: A Review. *Journal of Food Process Engineering*, 40, 1, e12308, 2017.
17. Laohavanich, J., and Wongpichet, S., Thin layer drying model for gas-fired infrared drying of paddy. *Songklanakarin Journal of Science & Technology*, 30, 3, 2008.
18. Jenkins, G. W., and Forth, M. W., Infrared drying of shelled corn. *Transactions of the ASAE*, 8, 4, 457-459, 1965.
19. Nindo, C. I., Kudo, Y., and Bekki, E., Test model for studying sun drying of rough rice using far-infrared radiation. *Drying Technology*, 13, 1–2, 225–238, 1995.
20. Sadeghi, E., Movagharnejad, K., and Haghighi Asl, A., Parameters optimization and quality evaluation of mechanical properties of infrared radiation

thin layer drying of pumpkin samples. *Journal of Food Process Engineering*, 43, 2, 2020.

21. Wu, B., Ma, H., Qu, W., Wang, B., Zhang, X., Wang, P., Wang, J., Atungulu, G. G., and Pan, Z., Catalytic Infrared and Hot Air Dehydration of Carrot Slices. *Journal of Food Process Engineering*, 37, 2, 111–121, 2014.
22. Shen, Y., Khir, R., Wood, D., McHugh, T. H., and Pan, Z., Pear peeling using infrared radiation heating technology. *Innovative Food Science and Emerging Technologies*, 65, 102474, 2020.
23. Wang, B., Venkitasamy, C., Zhang, F., Zhao, L., Khir, R., and Pan, Z., Feasibility of jujube peeling using novel infrared radiation heating technology. *LWT - Food Science and Technology*, 69, 458–467, 2016.
24. Li, X., Zhang, A., Atungulu, G. G., Delwiche, M., Milczarek, R., Wood, D., Williams, T., McHugh, T., and Pan, Z., Effects of infrared radiation heating on peeling performance and quality attributes of clingstone peaches. *LWT - Food Science and Technology*, 55, 1, 34–42, 2014.
25. Kate, A. E., and Sutar, P. P., Effluent free infrared radiation assisted dry-peeling of ginger rhizome: A feasibility and quality attributes. *Journal of Food Science*, 85, 2, 432–441, 2020.
26. Pan, Z., Li, X., Khir, R., El-Mashad, H. M., Atungulu, G. G., McHugh, T. H., and Delwiche, M., A pilot scale electrical infrared dry-peeling system for tomatoes: Design and performance evaluation. *Biosystems Engineering*, 137, 1–8, 2015.
27. Vishwanathan, K. H., Giwari, G. K., and Hebbar, H. U., Infrared assisted dry-blanching and hybrid drying of carrot. *Food and Bioproducts Processing*, 91, 2, 89–94, 2013.
28. Erdogdu, S. B., and Ekiz, H. I., Effect of ultraviolet and far infrared radiation on microbial decontamination and quality of cumin seeds. *Journal of Food Science*, 76, 5, M284-M292, 2011.
29. Staack, N., Ahrné, L., Borch, E., and Knorr, D., Effect of infrared heating on quality and microbial decontamination in paprika powder. *Journal of Food Engineering*, 86, 1, 17-24, 2008.
30. Shavandi, M., Taghdir, M., Abbaszadeh, S., Sepandi, M., and Parastouei, K., Modeling the inactivation of Bacillus cereus by infrared radiation in paprika powder (Capsicum annuum). *Journal of Food Safety*, e12797s, 2020.
31. Gande, N., and Muriana, P., Prepackage surface pasteurization of ready-to-eat meats with a radiant heat oven for reduction of *Listeria monocytogenes*. *Journal of Food Protection*, 66, 9, 1623-1630, 2003.
32. Huang, L., and Sites, J., Elimination of *Listeria monocytogenes* on hotdogs by infrared surface treatment. *Journal of Food Science*, 73, 1, M27-M31, 2008.
33. Aboud, S. A., Altemimi, A. B., Al-Hilphy, A. R., and Watson, D. G., Effect of batch infrared extraction pasteurizer (BIREP)-based processing on the quality preservation of dried lime juice. *Journal of Food Processing and Preservation*, 44, 10, e14759, 2020.

34. Ramaswamy, R., Krishnamurthy, K., and Jun, S., Microbial decontamination of food by infrared (IR) heating. In *Microbial Decontamination in the Food Industry* (pp. 450-471), Woodhead Publishing, 2012.
35. Rastogi, N. K., Infrared heating of fluid foods. In *Novel thermal and non-thermal technologies for fluid foods* (pp. 411-432), Academic Press, 2012.
36. Bingol, G., Yang, J., Brandl, M. T., Pan, Z., Wang, H., and McHugh, T. H., Infrared pasteurization of raw almonds. *Journal of Food Engineering*, 104, 3, 387-393, 2011.
37. Subramanyam b., Hot technology for killing insects. *Milling Journal*, 48–50, 2004.
38. Pei, Y., Tao, T., Yang, G., Wang, Y., Yan, W., and Ding, C., Lethal effects and mechanism of infrared radiation on Sitophilus zeamais and Tribolium castaneum in rough rice. *Food Control*, 88, 149-158, 2018.
39. Ratti, C., Hot air and freeze-drying of high-value foods: a review. *Journal of Food Engineering*, 49, 4, 311–319, 2001.
40. Flink, J. M., Energy analysis in [food] dehydration processes. *Food Technology (USA)*, 1977.
41. Lin, Y. P., Lee, T. Y., Tsen, J. H. and King, V. A. E., Dehydration of yam slices using FIR-assisted freeze drying. *Journal of Food Engineering*, 79, 4, 1295–1301, 2007.
42. Pan, Z., Shih, C., McHugh, T. H. and Hirschberg, E., Study of banana dehydration using sequential infrared radiation heating and freeze-drying. *LWT - Food Science and Technology*, 41, 10, 1944–1951, 2008.
43. Khampakool, A., Soisungwan, S. and Park, S. H., Potential application of infrared assisted freeze drying (IRAFD) for banana snacks: Drying kinetics, energy consumption, and texture. *LWT- Food Science and Technology*, 99, 355–363, 2019.
44. Wang, H. C., Zhang, M. and Adhikari, B., Drying of shiitake mushroom by combining freeze-drying and mid-infrared radiation. *Food Bioproducts Processing*, 94, 507–517, 2015.
45. Lin, Y. P., Tsen, J. H. and King, V. A. E., Effects of far-infrared radiation on the freeze-drying of sweet potato. *Journal of Food Engineering*, 68, 2, 249–255, 2005.
46. Antal, T., Comparative study of three drying methods: freeze, hot air-assisted freeze and infrared-assisted freeze modes. In *Agronomy Research*, 13, 4, 2015.
47. Shih, C., Pan, Z., Mchugh, T., Wood, D. and Hirschberg, E., Sequential infrared radiation and freeze-drying method for producing crispy strawberries. *Transactions of the ASABE*, 51, 1, 205-216, 2008.
48. Kumar, D. G. P., Hebbar, H. U., Sukumar, D. and Ramesh, M. N., Infrared and hot-air drying of onions. *Journal of Food Processing and Preservation*, 29, 2, 132-150, 2005.
49. Hebbar, H. U., Vishwanathan, K. H. and Ramesh, M. N., Development of combined infrared and hot air dryer for vegetables. *Journal of Food Engineering*, 65, 4, 557-563, 2004.

50. Afzal, T. M. and Abe, T., Some fundamental attributes of far infrared radiation drying of potato. *Drying Technology*, 17, 1-2, 138-155, 1999.
51. Nasiroglu, S. and Kocabiyik, H., Thin-layer infrared radiation drying of red pepper slices. *Journal of Food Process Engineering*, 32, 1, 1-16, 2009.
52. Das, I., Das, S. K., and Bal, S., Drying performance of a batch type vibration aided infrared dryer. *Journal of Food Engineering*, 64, 1, 129-133, 2004.
53. Fasina, O.O., Tyler, R.T., Pickard, M.D., Zheng, G. and Wang, N., Effect of infrared heating on the properties of legume seeds. *International Journal of Food Science and Technology*, 36, 79-90, 2001.
54. Konopka, I., Markowski, M., Tańska, M., Żmojda, M., Małkowski, M. and Białobrzewski, I., Image analysis and quality attributes of malting barley grain dried with infrared radiation and in a spouted bed. *International Journal of Food Science & Technology*, 43, 11, 2047-2055, 2008.
55. Wesołowski, A. and Głowacki, S., Shrinkage of apples during infrared drying. *Polish Journal of Food and Nutrition Sciences*, 12, 4, 9-12, 2003.
56. Das, I., Das, S. K. and Bal, S., Drying kinetics of high moisture paddy undergoing vibration-assisted infrared (IR) drying. *Journal of Food Engineering*, 95, 1, 166-171, 2009.
57. Nimmol, C., Devahastin, S., Swasdisevi, T. and Soponronnarit, S., Drying of banana slices using combined low-pressure superheated steam and far-infrared radiation. *Journal of Food Engineering*, 81, 3, 624-633, 2007.

13

Radiofrequency Heating

Chirasmita Panigrahi[1]*, Monalisha Sahoo[2], Vaishali Wankhade[3] and Siddharth Vishwakarma[1]

[1]Agricultural and Food Engineering Department, Indian Institute of Technology Kharagpur, West Bengal, India
[2]Centre for Rural Development and Technology, Indian Institute of Technology Delhi, New Delhi, India
[3]Maharashtra Institute of Technology, Aurangabad, India

Abstract
Radiofrequency (RF) heating with superior properties, like high energy efficiency due to volumetric heating, lower processing time, and greater energy saving has emerged as a potential alternative to the conventional method. RF technology has proven to process food materials with improved physical, chemical, and sensory properties, offering multiple applications in food processing. However, so far, the application of RF heating to the food industries is limited due to non-uniform temperature distribution and the complexity of food composition. Its use is restricted to lab-scale research and needs more systematic comprehensive research on a pilot scale and commercial scale. The application of computer simulation on the improvement of design for scale-up operations is an emerging area. Furthermore, health hazards arising out of the use of high-frequency radio waves during operations are the major concerns for the adoption of this technology. This chapter discusses the principle and mechanism of RF heating, application areas in food processing and preservation, technological constraints and associated safety hazards, commercialization aspects, and future scope of the technology.

Keywords: Radiofrequency, food applications, advantages, limitations, safety aspects, future scope

**Corresponding author*: chirasmitapanigrahi8@gmail.com

Nitin Kumar, Anil Panghal and M. K. Garg (eds.) Thermal Food Engineering Operations, (375–414)

13.1 Introduction

The radiofrequency (RF) band of the electromagnetic spectrum covers wide-ranging frequencies (1–300 MHz) that lie within the radar range for communication systems. Some of the selected frequencies (13.56, 27.12, and 40.68 MHz) are permitted by the US Federal Communications Commission (FCC) for domestic, scientific, medical, and industrial applications [1]. Radiofrequency heating is a non-ionizing radiation thermal process since insufficient energy (less than 10 eV) is produced to ionize biologically important atoms, and therefore it is also called dielectric heating. The dielectric-assisted heating methods offer volumetric heating, deeper penetration ability, and possibilities to heat packaged food products. Radiofrequency has many applications in the food processing sector. The most preferred technology for preserving the food is drying which decreases the water activity of food by removing moisture. The drying technologies should yield desirable product quality attributes, such as color, flavor, nutrients, texture, and rehydration capacity. Microwave energy has found greater interest in quick drying of food products along with the minimum change in quality attributes as compared to conventional methods. However, the industrial scalability of microwave technology is a big issue due to lower uniform heating and lower penetration depth. Hence, RF technology comes as an emerging technology that generates heat inside the food product in a short time due to molecular friction. The radio waves have properties of greater penetration depth, better heating uniformity, and highly stable product temperature control. Therefore, the RF technology has been found to have application in many areas like baking, roasting, cooking, disinfesting, pasteurization, sterilization, thawing, and many more [2].

The freezing process is also used for preserving food products by reducing the temperature to such an extent that the microbial growth gets stunted. For reusing the frozen products, thawing is usually practiced using air impingement and hot water. However, these conventional processes result in a longer thawing process, which provides the opportunity for microorganisms to grow on the surface. A rapid thawing method like RF technology has become a proven method. The RF technology can reduce the thawing time significantly, which results in the prevention of food quality deterioration, inhibition of microbial activities, and controlling water loss from dripping [3]. The study [4] found 85 times faster thawing of meat blends than the conventional air thawing when using an RF system of 27.12 MHz frequency.

Microbial inactivation is crucial for preventing denaturation of proteins, enzymes, nucleic acids, or other vital food components. This is generally done by using thermal techniques; however, high temperatures cause deterioration of nutrients and sensory attributes. This sometimes leads to the unacceptability of processed food from the consumer point of view. These problems can be prevented by using RF technology for microbial inactivation along with the preservation of quality attributes. Moreover, many studies have shown validated studies about the potential of RF heating in food pasteurization. The RF technology has proven to be used as an alternative method to fumigation which is generally used for inactivating pests in food products [5].

Recently, these alternatives to conventional processing methods have been widely explored for enhancing the quality of foods, especially value-added and heat-sensitive foods [6]. Besides, these may offer a plethora of opportunities for new product development. RF has a great potential to be used as an alternative pasteurization technology in the food industry since heating is rapid, volumetric, and can penetrate much deeper into most foods. Moreover, RF heating has become the preferred method for biscuit baking and holds a promising future for the snack food industry [7].

In this chapter, the brief history of RF technology is described to give a glimpse of its origin, evolution, and progress. One section explains the main principle and mechanism of heat generation. The working of RF technology in the food sector is elaborated with the effect on performance from various factors related to food. Further, lab-based research has been discussed along with the modern-time RF technology used in the industry. Some specific food applications were deliberated to provide information about the potential usefulness of using RF technology. At last, some potential hazards of using RF technology are discussed which can be prevented by using important safety aspects. The prospects are delineated for providing the area of research still within scope for the RF technology.

13.2 History of RF Heating

The existence of electromagnetic (EM) fields was postulated by Michael Faraday in 1832. After 41 years, radiowaves' existence and their properties were being predicted by James Clerk Maxwell in the mathematical form. This work was then experimentally proved by Heinrich Hertz in 1885. Later on, experiments were done on animals using low voltage alternating current and high frequency (500–1500 kHz) through Hertz's first high-frequency oscillator by Jacques Arsene dÁrsonval. This experiment led to the development of the first heat therapy unit using high-frequency current as the heat was found to get produced due to alternating current. In Paris, during 1895, the therapy unit was created in Hotel Dieu Hospital.

After World War II, the technology was used in the area of food processing during the 1940s. Sherman explained the effective use of RF energy as a source of 'electric heat' and the procedure to produce it [8]. Further, his work described the application of RF energy in food like melting of frozen foods, processing of food, preservation of meat products, etc. In 1947, RF energy was used for the cooking of processed meat products, bread heating, blanching, and drying of vegetables [9]. The problems of high operational costs of RF energy appeared to be the biggest hindrance for the commercial application of the above products. However, industrial installation was done in enough numbers due to the focused study of RF energy for the defrosting of frozen products [10]. Later, in the late 1980s, RF energy was used for the post-bake drying of cookies and snack foods in the food industry [7, 11].

The study of the use of RF energy in the pasteurization of foods was done to address energy efficiency and runaway heating problems [12, 13]. This resulted in the appropriate understanding of the dielectric properties of food at RF frequencies along with some factors affecting RF heating [1, 14]. Later on, many recent pieces of research were carried out on baking and roasting, cooking, disinfestation, drying, pasteurization, sterilization, defrosting, and thawing, which are discussed in later sections.

13.3 Principles and Equipment

13.3.1 Basic Mechanism of Dielectric Heating

Dielectric heating refers to the application of electromagnetic waves of certain frequencies to generate heat into food materials. Material composed of atoms is heated by producing dielectric motion in its molecules by passing it through alternating electric fields, [15]. Heating is based on the ability of materials to absorb electromagnetic waves and to convert them into heat [16]. Figure 13.1 shows the schematic circuit diagram of the arrangement for dielectric heating of food. It consists of two metal plates supplied with the electric field and in between these plates, the material being heated is fed.

Dielectric heating in food processing is performed in two ways, viz., radiofrequency heating and microwave heating, based on different frequency ranges. The basic principles and working of both dielectric heating techniques of food are explained in detail in the following section.

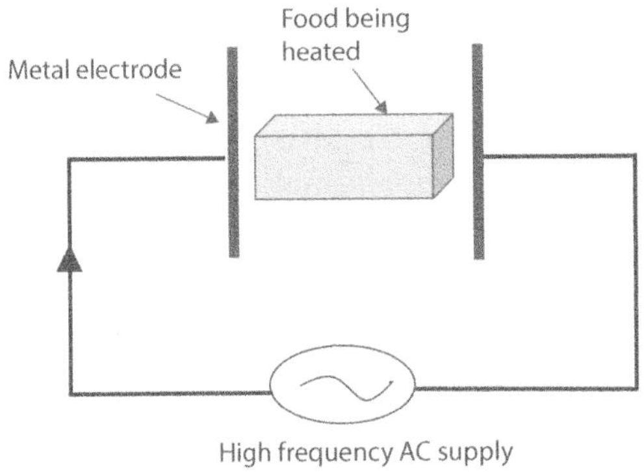

Figure 13.1 Schematic circuit diagram of arrangement for dielectric heating of food.

13.3.1.1 Basic Mechanism and Working of Radiofrequency Heating

Radiofrequency heating involves the heating of poor electrical conductors which is also characterized by freedom from electrical and mechanical contact. RF heating is carried out by application of high-voltage alternating electric field to a medium interpolated between two parallel electrodes among which one is grounded to setup capacitor structure. This causes the operation to resemble a condenser in which some energy loss occurs across the plates, causing a heating effect in the material spacing the plates. The materials like paper, air, or denser, or other insulating materials having the ability to separate the potential difference of the plates without breakdown at high voltage may be used as condensers. In general, an RF system generating an oscillating field of electromagnetic energy is used to generate heat. The system consists of an RF generator, hydraulic press, control circuit board, power supply, and parallel plate along with some supporting materials system. The plate walls aid the purpose of providing boundaries to the RF field generated by the RF generator. Commonly employed frequencies for RF heating are 13.56, 27.12, and 40.68 MHz [5].

The materials dielectric in nature are composed of electrons, atoms, molecules, and ions. When such materials are applied with an alternating electric field, the polar molecules try to line themselves up with the nearby molecules, which results in the lattice and frictional losses due to rotation, and thus, the electric resistance heating turns out due to movement of the dissolved ions. The RF heating is accomplished by different possible combinations of dipole rotations, like, electronic, atomic, molecular, and ionic, etc., as elements of materials may be locked into the systematic structure of crystals or free to wander over the structure [17].

13.3.1.2 Basic Mechanism and Working of Microwave Heating

In microwave (MW) heating, the range of frequency is higher compared to RF heating. It is generally operated at 915 to 2450 MHz, resulting in a smaller wavelength than the dimensions of sample material to be heated [18]. Heating occurs in the metal chamber with resonant electromagnetic standing wave mode. A major component of the MW heating system is an oscillator tube magnetron or klystron which generates microwaves from high voltage direct current, and the waves are then transmitted to an applicator or antenna through a waveguide or coaxial transmission line. Magnetrons give output in the range of 0.5 to 100 kW of MW radiations which are radiated into space in the oven or cavity. Waveguide assimilated in MW systems helps to streamline these microwaves to the stirrer which disperse them over the food being cooked in the cavity [19].

13.3.2 Factors of Food Affecting the Performance of RF Processing

In radiofrequency heating of foods, many factors of food affect the heating process. Among all, the dielectric properties of foods are the most influential factors as dielectric heating is based on the capacity of materials to engross electromagnetic waves and convert them into heat. The dielectric properties of food affecting the performance of RF processing are explained briefly below.

13.3.2.1 Permittivity and Loss Factor

Permittivity is the dielectric property of food which determines the dielectric constant, dielectric loss factor, and the loss angle. RF heating is highly influenced by these factors [20]. Dielectric constant or "capacitivity" is defined as the measure of the capacity of the material to store the electric energy as a measure of induced dispersion in a material. This dielectric property shows significant change over an offered range of frequencies which is generally notified as "dielectric dispersion," whereas it remains constant at constant frequency [16].

The absolute permittivity of vacuum is ε_0 and it is determined by the speed of light (c_0) and the magnetic constant (μ_0) which are combined as the following equation:

$$c_0^2 \mu_0 \varepsilon_0 = 1$$

The numerical value of ε_0 is about 8.854 x 10^{12} F/m. Permittivity shows higher value in other mediums and it is usually presented relative to the vacuum [21].

$$\varepsilon_{abs} = \varepsilon_r \varepsilon_a$$

Where, ε_{abs} = Absolute permittivity of a material, ε_r = Relative permittivity of the material

The high frequency and microwave fields are sinusoidal or time dependent and this time dependence is commonly expressed with the help of notations [21]. Therefore, the permittivity is also quantified as a complex quantity with imaginary components. The equation for complex permittivity is as follows [22],

$$\varepsilon = \varepsilon' - j\varepsilon''$$

Where ε = Relative complex permittivity

ε' = Relative real permittivity (Dielectric constant). For vacuum, $\varepsilon' = 1$.

ε'' = Relative dielectric loss factor

j = Imaginary unit

The real permittivity, ε', affects the electric field of a propagating wave, due to which the ratio between the electric and magnetic field strength which exists with the wave impedance gets changed. The dielectric loss factor (ε'') is correlated to several absorption mechanisms of energy dissipation and is always positive and generally considerably smaller than ε'. It is approximately proportional to the attenuation of the propagating wave. The substance is lossless if $\varepsilon'' = 0$ [21, 23]. The ratio of ε'' to ε' is called the dielectric loss tangent (tan δ), where $\tan \delta = \varepsilon''/\varepsilon'$.

13.3.2.2 *Power Density and Penetration Depth*

The rate of heating can be expressed by the power equation

$$P_v = 2\pi f \varepsilon_o \varepsilon'' |E|^2$$

Where P_v = Energy developed per unit volume (W/m^3),

f = Frequency (Hz),

| E | = Electric field strength inside the load (V/m).

The dielectric properties of the material, the geometry of the load, and oven configuration are the points considered to determine the electric field inside the load. Therefore, the equation above is not feasible and the determination of the electric field distribution is complex [24].

The calculation of penetration depth can be done with the help of the dielectric properties of the product. The penetration depth is also known as "power penetration depth" and is denoted by d_p. It is theoretically defined as the depth below a large plane surface of a substance at which the power density of a perpendicular impinging forward propagating plane electromagnetic wave has decayed by 1/e from the surface value (1/e ≈ 37%). If tan ε is smaller than about 0.5, the following formula gives 97-100% of the correct value [24],

$$d_p \frac{\lambda_o \sqrt{\varepsilon'}}{2\pi\varepsilon''}$$

Where λ_o is the free-space wavelength. In an infinite inhomogeneous slab, the absorbed power density is accordingly, approximately proportional to ε'' and ε' near the surface.

13.3.2.3 *Wave Impedance and Power Reflection*

The distribution of energy is determined by the transmission properties. This is related to the dielectric and thermal properties of the product medium [23]. In comparison with free space, the wavelength of waves will be shorter than the dielectric medium, since ε' decreases the speed of propagation. The reflection at the interface between two media with different ε' happens due to this change in wavelength [21]. The reflection phenomena can be analyzed in terms of characteristic wave impedance (η) [25]:

$$\eta = \frac{\eta_o}{\sqrt{\varepsilon}}$$

Where η_o is the wave impedance of free space (≈ 377Ω)

The reflection and transmission at a plane boundary are primarily related to √ε, and the principal determining factor for the magnitude of the reflection is the real permittivity (ε') of the material. Errors due to neglecting ε" are less than 5% for virtually all foods [24].

If different materials are to be heated simultaneously, special consideration is given to characteristic impedance as it is an important factor. The typical value of characteristic impedance is about 50 Ω for the average food. The reflection of about 50% of the microwave power falling on the food surface arises from a change in characteristic impedances at the food surface. Furthermost of this energy is reflected in the food through the metal cavity walls. For frozen food, the impedance matching demonstrates higher power utilization for thawing than for heating, hence, is better in the case of frozen food [25]. Dielectric properties are reliant on frequency, temperature, density, moisture, and composition of materials. Moreover, several kinds of fresh food materials come up with diverse dielectric properties. Food of similar kind also owns discrepant dielectric properties in its different parts [26].

13.3.3 Comparison of RF Heating With Other Methods

Some major points of comparison between RF heating and conventional methods as well as with other electromagnetic techniques are listed in Table 13.1.

In RF heating, the heat is generated within the product, providing more uniform heating than the traditional conventional heating in which heat is transferred from a hot medium to a food sample. Moreover, the conventional heating of low moisture food is very slow and inefficient due to their low thermal diffusivities. The rapid generation of heat in RF heating allows maintaining the product quality with a shorter heat treatment time [27]. The problems of overheating and earlier dehydration of the product's surface (mainly due to conduction) is easily solved using RF energy. Moreover, the RF heating can be applied in those foodstuffs which are already bottled or packed. Thus, it may be possible to process delicate foods like honey using RF energy [28]. The high energy efficiency due to volumetric heating, lower processing time, and greater energy densities make the RF application energy saving. Further, the apparatus for RF energy application takes less floor-space with an environmentally friendly nature and gets compatible with batch processing and/or continuous flow processing [13].

In comparison to conventional heating, RF technology has proven to process food materials with improved physical, chemical, and sensory properties and this nature is highly desirable in the area of food processing and preservation [29]. It was reported that during the removal of moisture, the efficiency of the RF system was 70% where the conventional oven stood out with only 10% efficiency in the post-baking applications [30]. The work of Demeczky showed better bacteriological and organoleptic qualities of

Table 13.1 Comparison of RF heating with conventional and other electromagnetic techniques.

Conventional and other electromagnetic techniques	RF heating	References
Excess heated areas of solid food have degraded quality	Uniform heating throughout the food surface	[32]
In the retort system, the air is exhausted using steam and this step is highly energy-intensive	The pro-automation and environment-friendly nature of EM waves in RF processing	[33]
Lower rate of heat transfer due to poor thermal conductivity of most of the foods. This leads to higher treatment time during processing.	Volumetric heating by EM waves are the best possible ways to heat these types of foods	[27]
Ohmic heating –Food product is placed in the path of AC having a low frequency	The food product is placed between directional electric field having a higher frequency	[10]
Microwave (MW) wavelengths are shorter than RF wavelength and thus provide relatively shorter penetration depth.	For heating of bulk material and products with big size, RF energy is comparatively the most suitable method. The construction of such a commercial scale RF system is much easier than the respective MW system. The scale-up cost is also lesser in the RF system due to the absence of magnetrons in it.	[27, 32, 34]
The possibility of occurrence of cold and hot spots in the food products due to microwaves is higher.	RF system results in well-distributed heating of food and also does not create any interference or negative results in the food items.	[1]

RF processed peach, quince and orange bottled juice in comparison to conventionally or thermally treated respective juices [31]. RF energy was found to be used for the sterilization process. A reduction in processing time was reported; 30 min as compared to 90 min in traditional heating producing a process lethality (F_0) of 10 min [32].

In the microwave oven, the food product is run over by the waves generated by the magnetrons, whereas in the case of an RF system, the food products are generally placed between the electrodes [13]. Moreover, the RF system produces better penetration depth and homogeneous heating as compared to microwaves. This property can be used in the pasteurization and sterilization of liquid food products [35]. However, for penetrating inside the food having a bigger size, the microwave is used, though the penetrating ability is still limited. For instance, milk or yogurt products were passed through microwaves having a frequency of 2.35 kHz and RF systems having a frequency of 27.12 MHz. The results showed only 1 cm penetration depth of microwaves, whereas RF has resulted in 20 cm of penetration depth. Moreover, the higher affected diameter of an RF system produced more uniform and consistent heating [36]. Therefore, the limited ability of microwaves to penetrate the food products having high volume provides an opportunity to explore the potential of RF energy in commercial high-volume heating operations.

13.3.4 Lab Scale and Commercial Scale of RF Equipment

The principle of rapid heating and large depth penetration quality of radiofrequency can be used in several fields of food processing. Many types of research were done on the microorganism reduction, disinfection, thawing, and blanching of food products using RF. Short processing time, preserving texture, flavor, color, and nutrients of food products like vegetables, fruits, meats, and aquatic products are some advantages of radiofrequency. Generally, high-quality food products are the results of RF processing, along with the advantage of short processing time and low energy consumption. However, most of the research in RF processing of food is done on a laboratory scale which may or may not be scalable for commercial operations [26]. This is because some significant challenges occurred in the industrial application of RF processing in foods, such as non-uniform heating of food, which occurred at large volumes, high equipment cost, more energy consumption, etc. [26].

13.3.4.1 Radiofrequency Processing of Food at Lab Scale

The radiofrequency processing of foods has been researched since the 1960s. In the early time, the effect of RF processing on defrosting of the frozen products was searched, and its effect on quality was observed. Moreover, the potential use of RF processing in assisting food freezing was observed and found well-accepted results. A decrease in thawing loss of RF cryogenic frozen meat as compared to air cryogenic frozen meat was found [37]. RF processed cryogenic frozen meat showed better cellular structure, fewer intercellular voids, lower cell disruption, and smaller ice crystals at the intracellular level. RF was also applied in the cooking and pasteurization of the food, which produced better texture, color, and flavor than the conventional method [18]. Pasteurization of ready-to-eat aquatic food using RF at 25 MHz and 65°C showed complete inactivation of *innocua cells microorganisms* with excellent and acceptable product quality [38]. The combination of RF and convection heating showed remarkable 86% of cooking loss along with the similar or better inactivation of selected foodborne pathogens (*Escherichia coli, Listeria monocytogenes,* and *Salmonella Typhimurium* as well as spores of *Bacillus thuringiensis* and *Bacillus cereus*) as compared to convection only [39]. The RF can be easily applied in the drying of food products, especially for post-bake drying of cookies, crackers, and pasta. The problem of removing the non-uniform distributed moisture can be quickly sorted out by the application of RF processing [18]. Among the four drying techniques viz. hot air drying, infrared drying, hot-air-assisted radiofrequency drying, and microwave-assisted hot air drying, stem lettuce slices (*Lactuca sativa L.*) were dried uniformly and quickly by hot-air-assisted radiofrequency drying with most optimum quality retention [40]. Apart from drying, the heating principle of RF processing is also utilized in the disinfestation of the grains like chickpea, green pea, lentil, etc. The experiment of RF heating of these grains showed no significant differences in weight loss, color, moisture content, or germination when compared to unheated controls. However, lower heating time and increased heating rate were observed in comparison to hot air-based heating [41]. The disinfestation by RF is mainly due to the variation in the dielectric properties of grains and target insects. During RF heating, the lethal temperature for insects is achieved without any significant effect on the grains, thus resulting in the killing of insects. Moreover, the RF heating can reach the corners and edges effectively, which makes them an alternative option for disinfestation rather than fumigation [26].

13.3.4.2 *Radiofrequency Processing of Food at Industrial Scale*

A vast amount of research work hads been conducted in RF processing of food; however, there are some limitations in technology transfer to industrial scale like the higher capital cost of the equipment, non-uniform heating particularly in heterogeneous food material with an irregular shape, complex mechanism, and troubleshooting of RF equipment, etc. [5]. Nevertheless, in some areas like thawing, post-baking, and drying of selected food products, RF technology is used at the industry level.

During the post-baking process of biscuit, the most common problem of cracking of biscuit product arises due to the generation of differential stresses at varied locations. The stress developed by the removal of moisture at various uneven locations results in varied shrinkage within a single biscuit during the baking process. This process is generally called 'checking' in the biscuit industry. The checking was significantly reduced from 50% to almost 0% due to the incorporation of RF heating in the post-baking treatment of cracker or biscuit production line. The RF heating had the advantage of removing the inner moisture (generally present in the central zone) by penetrating through the insulating outer crust of the product. The removal of moisture was quick along with the increase in throughput by 40% [5].

Normal circulation of air upon the frozen products inside a cooling room (4°C) is the general practice of thawing or tempering. The process takes too much time for thawing, which causes substantial drip loss, an increase in bacteria growth, and product quality deterioration [5]. The utilization of RF heater (27.12 MHz, 500 W) for tempering 4-kg beef block from −20°C to −3.6°C was done [42]. The results showed a tremendous decrease in thawing time from 50 h (traditional method) to 35 min (RF heating). In the frozen tuna block, the RF tempering method was applied and found a threefold reduction in thawing time as compared to traditional thawing. The experiment showed that the use of a similar electrode size as of sample surface provided the best heating uniformity [42]. Many companies of France, Italy, the United States, the United Kingdom, and Japan have produced the defrosting/tempering RF system on a commercial scale. Italy-based company Stalam used an RF system having a tempering capacity of 2,500 kg/h (from −25°C to −5°C) with the power requirement of approximately 3-85 kW for meat, fish, and vegetables. Yamamoto Vinita (Japan) manufactures thawed up to 4,000 kg/h of thick meat blocks by RF equipment having the power of approximately 5–120 kW with the frequency of 13.56 MHz [5].

In the drying process using RF heating, the most challenging task to tackle is the uniform heating of the product like in the case of walnuts. Due to variations in the moisture content of walnuts, the non-uniform electromagnetic field is formed, which causes non-uniform heating, thus resulting in temperature variations among walnuts. Some other factors like walnut size, orientation, and location also affect the uniform heating. This can effectively be addressed by mixing walnuts between RF exposures [44]. One industrial-scale RF drying system of 1561.7 kg/h of capacity with almost 80% of average heating efficiency was developed with the power of 25 kW. The system utilizes the intermediate mixing of walnuts between RF exposures. However, the initial installation cost was significantly high, but the cost of energy for RF treatment of one kg of the walnut product was comparable to the cost of methyl bromide fumigation. The industrial-scale RF drying system for walnuts was effective; however, further, development is still required [45].

13.4 Applications in Food Processing

Radiofrequency heating has many promising applications in the food industry. Over the last two centuries, research has been carried out in the field of agriculture for postharvest pest control. Several applications of RF heating are described in the following sections and are summarized in Table 13.2.

13.4.1 Drying

RF heating system is used mostly during the final stage of drying and it has been used for several drying applications such as drying crackers, cookies, and snack foods (post-baking drying) [7]. It is more advantageous than conventional drying and microwave drying as it removes moisture more efficiently from solid or semi-solid foods, and prevents cracking and hardness of the shell. RF drying has the unique property of volumetric heating. During RF heating, water molecules absorb more energy compared to solid materials due to differences in their dielectric properties. Hence, the quality of the food products is protected by evaporating the water with minimal heating of food. However, the high operating cost is one of the major constraints for solely RF drying which is a key drawback to the food industry. Therefore, a combination of RF with other drying methods has been explored and utilized which maximizes the strength of technology by reducing the limitations. A combination of RF drying and vacuum drying

Table 13.2 Application of radiofrequency for disinfestations, thawing, drying and enzyme inactivation.

Purpose	Sample	Power/ frequency	Treatment conditions	Salient findings	References
Disinfestation	Walnuts (in-shell)	12 kW, 27.12 MHz	Heating to 55 °C at 5–6 °C. min^{-1}	Resulted in 100% mortality of fifth instar navel orangeworm. A combined system of RF heating with hot air has the potential to accelerate or even replace batch drying of walnuts.	[51]
Disinfestation	Walnuts (in-shell)	6 kW, 27.12 MHz	3 min heat to 54°C, hold for 5 min	Codling moth larvae mortality reached 100% after 3 min heating. The peroxide value of RF-treated walnuts was less than 1.0 meq/kg (the upper limit for good quality walnuts), after 20 days of storage at 35°C that offered 2-year storage at 4°C.	[69]

(*Continued*)

Table 13.2 Application of radiofrequency for disinfestations, thawing, drying and enzyme inactivation. (*Continued*)

Purpose	Sample	Power/ frequency	Treatment conditions	Salient findings	References
Disinfestation	Lentils	6 kW, 27.12 MHz	5-7 min heat to 60°C, hold for 10 min	RF treatments can be practical, effective, and environmentally friendly method for disinfestation of postharvest legumes while maintaining product quality.	[41]
Disinfestation	Rice (milled)	6 kW, 27.12 MHz	4.3 min heat to 50°C, hold for 5 min	RF treatments had no negative effects on the quality parameters and may provide a practical, effective, and environmentally friendly method for disinfesting milled rice.	[64]
Disinfestation	Lentils (Black-eyed peas and mungbeans)	6 kW, 27.12 MHz	5.5 min heat to 60°C, hold for 10 min	Industrial RF heating systems could be used to disinfect large quantities of these products.	[54]

(*Continued*)

Table 13.2 Application of radiofrequency for disinfestations, thawing, drying and enzyme inactivation. (*Continued*)

Purpose	Sample	Power/ frequency	Treatment conditions	Salient findings	References
Thawing	Lean beef meat	0.6 kW, 27.12 MHz	20–45 min heat to 0°C	Micronutrient loss and dip loss were lower and the process was rapid in RF thawed samples compared to the conventional air method.	[48]
Thawing	Tuna muscle	1 kW, 13.56 MHz	39.8, 32.4, 31.6, 30.5, 30.0 min heat to −3°C	Greater uniformity of end-point temperature distribution and a 3-fold reduction in thawing time was obtained with the RF system compared with the conventional process.	[43]
Drying	Macadamia nuts	12 kW, 27.12 MHz	20–120 min heat to 50–80°C	Hot air acted as a mixing medium and improved RF heating and drying uniformity in nuts.	

(*Continued*)

Table 13.2 Application of radiofrequency for disinfestations, thawing, drying and enzyme inactivation. (*Continued*)

Purpose	Sample	Power/ frequency	Treatment conditions	Salient findings	References
Enzyme inactivation	Apples	3.5 kW, 27.12 MHz	4-6.8 kV, 0-16 min Heat to 98°C, hold 30 min at room temperature	Complete inactivation of polyphenol oxidase and lipoxygenase with an electric field of 6.8 kV for 5 and 3 min, respectively. A slightly higher perceivable sweetness of the puree obtained from RF blanched samples was found in contrast to the water blanched sample.	[35]
Enzyme inactivation	Mustard seed flour	Power not given, 13.5 MHz	Heat to 112°C	The pungent flavor was eliminated by RF treatment as a result of an 85–90% reduction in myrosinase activity.	[70]

is used in textile, paper, and wood industries. RF drying of peanuts was studied to compare the drying rates and influential factors on the power absorption [45]. They found that power absorption is influenced by moisture content, bulk volume, and electrode distance while drying efficiency is increased by internal heating characteristics while combined with hot air drying.

Computer simulation of an RF-assisted fluidized bed drying was conducted by [46]. The author studied the key operating parameters and their effect on the time-varying RF heating system. The developed model predicted the drying curves and recommended that the product quality of heat-sensitive material could be improved by the use of an intermittent RF-assisted fluidized bed dryer. The effect of RF assisted with hot air oven on macadamia nuts was experimented [47]. The authors found that RF-assisted hot air drying reduced the drying time by half. But at the same time, they observed that there was no significant difference in the product quality. In another set of experiments, they evaluated the heating uniformity of nuts during hot-air-assisted RF and RF drying. It was observed that the intensity of the electric field was higher at the edges than the center. Therefore, the rate of heating at the corner was higher than at the center of the container. No significant improvement in uniform heating was observed while moving samples on a conveyor belt.

Although RF drying is extensively industrialized in textile, wood, and paper industries, as well as in post-baking of crackers and biscuits, few kinds of research have been carried out on drying of foods. This might be due to the thin layer drying of agricultural products that are mainly required and hot air drying is cost-effective for this, which too produces a reasonable quality of products. On the other hand, RF drying systems are too costly for drying bulk agricultural materials, such as cereals and grains. RF drying can be beneficial for achieving high-quality end products where no conventional means work out. It can also be helpful and cost-effective where space is limited. For example, throughput is increased up to 40% in the cracker/cookies production line using RF post-baking. Although the combined effect of RF and hot air drying is effective, many manufacturers have found that combining equipment needs many conflicting requirements which gives rise to many problems in different applications, making it unreliable and high cost.

13.4.2 Thawing

Thawing or tempering is a traditional process that is normally carried by still or circulating air in a cool room (4°C). This process is slow and time-consuming as it takes several days depending upon the size of the frozen products. Growth of microbes, drip loss, and degradation of product quality are the main problems of the traditional lengthy thawing process. Commercial use of RF-based thawing or tempering has been explored for decades, but in recent years, research on fish and meat thawing has been published with an experimental set scale of 50 Ω RF to provide high-quality products, reduce food waste, and meet the rising demands for food safety in consumer markets [5]. The tempering effect of RF heating on beef was studied and it was concluded that there was a reduction in thawing time from 50 h to 35 min when compared to traditional methods [42]. It was also observed that the homogeneity in sample composition could improve the uniform thawing. In a quality evaluation of defrosted beef, results showed a significant reduction of micronutrient loss and drip loss in RF defrosted products compared to conventional air-thawed meat [48]. A mathematical model was developed to predict the heating patterns during RF thawing of meat blocks with different electrode gaps [49]. This model was developed based on the experimental results of the previous literature [42]. The developed model was able to predict the temperature distribution in products during RF processing. Dielectric properties and efficacy of RF tempering of frozen tuna over a temperature range were determined [43]. The authors found that thawing time was reduced by threefold compared to the traditional thawing processes. Moreover, it was noticed that the use of an electrode comparable to the sample size provides the best heating uniformity. Nowadays, RF thawing and tempering methods are commercially used in processing industries but most are limited to regular shaped fish and meat products. There are still many challenges in the thawing of irregular food products. Extensive research may be carried out to solve these issues for large-scale applications.

13.4.3 Roasting

The effect of hot-air-assisted RF roasting on salted peanuts was studied [50]. An experimental setup with a 12-kW, 27.12-MHz was used for roasting at 110–130°C for 45 min. The volatile components and oxidation values of both the treated and untreated samples were analyzed. High roasting efficiency, improvement in product quality, and shelf life was observed during storage for 13-15 weeks.

13.4.4 Baking

For more than 50 years, commercial application of RF has been carried out in biscuit industries for baking. Checking is a term that refers to the cracks developed in the biscuits due to the stress differences at different locations, which is caused by uneven moisture removal and consequent shrinkage. Application of RF heating for baking of cracker or biscuit production reduces the checking approximately from 50% to 0%. Meanwhile, throughput is increased up to 40% because of the removal of water from the high moisture center zone by RF energy penetration to the insulating outer crust. By 1960, commercial application of RF post-baking was fully utilized; therefore very few recent publications can be found [5].

13.4.5 Disinfestation

Pest invasion and infestation are the major concerns in many countries due to the import and export of agricultural commodities. Cheap and effective control measures are the chemical methods (fumigation) to control infestation but are restricted in several countries. Various physical and biological methods are under development for more efficient and environmentally friendly use of RF heating that can kill the pest without damaging the product with its distinctive heating mechanism.

Several researchers have conducted extensive research to develop RF disinfestation systems by bridging the knowledge gap [51–53]. Their studies have led over a wide range of pests and host agricultural commodities for dielectric properties over a temperature range [54, 55], understanding the thermal death lethality of pests at different life stages [56, 57], the effectiveness of RF disinfestations, and pilot-scale validation in industrial level [44, 58], and computer simulation for improvement in uniformity of RF disinfestation system [59]. The RF disinfestation method could be applied to nuts, grains, and dried fruits because of the low heat sensitivity of dry foods compared to fresh produces [32, 60].

Many recent researches have been carried out on the inactivation of pests in coffee beans [61], legumes [41, 62], fish meal [63], milled rice [64], stored wheat [65], and nuts [44, 58, 66]. To improve the uniformity in temperature, mixing, moving and hot-air-assisted methods were applied within the food volume. The research reported no significant degradation in product quality at 100% insect lethality after RF treatment. This exhibits the potential of RF disinfestations for industrialization.

13.4.6 Blanching

Blanching or enzyme inactivation is a vital preservation storage step in fruits and vegetable processing industries. In 2006, the effect of RF heating (at 112°C in a 13.5-MHz RF machine) for inactivation of myrosinase enzyme in white mustard for removal of spicy flavor was studied [67].

They concluded that a low-cost and environmentally friendly RF technology could be developed for the thermal inactivation of enzymes. This will produce products with acceptable sensory attributes.

RF blanching of whole apples was studied at 6.2 kV for 3 min [35]. The effect of enzyme inactivation was compared with water blanched apple cubes. It was observed that RF efficiently inactivates the oxidizing enzymes and preserves more sweetness in apples under proper operating conditions. Waste management cost is reduced, and time and energy are saved during RF blanching. RF heating of miso paste could completely inactivate enzymes at 72°C, which is 12°C lower than the traditional methods [68]. This RF treatment also reduces two-thirds processing time of the conventional method. Although RF heating or blanching has many benefits, such as less water usage, shorter processing time, high capital cost, and lack of organized research methodologies in experiment development are the main constraints for commercial application of RF heating in blanching.

13.4.7 Pasteurization/Sterilization

To extend shelf life and ensure food safety, RF pasteurization or sterilization technologies are developed. RF heating of food elevates the food temperature much faster than traditional pasteurization or sterilization by directly penetrating the food packages. It further prevents cross-contamination and allows in-package pasteurization or sterilization. Food quality improves by a significant reduction in heating time. However, to assure the inactivation of microbes, the temperature must be monitored all the time at cold spots to guarantee lethality.

RF heating was used to destroy three different microbes in alfalfa seeds. However, the 5-log reduction of the pathogen was not achieved before affecting germination [71]. In a US-funded research project, a pilot-scale sterilization system was developed to process a six-pound tray containing macaroni and cheese for military group rations [32]. To reveal the heating pattern and simulate macaroni and cheese, a model food was developed with a chemical marker. It was observed that F_0=10 min lethality could be achieved in a 30 min RF heating instead of a 150 min retort process. Microbial efficacy of RF sterilized smashed potato packed in a six-pound tray using *Clostridium sporogenes* was verified [72]. The RF applications for bacterial inactivation are listed in Table 13.3.

Several researches have been carried out for microbial inactivation of food products, such as packaged vegetables [83], peaches/nectarines/stone fruits [79, 84, 85], peanut butter cracker sandwiches [86], and ground beef [39]. RF for pasteurization of liquid foods, e.g., soy milk [78], apple juice/

Table 13.3 Applications of radiofrequency in industrial application for bacterial inactivation.

Material	RF system	Power/ frequency	Treatment conditions	Log reduction/ species	References
Wheat flour	Free-running	6 kW, 27.12 MHz	15 min heat to 85 °C, hold for 25 min	3.7–5 log, *Enterococcus faecium*, *Salmonella Enteritidis*	[73]
Wheat flour	Free-running	500 W, 27.12 MHz	8.5–9 min heat to 75°C	3–7 log, *S. Enteritidis*, *E. faecium*	[74]
Non-fat dry milk	Free-running	3 kW, 27.12 MHz	4.3–5.5 min heat to 75–90 °C, hold for 13.6–75 min	3 log, *Cronobacter sakazakii* and *Salmonella*	[75]
Apple juice	Not provided	25 kV/cm, 20-60 kHz	1.6 s to 25, 55, and 75°C (depending on the voltage supply)	0.8, 7.3, and 6.6 log, *E. coli*	[76]
Orange juice	Not provided	80 kW, 20-40 kHz	270 µs heat to 60°C and 65°C, hold for 3 s	2.1 and 3.3 log, *E. coli* K12	[77]
Soymilk	Not provided	1 kW, 28 MHz	0.4 s heat to 115°C, hold for 10 s	4 log, *Bacillus subtilis*	[78]

(Continued)

Table 13.3 Applications of radiofrequency in industrial application for bacterial inactivation. (*Continued*)

Material	RF system	Power/ frequency	Treatment conditions	Log reduction/species	References
Peach and nectarines	50 Ω	15 kW, 27.12 MHz	18 min heat to 62.5°C	Not provided, *Monilinia* spp.	[79]
Meat (comminuted)	Not provided	0.5 kW, frequency not given	33 min heat to 67.6–76 °C, hold for 2 min (Sample in water)	5.4 and 1.8 log, *Bacillus cereus* and 6.8 and 4.1 log, *Clostridium perfringens*	[80]
Ground beef	Not provided	1.5 kW, 27.12 MHz	4.25 min heat to 72°C	7 log, *E. coli* K12	[81]
Fish meal	Free-running	0.17-1.2 kW, 6-14 MHz	1–2 min heat to 70–90 °C, hold for 1 min	3.3–6.2 log, *Salmonella* spp. and 3.1–5.3 log, *E. coli* O157:H7	[63]
Almonds	Free-running	6 kW, 27.12 MHz	2–4 min heat to 75°C, hold for 1.5 min	5 log, *Salmonella*	[82]
Alfalfa seed	Not provided	3 kW, 39 MHz	0–28 s heat to 50–126°C	0–1.5 log, *Salmonella, E. coli* O157:H7	[71]

cider [77, 87, 88], orange juice [89], and milk [6] has been carried out and found to enhance product quality with efficient processing.

In recent years, several researchers have investigated the potential application of RF in low moisture bulk foods and ingredients for pathogen control. Conventional heating of low moisture foods in bulk heating is slow and inefficient. This is due to the low thermal conductivity of foods with low moisture. But RF has unique properties of penetration depth and volumetric heating which can replace the traditional methods of pathogen control. Many researchers have studied the combined effect of thermal kinetics and water activity in low moisture foods [90–92] in developing RF pasteurization systems for spices, almonds, and peanut butter crackers [82, 86, 93–95]. Some research studies also indicated the issues of non-uniform heating during pathogen control in low moisture food [27, 92].

The main issue in the development of thermal treatments for pathogen control is that low water activities in low moisture food make extremely heat tolerant to the bacterial pathogens. To understand the relationship between pathogen kinetics and water activity during temperature elevation, a special water activity (WA) measuring apparatus was developed to measure the WA changes at different temperatures [90, 97]. A 5-7 log reduction of salmonella could be achieved after 8.5-9 min RF treatment of wheat flour (water activity 0.25, 0.45, and 0.65) [74]. Moreover, the efficacy of RF (Pilot-scale RF system) to inactivate *Enterococcus faecium* in wheat flour with 0.45 WA was evaluated. It was found that an RF treatment (kW, 27 MHz) for 39 min was sufficient to inactivate pathogens in wheat flour samples (3 kg). This concludes that RF heating can effectively destroy pathogens in a low moisture environment, though it is difficult to inactivate pathogens in low moisture foods than high moisture foods [73].

However, so far, the application of RF heating to the food industries is limited due to the complexity of food composition. The use of RF in food processing is restricted to lab-scale research and needs more systematic, comprehensive research on a pilot scale and commercial scale. Simultaneously, electric power consumption, capita cost, training, maintenance, and operations are the major concerns for the adoption of technology. Other constraints are discussed in detail in the subsequent section.

13.5 Technological Constraints, Health Hazards, and Safety Aspects

Radiofrequency finds many applications in the food processing sector and has become an emerging technology for rapid cooking, drying, disinfestation, etc. However, there are still potential areas in food processing where the advantages of RF energy are not yet utilized enough. There are some drawbacks of using an RF system which are mainly due to inconsistent product characteristics. If the food products having non-homogeneous dielectric properties are used in the RF system, then it can lead to possible non-uniform distribution of electric field in the product. This can cause non-uniform temperature distribution which ultimately leads to the occurrence of hot spots in several zones based on the product geometry, shape, size, etc. [97, 98]. This non-homogeneous temperature distribution was seen to be occurred in many food products, like fresh fruits [98], dried fruits [98], dry nuts [47], eggs [97], legumes [41, 100], and meat [101, 102]. This leads to serious problems of RF processed products, like, the survival of pathogen and/or insects, issues of overheating, concern regarding heating uniformity, microbial safety, and poor end quality [62, 91]. In this regard, different RF systems were developed to increase heating uniformity with the help of water immersion, hot air circulation, mixing, preconditioning the sample, rotating thc sample along its central axis, intermittent stirring/shaking, conveyor belt movement, electrode modification, adjusting the electrode gap during heating, and surrounding the container with polyetherimide (PEI) [103, 104].

If the electrical field of high strength or power is used across the sample packed food products processed using RF energy, then the possibility of dielectric breakdown (arcing) or damages can occur, resulting in packaging failure or product destruction [105]. This problem can be solved by using appropriate packaging material for the food products during RF pasteurization and sterilization of foods using RF energy [34]. Apart from this, the cost of RF-H equipment is more than the traditional heating systems due to the demand for high efficiency and output quality from RF systems [106].

As discussed above, the property of rapid heating with additional large depth penetration of radio waves is the major property due to which RF is beneficial in the food processing sector. However, there are some significant health hazards when the high frequency of radio waves is used during working. The primary one is the biological effects on humans when exposed to RF energy. This exposure results in heating of tissues and an increase in body temperature; the effect is often known as the "thermal" effect. This happens due to the inability of the human body to dissipate the excessive heat generated by the high-frequency radio waves. Particularly, the parts of the body where the blood circulation is low are much vulnerable to the RF heating phenomenon because of the lower rate of heat dissipation. However, low levels of RF do not produce significant harmful biological effects on the body as a result of heating. But if any such effect comes, then that will be referred to as "non-thermal" effects [107]. Specific absorption rate (SAR) is generally used to test the amount of RF exposures to the body. It shows the average rate of absorption of RF energy by each kilogram of tissue. The unit of SAR is watts per kg. Most of the time, this measurement criterion is used for checking the following safety guidelines by mobile phones. The exposure limit is set well below the levels known to show biological effects, i.e., the dissipation of heat by the tissues is good [108].

Among most of the concerns regarding RF exposure, one is the occurrence of cancer and tumor in the body. However, none of the studies has scientifically proven the statement. Some studies on the tumor formation in the animals due to RF exposures were not able to replicate their results independently. Most studies have failed to get any link of RF exposure to cancer [109]. Radiofrequency radiation can penetrate the thin skull's bone marrow of children more easily than the adult male. The absorption of radiation in doses per unit volume is roughly ten times higher in children than adults. However, most of these outcomes are based on the RF radiation of mobile phones [110]. Nevertheless, if the same frequency is being used in the RF processing of food, then unexpected exposure can cause similar results with the operators working with the RF equipment. The carcinogenic effect of RF on the food was not being noticed during the study of many literary works. However, there are some adverse and undesirable effects on the food like non-uniform heating of food products, and high temperatures on the surfaces and corners of the products [18, 49, 111]. Moreover, the RF equipment for heating is expensive as compared to equipment used in traditional heating systems [29].

Some safety aspects should always be maintained while working with RF equipment. Proper distance with the RF equipment must be maintained during the processing of food. As electromagnetic waves are invisible, any leakage of high-intensity radio waves can unknowingly cause some harm, which probably would not be detected until much later. During the operations, the workers can use an RF field monitor to check the presence of radio waves outside the equipment. These devices must have some sounding or alarming system to tell the operator of potential exposures of RF over the Maximum Permissible Exposure (MPE) limit. However, proper training must be given to use the monitors properly as per the manufacturer's instructions. In some cases, the RF protective clothing can be used for considering the extra safety of the operators using RF equipment where high-intensity radio waves are being used (CPWR - The Center for Construction Research and Training, 2016) [112].

13.6 Commercialization Aspects and Future Trends

Radiofrequency technology has considerable potential to be applied in various sectors of food processing like heating, drying, thawing, pasteurization, preservation, and many more. This has become possible due to the advantage of penetrating food products greater than 20 cm under processing. As a result, the heating becomes relatively efficient and uniform with the least adverse effect on the food quality and sensory attributes. As compared to MW-based processing, the RF units are typically less expensive and require less maintenance cost when scaling up to commercial plant applications because of the absence of magnetrons [29]. Laboratory application can be scaled up to a processing plant level owing to the greater economy at higher power levels and the incurrence of lower maintenance costs. Post-baking of biscuits and thawing/tempering of the meat are two RF applications that have been successfully commercialized in the food industry so far. RF drying and disinfestation are the processes that are partially commercialized in the food industry. The combination of hot air circulation with RF treatment has been found effective for drying and is commonly used to minimize the operational cost. Recently, RF postharvest disinfestation of agricultural samples has been commercialized in the food industry [5]. Furthermore, RF ovens are expected to appear as a replacement for microwave ovens for daily use.

However, most of the studies on RF processing still are lab-based, and it is required to transfer or scale them up to be utilized in commercial-scale operations. If the scale-up is made as a continuous operation, then the cost of mass production will be less, and the quality of the products will be higher as compared to batch operation. These developments can be carried out by experimenting on the effect of movement of the food products, the gap between the two electrodes, heat dissipation to the surrounding, cost per unit production, etc., on the pilot-scale production. The designed process parameters can be further optimized or maybe scaled-up by using modeling and simulations [3].

The commercial scale-up of the lab-based developed RF technology is the most futuristic area to be worked upon. The major challenges during the scale-up are heating uniformity, energy consumption, and equipment cost [26]. The RF heating technology can be combined with the other conventional drying methods like hot air and vacuum drying, which can provide improved heating, drying uniformity, high energy efficiency, and better product quality. Thus, further studies can be done on the development of different RF-conventional combined drying technologies along with their optimization. The application of computer simulation on the improvement of design for scale-up operations is another emerging area. Computer simulation has a great ability to provide an understanding of the heating mechanism, the test strategy, parameters optimization, and identification of the best conditions of RF treatments for target food products by solving the governing equations demonstrating the 3D distribution of the desired parameters including temperature and electric field distribution with minimum time, labor and experimental work [113, 114]. The simulations of the processing conditions are rapid, cheap, and flexible, which can be used for making optimal process design. Moreover, the validated simulations can predict important operational parameters, such as the distributions of RF power, moisture, and temperature in specific materials. The development of mathematical models backed by experimental results on the effect of different factors and methods on the electromagnetic interactions between RF and drying materials can be used for improving heating uniformity. These models can be further utilized for computer simulations. However, more focus should be done on developing multi-phases transport models and 3D coupled electromagnetics with consideration on sample shape changes for detailed knowledge of the RF drying process [2]. The application of RF technology in the preservation process like pasteurization, blanching, and disinfestation is still an important area of research. The interaction of RF electromagnetic field and the cells of food materials, and microorganisms as well as the food macromolecules, such as protein, carbohydrate, and enzymes still need to be clarified [26].

Apart from the industrial scalability, more research work is needed to be carried out on the practical applicability of RF technology through laboratory experiments. Reliable information about the dielectric properties of many experimented foods needs to be explored along with the effect of the temperature, moisture content, and frequency of radio waves. These data of dielectric properties can be used for improving the computer model prediction precision. The effect of pre-treatments of food products before RF drying should also be explored. This not only would improve drying rates but also affect the dielectric property of food products. Apart from salt or sucrose-based pre-treatment, new methods, like, blanching, dipping, osmotic dehydration, ultrasound, and pulsed electric fields should be researched upon due to the ability to prevent nutritional and sensory properties with improved drying rates [2]. Food quality and sensory perception of the RF heated powders can be evaluated to design an effective RF system by optimizing the factors influencing heating for the food powders. More studies need to be conducted on the microbial validation of RF heated food powders. The study on the lethal kinetics of various pathogens including their spores can be studied to develop an effective RF pasteurization/sterilization system [115].

Radiofrequency-based devices are common these days, like mobiles, laptops, cellphones, etc., and their RF-band designations are fixed in each country. If the RF is used in food applications and these frequencies are required, then there might be some chance of interference that should be avoided. This can be done by shielding the RF electromagnetic field used in the food processing equipment from other devices. Fortunately, radio waves have longer wavelengths, which makes them easier to shield than microwaves. Moreover, different foods need different RF for optimum processing [26, 29]. Thus, due to better shielding of RF than microwaves, there are future opportunities to design RF equipment for heating or processing of food products at home to replace the microwaves.

13.7 Conclusions

The use of radiofrequency (10-50 MHz) is one of the most important and promising modern heating techniques and has immense potential for achieving rapid and volumetric heating in foods. It provides safe and good-quality food products owing to deep penetration depth and moisture self-balance effects with the generation of no chemical residues. However, this technology is still at a preliminary stage in various food processing domains, and a lot of research would have to be carried out to come up with the industrial applications and consequent marketing. Food scientists and engineers can anticipate determining suitable RF frequencies, exposure time, and heating configuration for RF equipment. At the same time, the relative impact of RF on quality and organoleptic perception of food can be investigated to design an RF unit for a particular food or group of foods. Further improvements in RF heating uniformity with the combination of experimental, computer simulation, and microbiological validation studies will help RF heating to be scaled-up for the commercialized application of RF pasteurization in the food industry.

References

1. Piyasena, P., Dussault, C., Koutchma, T., Ramaswamy, H. S., Awuah, G. B. (2003). Radio frequency heating of foods: principles, applications and related properties–a review. *Critical Reviews in Food Science and Nutrition,* 43: 587–606.

2. Zhou, X., & Wang, S. (2019). Recent developments in radio frequency drying of food and agricultural products: A review. *Drying Technology, 37*(3), 271-286.
3. Bedane, T. F., Chen, L., Marra, F., & Wang, S. (2017). Experimental study of radio frequency (RF) thawing of foods with movement on conveyor belt. *Journal of Food Engineering, 201*, 17-25.
4. Farag, K. W., Lyng, J. G., Morgan, D. J., & Cronin, D. A. (2011). A comparison of conventional and radio frequency thawing of beef meats: effects on product temperature distribution. *Food and Bioprocess Technology, 4*(7), 1128-1136.
5. Jiao, Y., Tang, J., Wang, Y., & Koral, T. L. (2018). Radio-frequency applications for food processing and safety. *Annual Review of Food Science and Technology, 9*, 105-127.
6. Awuah, G. B., Ramaswamy, H. S., Economides, A., & Mallikarjunan, K. (2005). Inactivation of *Escherichia coli* K-12 and *Listeria innocua* in milk using radio frequency (RF) heating. *Innovative Food Science & Emerging Technologies, 6*(4), 396-402.
7. Mermelstein, N.H. (1998). Microwave and radio frequency drying. *Food Technology*, 52(11):84–86.
8. Sherman, V.W., 1946. *Food Industry* 18, 506–509. 628.
9. Moyer, J., and Stotz, E. (1947). The blanching of vegetables by electronics. *Food Technol.* **1**: 252-257.
10. Marra F, Zhang L, Lyng JG (2009) Radio frequency treatment of foods: review of recent advances. *J Food Eng* 91(4):497–508. https ://doi.org/10.1016/j.jfood eng.2008.10.015
11. Rice J. 1993. RF technology sharpens bakery's competitive edge. *Food Process.* 6:18–24.
12. Houben, J., Schoenmakers, L., Putten, E. V., Roon, P. V., & Krol, B. (1991). Radio-frequency pasteurization of sausage emulsions as a continuous process. *Journal of Microwave Power and Electromagnetic Energy, 26*(4), 202-205.
13. Zhao, Y., Flugstad, B., Kolbe, E., Park, J. W., and Wells, J. H. (2000). Using capacitive (radio frequency) dielectric heating in food processing and preservation-a review. *J. Food Proc. Eng.* **23**: 25-55.
14. Birla, S., Wang, S., Tang, J., Fellman, J., Mattinson, D., and Lurie, S. (2005). Quality of oranges as influenced by potential radio frequency heat treatments against Mediterranean fruit flies. *Postharvest Biol. Technol.* **38**: 66-79.
15. Datta, A. K., & Davidson, P. M. (2000). Microwave and radio frequency processing. *Journal of Food Science, 65*, 32-41.
16. Ryynänen, S. The electromagnetic properties of food materials: a review of the basic principles. *Journal of Food Engineering*, 26, 4, 409-429, 1995.
17. Hartshorn, Leslie. Radio-frequency heating, 607, 1949.
18. Jojo, S., & Mahendran, R. (2013). Radio frequency heating and its application in food processing: A review. *International Journal of Current Agricultural Research (IJCAR), 1*(9), 042-046.

19. Bhatt, Kanchan, Devina Vaidya, Manisha Kaushal, Anil Gupta, Pooja Soni, Priyana Arya, Anjali Gautam, and Chetna Sharma. Microwaves and Radio waves: In Food Processing and Preservation. *International Journal of Current Microbiology and Applied Science*, 9, 9, 118-131, 2020.
20. Nelson, S.O. Electrical properties of agricultural products—a critical review. *Transactions of the ASAE*, 16, 1, 384–400, 1973.
21. Nyfors, E. & Vainikainen, P. *Industrial Microwave Sensors,* Chapter 2. Artech House, Norwood., 1989.
22. Risman, P. O. Terminology and notation of microwave power and electromagnetic energy. *I. Microw. Power Electromagn. Energy*, 26, 243-50, 1991.
23. Mudgett, R.E. Electrical properties of foods. In *Engineering Properties of Foods*, Ed. M. A. Rao & S.S.H. Rizvi; Marcel Dekker, New York, 329–390, 1986.
24. Buffler, C. R. "Dielectric properties of foods and microwave materials." *Microwave Cooking and Processing*, 46-69, 1993.
25. Ohlsson, T. Dielectric properties and microwave processing. In *Food Properties and Computer-aided Engineering of Food Processing Systems*, eds. R. P. Singh & A. G. Medina. Kluwer Academic Publishers, Amsterdam, 73-92, 1989.
26. Guo, C., Mujumdar, A. S., & Zhang, M. (2019). New development in radio frequency heating for fresh food processing: A review. *Food Engineering Reviews, 11*(1), 29-43.
27. Jiao S, Luan D, Tang J (2014) Principles of radio-frequency and microwave heating. In: Awuah GB, Ramaswamy HS, Tang J (eds.) *Radio-frequency heating in food processing: principles and applications.* CRC Press, Boca Raton.
28. Di Rosa AR, Leone F, Cheli F, Chiofalo V (2019) Novel approach for the characterisation of sicilian honeys based on the correlation of physicochemical parameters and artificial senses. *Italian J Anim Sci* 18(1):389–397. https://doi.org/10.1080/1828051X.2018.15309 62
29. Altemimi, A., Aziz, S. N., Al-Hilphy, A. R., Lakhssassi, N., Watson, D. G., & Ibrahim, S. A. (2019). Critical review of radio-frequency (RF) heating applications in food processing. *Food Quality and Safety*, *3*(2), 81-91.
30. Mermelstein, N. (1997). Interest in radio frequency heating heats up. *Food Technol.* **50**: 94-95.
31. Demeczky, M., 1974. Continuous pasteurisation of bottled fruit juices by high frequency energy. *Proceedings of IV International Congress on Food Science and Technology IV*, 11–20.
32. Wang, Y., Wig, T. D., Tang, J., & Hallberg, L. M. (2003a). Sterilization of foodstuffs using radio frequency heating. *Journal of Food Science, 68*(2), 539-544.
33. Tang J, Chan TVCT (2007) Microwave and radio frequency in sterilization and pasteurization applications. In: Yanniotis S (ed.) *Heat transfer in food processing*, Ashurst Lodge. WIT Press, Ashurst, Southampton.

34. Hassan HF, Tola Y, Ramaswamy HS (2014) Radio-frequency and microwave applications. In: Awuah GB, Ramaswamy HS, Tang J (eds.) *Radio-frequency heating in food processing: principles and applications*. CRC Press, Boca Raton.
35. Manzocco, L., Anese, M., & Nicoli, M. C. (2008). Radiofrequency inactivation of oxidative food enzymes in model systems and apple derivatives. *Food Research International, 41*(10), 1044-1049.
36. Felke, K., Pfeiffer, T., Eisner, P. (2009). Neues Verfahren zur schnellenund schonendenErhitzungvonverpacktenLebensmitteln: Hochfrequenzerhitzung im Wasserbad. *Chemie Ingenieur Technik*, 81(11): 1815–1821.
37. Anese, M., Manzocco, L., Panozzo, A., Beraldo, P., Foschia, M., & Nicoli, M. C. (2012). Effect of radiofrequency assisted freezing on meat microstructure and quality. *Food Research International, 46*(1), 50-54.
38. Al-Holy, M., Ruiter, J., Lin, M., Kang, D. H., & Rasco, B. (2004). Inactivation of *Listeria innocua* in nisin-treated salmon (*Oncorhynchus keta*) and sturgeon (*Acipenser transmontanus*) caviar heated by radio frequency. *Journal of food protection, 67*(9), 1848-1854.
39. Schlisselberg, D. B., Kler, E., Kalily, E., Kisluk, G., Karniel, O., & Yaron, S. (2013). Inactivation of foodborne pathogens in ground beef by cooking with highly controlled radio frequency energy. *International Journal of Food Microbiology, 160*(3), 219-226.
40. Roknul, A. S., Zhang, M., Mujumdar, A. S., & Wang, Y. (2014). A comparative study of four drying methods on drying time and quality characteristics of stem lettuce slices (Lactuca sativa L.). *Drying Technology, 32*(6), 657-666.
41. Wang, S., Tiwari, G., Jiao, S., Johnson, J. A., & Tang, J. (2010). Developing postharvest disinfestation treatments for legumes using radio frequency energy. *Biosystems Engineering, 105*(3), 341-349.
42. Farag, K., Lyng, J.G., Morgan, D.J., Cronin, D.A., 2008. Dielectric and thermophysical properties of different beef meat blends over a temperature range of -18 to +10 °C. *Meat Science* 79 (4), 740–747.
43. Llave, Y., Terada, Y., Fukuoka, M., & Sakai, N. (2014). Dielectric properties of frozen tuna and analysis of defrosting using a radio-frequency system at low frequencies. *Journal of Food Engineering, 139*, 1-9.
44. Wang, S., Monzon, M., Johnson, J. A., Mitcham, E. J., & Tang, J. (2007). Industrial-scale radio frequency treatments for insect control in walnuts: I: Heating uniformity and energy efficiency. *Postharvest Biology and Technology, 45*(2), 240-246.
45. Porterfield JG, Wright ME. (1971). Heating and drying peanuts with radio-frequency energy. *Trans. ASAE* 14(4):629–33.
46. Jumah R. 2005. Modelling and simulation of continuous and intermittent radio frequency–assisted fluidized bed drying of grains. *Food Bioprod. Process.* 83(C3):203–10.
47. Wang, Y., Zhang, L., Gao, M., Tang, J., & Wang, S. (2014). Pilot-scale radio frequency drying of macadamia nuts: heating and drying uniformity. *Drying Technology, 32*(9), 1052-1059.

48. Farag KW, Duggan E, Morgan DJ, Cronin DA, Lyng JG. 2009. A comparison of conventional and radio frequency defrosting of lean beef meats: effects on water binding characteristics. *Meat Sci.* 83(2):278–84.
49. Uyar, R., Bedane, T. F., Erdogdu, F., Palazoglu, T. K., Farag, K. W., & Marra, F. (2015). Radio-frequency thawing of food products–A computational study. *Journal of Food Engineering, 146*, 163-171.
50. Jiao, S., Zhu, D., Deng, Y., & Zhao, Y. (2016). Effects of hot air-assisted radio frequency heating on quality and shelf-life of roasted peanuts. *Food and Bioprocess Technology*, *9*(2), 308-319.
51. Mitcham, E. J., Veltman, R. H., Feng, X., de Castro, E. D., Johnson, J. A., Simpson, T. L., ... & Tang, J. (2004). Application of radio frequency treatments to control insects in in-shell walnuts. *Postharvest Biology and Technology*, *33*(1), 93-100.
52. Tang, J., Ikediala, J. N., Wang, S., Hansen, J. D., & Cavalieri, R. P. (2000). High-temperature-short-time thermal quarantine methods. *Postharvest Biology and Technology*, *21*(1), 129-145.
53. Wang, S., Tang, J., Cavalieri, R. P., & Davis, D. C. (2003b). Differential heating of insects in dried nuts and fruits associated with radio frequency and microwave treatments. *Transactions of the ASAE*, *46*(4), 1175.
54. Jiao, S., Johnson, J. A., Tang, J., Tiwari, G., & Wang, S. (2011). Dielectric properties of cowpea weevil, black-eyed peas and mung beans with respect to the development of radio frequency heat treatments. *Biosystems Engineering*, *108*(3), 280-291.
55. Guo, W., Wang, S., Tiwari, G., Johnson, J. A., & Tang, J. (2010). Temperature and moisture dependent dielectric properties of legume flour associated with dielectric heating. *LWT-Food Science and Technology*, *43*(2), 193-201.
56. Wang, S., Johnson, J. A., Hansen, J. D., & Tang, J. (2009). Determining thermotolerance of fifth-instar Cydia pomonella (L.)(Lepidoptera: Tortricidae) and Amyelois transitella (Walker)(Lepidoptera: Pyralidae) by three different methods. *Journal of Stored Products Research*, *45*(3), 184-189.
57. Armstrong, J. W., Tang, J., & Wang, S. (2009). Thermal death kinetics of Mediterranean, Malaysian, melon, and oriental fruit fly (Diptera: Tephritidae) eggs and third instars. *Journal of Economic Entomology*, *102*(2), 522-532.
58. Wang, S., Monzon, M., Johnson, J. A., Mitcham, E. J., & Tang, J. (2007b). Industrial-scale radio frequency treatments for insect control in walnuts: II: Insect mortality and product quality. *Postharvest Biology and Technology*, *45*(2), 247-253.
59. Tiwari, G., Wang, S., Tang, J., & Birla, S. L. (2011). Computer simulation model development and validation for radio frequency (RF) heating of dry food materials. *Journal of Food Engineering*, *105*(1), 48-55.
60. Hou, L., Johnson, J. A., & Wang, S. (2016). Radio frequency heating for postharvest control of pests in agricultural products: A review. *Postharvest Biology and Technology*, *113*, 106-118.

61. Pan, L., Jiao, S., Gautz, L., Tu, K., & Wang, S. (2012). Coffee bean heating uniformity and quality as influenced by radio frequency treatments for postharvest disinfestations. *Transactions of the ASABE, 55*(6), 2293-2300.
62. Jiao, S., Johnson, J. A., Tang, J., & Wang, S. (2012). Industrial-scale radio frequency treatments for insect control in lentils. *Journal of Stored Products Research, 48*, 143-148.
63. Lagunas-Solar, M. C., Zeng, N. X., Essert, T. K., Truong, T. D., Pina, C., Cullor, J. S., ... & Larraín, R. (2005). Disinfection of fishmeal with radiofrequency heating for improved quality and energy efficiency. *Journal of the Science of Food and Agriculture, 85*(13), 2273-2280.
64. Zhou, L., Ling, B., Zheng, A., Zhang, B., & Wang, S. (2015). Developing radio frequency technology for postharvest insect control in milled rice. *Journal of Stored Products Research, 62*, 22-31.
65. Shrestha, B., Yu, D., & Baik, O. D. (2013). Elimination of cruptolestes ferrungineus s. in wheat by radio frequency dielectric heating at different moisture contents. *Progress In Electromagnetics Research, 139*, 517-538.
66. Hou, L., Hou, J., Li, Z., Johnson, J. A., & Wang, S. (2015). Validation of radio frequency treatments as alternative non-chemical methods for disinfesting chestnuts. *Journal of Stored Products Research, 63*, 75-79.
67. Ildikó, S. G., Klára, K. A., Marianna, T. M., Ágnes, B., Zsuzsanna, M. B., & Bálint, C. (2006). The effect of radio frequency heat treatment on nutritional and colloid-chemical properties of different white mustard (Sinapis alba L.) varieties. *Innovative Food Science & Emerging Technologies, 7*(1-2), 74-79.
68. Uemura K, Takahashi C, Kobayashi I. (2014). Inactivation of enzymes in packed miso paste by radio-frequency heating. *J. Jpn. Soc. Food Sci. Technol.* 61(2), 95–99.
69. Wang S, Ikediala JN, Tang J, Hansen JD, Mitcham E, Mao R. 2001. Radio frequency treatments to control codling moth in in-shell walnuts. *Postharvest Biol. Technol.* 22:29–38.
70. Schuster-Gajzago I, Kiszter AK,Toth-Markus M, Bardth A, Markus-BednarikA, CzukorB. 2006. The effect of radio frequency heat treatment on nutritional and colloid-chemical properties of different white mustard (*Sinapis alba* L.) varieties. *Innov. Food Sci. Emerg. Technol.* 7(1-2):74-79.
71. Nelson, S. O., Lu, C. Y., Beuchat, L. R., & Harrison, M. A. (2002). Radio–frequency heating of alfalfa seed for reducing human pathogens. *Transactions of the ASAE, 45*(6), 1937.
72. Luechapattanaporn, K., Wang, Y., Wang, J., Al-Holy, M., Kang, D. H., Tang, J., & Hallberg, L. M. (2004). Microbial safety in radio-frequency processing of packaged foods. *Journal of Food Science, 69*(7), 201-206.
73. Liu, S., Ozturk, S., Xu, J., Kong, F., Gray, P., Zhu, M. J., ... & Tang, J. (2018). Microbial validation of radio frequency pasteurization of wheat flour by inoculated pack studies. *Journal of Food Engineering, 217*, 68-74.

74. Villa-Rojas, R., Zhu, M. J., Marks, B. P., & Tang, J. (2017). Radiofrequency inactivation of *Salmonella enteritidis* PT 30 and *Enterococcus faecium* in wheat flour at different water activities. *Biosystems engineering, 156*, 7-16.
75. Michael M, Phebus R K, Thippareddi H, Subbiah J, Birla SL, Schmidt, K. A. 2014. Validation of radio-frequency dielectric heating system for destruction of *Cronobacter sakazakii* and *Salmonella* species in nonfat dry milk. *J. Dairy Sci.* 97(12):7316–24
76. Ukuku DO, Geveke DJ. 2012. Effect of thermal and radio frequency electric fields treatments on *Escherichia coli* bacteria in apple juice. *J. Microb. Biochem. Technol.* 4(3):76–81
77. Geveke, D. J., Gurtler, J., & Zhang, H. Q. (2009). Inactivation of *Lactobacillus plantarum* in apple cider, using radio frequency electric fields. *Journal of Food Protection, 72*(3), 656-661.
78. Uemura, K., Takahashi, C., & Kobayashi, I. (2010). Inactivation of Bacillus subtilis spores in soybean milk by radio-frequency flash heating. *Journal of Food Engineering, 100*(4), 622-626.
79. Casals, C., Viñas, I., Landl, A., Picouet, P., Torres, R., & Usall, J. (2010). Application of radio frequency heating to control brown rot on peaches and nectarines. *Postharvest Biology and Technology, 58*(3), 218-224.
80. Byrne B, Lyng JG, Dunne G, Bolton DJ. 2010. Radio frequency heating of comminuted meats—considerations in relation to microbial challenge studies. *Food Control* 21(2):125–31
81. Guo Q, Piyasena P, Mittal GS, Si W, Gong J. 2006. Efficacy of radio frequency cooking in the reduction of *Escherichia coli* and shelf stability of ground beef. *Food Microbiol.* 23(2):112–18
82. Gao, M., Tang, J., Villa-Rojas, R., Wang, Y., & Wang, S. (2011). Pasteurization process development for controlling Salmonella in in-shell almonds using radio frequency energy. *Journal of Food Engineering, 104*(2), 299-306.
83. Liu, Q., Zhang, M., Xu, B., Fang, Z., & Zheng, D. (2015). Effect of radio frequency heating on the sterilization and product quality of vacuum packaged Caixin. *Food and Bioproducts Processing, 95*, 47-54.
84. Sisquella, M., Casals, C., Picouet, P., Vinas, I., Torres, R., & Usall, J. (2013). Immersion of fruit in water to improve radio frequency treatment to control brown rot in stone fruit. *Postharvest Biology and Technology, 80*, 31-36.
85. Sisquella, M., Viñas, I., Picouet, P., Torres, R., & Usall, J. (2014). Effect of host and Monilinia spp. variables on the efficacy of radio frequency treatment on peaches. *Postharvest Biology and Technology, 87*, 6-12.
86. Ha, J. W., Kim, S. Y., Ryu, S. R., & Kang, D. H. (2013). Inactivation of *Salmonella enterica serovar Typhimurium* and *Escherichia coli* O157: H7 in peanut butter cracker sandwiches by radio-frequency heating. *Food Microbiology, 34*(1), 145-150.
87. Geveke, D. G., & Brunkhorst, C. (2004). Inactivation of in apple juice by radio frequency electric fields. *Journal of Food Science, 69*(3), FEP134-FEP0138.

88. Geveke, D. J., & Brunkhorst, C. (2008). Radio frequency electric fields inactivation of *Escherichia coli* in apple cider. *Journal of Food Engineering, 85*(2), 215-221.
89. Geveke, D. J., Brunkhorst, C., & Fan, X. (2007). Radio frequency electric fields processing of orange juice. *Innovative Food Science & Emerging Technologies, 8*(4), 549-554.
90. Syamaladevi, R. M., Tadapaneni, R. K., Xu, J., Villa-Rojas, R., Tang, J., Carter, B., ... & Marks, B. (2016a). Water activity change at elevated temperatures and thermal resistance of Salmonella in all purpose wheat flour and peanut butter. *Food Research International, 81*, 163-170.
91. Syamaladevi, R. M., Tang, J., Villa-Rojas, R., Sablani, S., Carter, B., & Campbell, G. (2016b). Influence of water activity on thermal resistance of microorganisms in low-moisture foods: a review. *Comprehensive Reviews in Food Science and Food Safety, 15*(2), 353-370.
92. Villa-Rojas, R., Tang, J., Wang, S., Gao, M., Kang, D. H., Mah, J. H., ... & Lopez-Malo, A. (2013). Thermal inactivation of Salmonella Enteritidis PT 30 in almond kernels as influenced by water activity. *Journal of Food Protection, 76*(1), 26-32.
93. Gao, M., Tang, J., Johnson, J. A., & Wang, S. (2012). Dielectric properties of ground almond shells in the development of radio frequency and microwave pasteurization. *Journal of Food Engineering, 112*(4), 282-287.
94. Jeong, S. G., & Kang, D. H. (2014). Influence of moisture content on inactivation of *Escherichia coli* O157: H7 and *Salmonella enterica* serovar Typhimurium in powdered red and black pepper spices by radio-frequency heating. *International Journal of Food Microbiology, 176*, 15-22.
95. Kim, S. Y., Sagong, H. G., Choi, S. H., Ryu, S., & Kang, D. H. (2012). Radio-frequency heating to inactivate *Salmonella Typhimurium* and *Escherichia coli* O157: H7 on black and red pepper spice. *International Journal of Food Microbiology, 153*(1-2), 171-175.
96. Tadapaneni, R. K., Syamaladevi, R. M., Villa-Rojas, R., & Tang, J. (2017). Design of a novel test cell to study the influence of water activity on the thermal resistance of Salmonella in low-moisture foods. *Journal of Food Engineering, 208*, 48-56.
97. Alfaifi, B., Tang, J., Jiao, Y., Wang, S., Rasco, B., Jiao, S., and Sablani, S. (2014). Radio frequency disinfestation treatments for dried fruit: model development and validation. *J. Food Eng.* 120: 268-276.
98. Birla, S., Wang, S., Tang, J., and Hallman, G. (2004). Improving heating uniformity of fresh fruit in radio frequency treatments for pest control. *Postharvest Biol. Technol.* 33: 205-217.
99. Lau, S. K., Thippareddi, H., Jones, D., Negahban, M., and Subbiah, J. (2016). Challenges in radio frequency pasteurization of shell eggs: coagulation rings. *J. Food Sci.* 81: 2492-2502.

100. Huang, Z., Chen, L., and Wang, S. (2015). Computer simulation of radio frequency selective heating of insects in soybeans. *Int. J. Heat Mass Transfer.* 90: 406-417.
101. Llave, Y., Liu, S., Fukuoka, M., and Sakai, N. (2015). Computer simulation of radiofrequency defrosting of frozen foods. *J. Food Eng.* 152: 32-42.
102. Uyar, R., Erdogdu, F., Sarghini, F., and Marra, F. (2016). Computer simulation of radio-frequency heating applied to block-shaped foods: Analysis on the role of geometrical parameters. *Food Bioprod. Process.* 98: 310-319.
103. Boreddy, S. R.; Thippareddi, H.; Froning, G.; Subbiah, J. Novel Radiofrequency-assisted Thermal Processing Improves the Gelling Properties of Standard Egg White Powder. *J. Food Sci.* 2016, *81*, E665–E671. DOI: 10.1111/1750-3841.13239.
104. Choi, E. J.; Park, H. W.; Yang, H. S.; Kim, J. S.; Chun, H. H. Effects of 27.12 MHz Radio Frequency on the Rapid and Uniform Tempering of Cylindrical Frozen Pork Loin (Longissimus Thoracis Et Lumborum). *Korean J. Food Sci. An.* 2017, *37*, 518–528. DOI: 10.5851/kosfa.2017.37.4.518.
105. Zhao, Y. (2006). Radio frequency dielectric heating. In: Sun D-W (ed.) *Thermal food processing: new technologies and quality issues*. CRC Press, Boca Raton.
106. Jones, P. L., Rowley, A. (1997). *Dielectric Dryers in Industrial Drying of Foods*. London: Blackie Academic and Professional.
107. Fields, E., Cleveland, R. F., & Ulcek, J. L. (1999). Questions and answers about biological effects and potential hazards of radiofrequency electromagnetic fields. In *Oet Bulletin*.
108. Kwan-Hoong, N. (2003). *Radiation, mobile phones, base stations and your health*. Malaysian Communications and Multimedia Commission.
109. Meena, J. K., Verma, A., Kohli, C., & Ingle, G. K. (2016). Mobile phone use and possible cancer risk: Current perspectives in India. *Indian Journal of Occupational and Environmental Medicine*, *20*(1), 5.
110. Miller, A. B., Sears, M., Hardell, L., Oremus, M., & Soskolne, C. L. (2019). Risks to health and well-being from radio-frequency radiation emitted by cell phones and other wireless devices. *Frontiers in Public Health*, *7*, 223.
111. Wang, J., Luechapattanaporn, K., Wang, Y., & Tang, J. (2012). Radio-frequency heating of heterogeneous food–meat lasagna. *Journal of Food Engineering*, *108*(1), 183-193.
112. CPWR - The Center for Construction Research and Training (2016). Radiofrequency (RF) Radiation Awareness Guide for the Construction Industry. North America's Building Trades Unions (NABTU). https://www.cpwr.com/wp-content/uploads/publications/RF_Radiation_Awareness_Program_Guide_8_2016.pdf
113. Huang, Z., Marra, F. Wang, S. (2016). A novel strategy for improving radio frequency heating uniformity of dry food products using computational modeling. *Innovative Food Science & Emerging Technologies*, 34: 100–111.

114. Li, Y.; Li, F.; Tang, J.; Zhang, R.; Wang, Y.; Koral, T.; Jiao, Y. Radio Frequency Tempering Uniformity Investigation of Frozen Beef with Various Shapes and Sizes. *Innov. Food Sci. Emerg. Technol.* 2018, *48*, 42–55. DOI: 10.1016/j.ifset.2018.05.008.
115. Dag, D., Singh, R. K., & Kong, F. (2020). Developments in Radio Frequency Pasteurization of Food Powders. *Food Reviews International*, 1-18.

14

Quality, Food Safety and Role of Technology in Food Industry

Nartaj Singh[1*] **and Prashant Bagade**[2]

[1]*Scientific & Technical Advisor, KFPL, Jammu, India*
[2]*Head – R&D, NCML, Hyderabad, India*

Abstract

Food has achieved global trading status, leading to a host of quality and safety issues. From the innocuous beginning of basic quality regulations implemented decades ago, there are a plethora of global quality and safety standards in existence now. However, despite the implementation of such standards, there have been many instances of compromises in food quality and safety. As expected, the effects of such compromises are broadly proportional to the socio-economic development of the countries. Traditionally, food quality and safety evaluations were tedious and time-consuming. But with recent technological advances, it has become easier and quicker to monitor such issues at various points across the food value chain. A few such technologies include the use of newer IoT-based technologies such as Artificial Intelligence (AI), Machine Learning (ML), Sensors, Hyperspectral Imaging, Blockchain, etc., which are being used across the value chain from production/processing to retail outlets. AI and ML are also being widely used in determining consumer preferences for product development. With the food processing industry projected to grow significantly across the globe during years to come, it is imperative that these technologies are widely deployed across processes to monitor and improve food quality and safety. This chapter gives an overview of developments in food standardization, the world situation concerning food safety and preventing food contamination, methods of quality control, and organizations dealing with food quality and safety.

***Keywords*:** Food contamination, machine learning, artificial intelligence, processing

**Corresponding author*: nartaj@gmail.com

Nitin Kumar, Anil Panghal and M. K. Garg (eds.) Thermal Food Engineering Operations, (415–454)

14.1 Introduction

Food quality and safety are the areas we bet on our businesses each and every day. It is more relevant in today's world considering the extent of quality and safety issues being witnessed regularly. The challenge is to ensure that the food being consumed is safe and is of claimed quality. Across the globe, various regulations have been promulgated to ensure food safety and quality. Despite the regulations, there have been many instances of compromises, either intentional or otherwise. Irrespective of the intent, they have been causing health hazards to consumers and costing the national exchequer in addressing the consequences. The challenge can be addressed by a unified and determined effort to address food quality and safety on the part of all the stakeholders - producers, wholesalers, retailers, processors, packers, regulatory agencies, and consumers.

Though food quality and food safety are two different terms and have their own meaning, they are intertwined to such an extent that it will be difficult to discuss one without talking about the other. In general, food quality broadly refers to both internal and external quality parameters of the product such as appearance, texture, size, shape, flavor, nutritional composition, and presence or absence of various chemicals such as pesticides, antibiotics, dyes, etc. It also involves determining pathogens responsible for having deleterious effects on human health and/or the product itself. However, food safety primarily refers to processes such as preparing (manufacturing), handling, storing food products that result in preventing health hazards when consumed within the recommended duration. Hence, food quality and safety can be defined as "a set of practices and conditions that ensure food is maintained in a state that is safe to consume by those for whom it is intended for and retains all the nutrients that it is intended to during its shelf life under the defined storage conditions". Though food safety has been practiced over the years by many food industries across the globe, it has gained a lot of importance in the recent past due to increased public awareness, stringent food regulations, and global trade requirements as mandated by importing countries. Despite such practices, there have been numerous instances of food safety issues not just in developing nations but in developed countries too. However, the effects are not uniform across the globe but are broadly in agreement with the socio-economic status of the region, as shown in Table 14.1. According to WHO [1], unsafe food causes 600 million cases of foodborne diseases and results in 420,000 deaths across the globe, annually. As per a report of the World Bank released in 2018, it costs low- and middle-income countries

Table 14.1 Regional effects of food quality and safety (Source: WHO).

Region	No. of people (in Millions) falling ill annually	No. of deaths
Eastern Mediterranean	> 100	37,000
European	23	5,000
Western Pacific	125	> 50,000
American	77	> 9,000
South-Eastern Asia	> 150	> 175,000
African	> 91	137,000

about USD 110 billion per year in lost productivity and for treating food-borne illnesses [2].

The twentieth century saw remarkable advances in all areas of food science and technology and also in nutrition science. These changes have in turn required greater flexibility in legal controls, to adequately protect the consumer from newly emerging hazards and to assist the food trade in its development. Many developed nations and some emerging nations have recently either completely reviewed and updated their food laws or provided new laws to meet the new situation.

14.1.1 Food Quality

The notion of food quality is multi-dimensional since it involves a wide range of characteristics that are unique to a product (Figure 14.1). It is one of the most important aspects considered by a consumer prior to purchase of food products and has assumed greater significance in the recent past due to increasing food frauds. In earlier days, quality referred primarily to the absence of defects, which for food products refers to the absence of extraneous matter. However, with time, it has been extended to include other parameters describing various aspects of the food including production practices, processing information, and specialty characters such as the absence of allergens, gluten-free, etc. Yet there is no clear and acceptable definition of food quality. It assumes many different forms depending on the person and situation. However, according to International Standards Organization [3], quality is defined as "the degree to which a set of inherent characteristics of an object fulfills requirements". Since this is a very

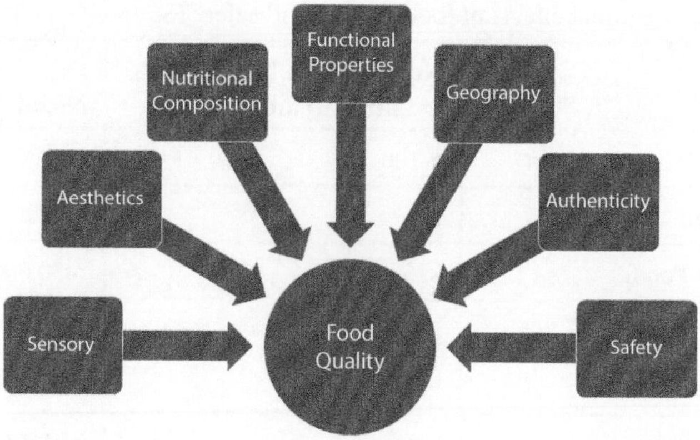

Figure 14.1 Contributors to food quality.

generic definition and is applied across segments and products, it will be useful to define quality in relation to the food. Keeping these in view, food quality can be defined as a set of characteristics, visual and determined, which when applied together result in a product that is safe to consume and delivers its intended benefits to the consumer. This definition implies that food quality is judged by its appearance, absence of undesirable chemicals such as contaminants, toxins, antibiotics, dyes, etc., no loss of desirable nutritional components including functional and nutraceutical properties, and free from microbial contamination.

According to the Glossary of the European Organization for Quality Control, quality may be defined as: "The totality of features and characteristics of a product or service that bear on its ability to satisfy a given need." More simply, quality may be defined as fitness for purpose. In the case of food, quality primarily involves safety, nutritive value, and acceptance.

Food quality is influenced by various factors like agricultural practices followed, such as application of pesticides, harvesting, cleaning, transport, storage, conditions existing at food processing facilities, etc. Critical points must be identified at each stage of the process and there must also be checks for ensuring that the food quality is not compromised. As a rule of thumb, checks should be implemented at a stage if it cannot be controlled at the subsequent stage. Such a strategy will be helpful in the production of safe food, apart from keeping the associated costs to a minimum. Due to economic and social changes, the quality of food is dynamic in nature, even in the same country, not to mention big differences between countries with different climatic conditions and levels of industrial development.

14.1.1.1 Primary and Secondary Food Processing

Food processing has been in existence since prehistoric days by adopting crude techniques such as treating with salt, fermentation, sun-drying, etc. With the advent of technology, significant progress has been made in terms of product development for different age groups such as baby foods, geriatric foods, fortified foods, allergen-free foods, diabetic-friendly foods, etc.; enhancing nutrient retention; increasing the shelf life of the product, reducing anti-nutritional factors such as phytates, tannins, etc. Though the primary processing industry has shown technological advances, it is the secondary processing industry that has grown beyond imagination, leading to significant improvements in the economies of certain countries. The food industry is a subsidiary of the agricultural sector. Without agriculture, practically, there is no food industry. Agricultural commodities form the primary food. However, they cannot be consumed in the form in which they are produced. For example, wheat cannot be consumed in the form it is produced. At a minimum, it needs to be converted into flour to prepare basic food. This is known as primary processing. A step further – secondary processing – involves converting the flour into other food products such as bread, pasta, vermicelli, etc. – an endless list. Therefore, food processing, in simple terms, can be said to be transforming raw ingredients into a form that can be consumed. It can be as simple as removing undesirable components from commodities to the production of frozen foods that are ready to cook and/or ready to eat. Simply put, it adds value to the raw ingredients, thereby increasing their marketability. With lifestyle changes, secondary processed foods have gained more popularity and the secondary processing industry has been growing by leaps and bounds. It has become so common that the word "food processing" by default means secondary processing. The food processing market is expected to reach USD 4.1 trillion by 2024 with a CAGR of 4.3% from 2019. However, the main challenge to the food processing industry is to ensure the delivery of safe and quality food. It requires ensuring quality and safety at almost every step during the supply chain, which is not only tedious but requires significant resources – human, infrastructure, logistics, financial, etc. Despite such challenges, the food and beverage (F&B) market is rapidly growing across the globe with Asia-Pacific growing fastest during 2006–16 and expected to show better growth than the rest of the world during 2017–26 (Table 14.2).

Possible reasons for such growth are increased disposable incomes, changing lifestyles, dual-income families, westernization of local cultures, increased urbanization, etc. The average annual growth rate of consumer spending on eating out varied in different regions of the world.

Table 14.2 Region-wise average annual growth rate of consumer spending on eating out during 2006 to 2026 (Source: Oxford Economics).

Region	Average annual growth rate of consumer spending on eating out	
	2006 – 16	**2017 – 26 (Expected)**
Asia Pacific	9.8%	7.5%
Middle East & Africa	7.4%	7.3%
The Americas	6.1%	5.5%
Europe	4.2%	4.9%

Table 14.3 Ranking of countries with largest food and beverages market (Source: Oxford Economics).

Country	Rank during		
	2006	**2016**	**2026**
United States	1	– 1	– 1
China	3	↑ 2	– 2
India	14	↑ 3	– 3
Spain	2	↓ 4	↓ 5
Japan	4	↓ 5	↓ 7
UK	5	↓ 6	↑ 4
Brazil	6	↓ 7	↓ 8
Italy	7	↓ 8	↓ 9
Thailand	12	↑ 9	↑ 6
Germany	8	↓ 10	↓ 11

Symbols preceding the rank indicate change in ranking: – no change; ↑ increased ranking; ↓ decreased ranking.

The Americas, which primarily include the USA and Canada, showed a lower growth rate. It doesn't mean that these countries are spending less. Rather, it just means that the rate of growth is lower. In fact, the USA has been the largest F&B market and tops the world and is expected to stay that way even up to 2026. The top 10 countries in terms of the F&B market, along with their previous, current, and expected rankings are shown in Table 14.3.

From various sources, it is clear that the food processing industry will inevitably make significant inroads into the everyday life of the majority of the population irrespective of the geographic region. Its contribution to world trade is going to increase and form a core sector in economic development.

14.1.1.2 Historical Trends in Food Quality

To determine food quality, it is important to follow good laboratory practices (GLP) to maintain consistency in results. The concept of GLP, which helps laboratories in ensuring quality, originated during the 1970s in the USA to address the malpractices in R&D activities of pharmaceutical companies [4]. The US Food and Drug Administration (FDA) after a series of thorough investigations, concluded that the problems could be addressed only by developing and imposing binding regulations, which became GLP regulations. These were first implemented by the FDA followed by the US Environmental Protection Agency (EPA). Subsequently, they were followed by various nations as they were involved in trade with the US. In 1981, the Organisation for Economic Co-operation and Development (OECD) also published GLP Principles, which are currently in use. The primary intent of GLP principles is to define, implement and follow practices that are verifiable in nature, leading to the generation of reliable data ensuring complete traceability.

Various aspects of food quality determination, preservation, dyeing, etc., have been practiced for a very long time and have been documented sporadically in ancient scriptures. However, with advances in technology, the same activities are being performed with more precision, accuracy, and speed. Currently, major emphasis is being laid on the determination of pesticide residues and heavy metal residues in agricultural products to determine food quality due to the indiscriminate use of a wide range of chemicals during crop production. For a long period, pesticides were quantified using simple chromatographic techniques. However, with the implementation of stringent regulations, there has been a need to quantify them at very low levels (parts per billion or μg/kg). In certain cases, the requirements are even in the range of nanograms, depending upon the compound of interest and the requirements of the national regulations. The most commonly used equipment for such low-level quantification is LC-MS/MS and GC-MS/MS, which are being touted as the "gold standard" for analytical needs. Though they can be used for very low-level quantification of a wide range of pesticides, analytical requirements seem to be moving a step further where the requirement is to determine a wide array of compounds at least at a qualitative level leading to the development and use of "Time of Flight Mass Spectrometry (TOFMS)" technology. It is based on the principle of identifying an ion based on the mass-to-charge ratio and time taken by the ion to reach the detector, which is at a known distance. Such technologies can be deployed to identify the presence or absence of a wide range of compounds at a faster rate. Upon identifying the presence of compounds of interest, further quantification can be taken up.

Similarly, the determination of heavy metal residues has come a long way in terms of technology. In earlier days there was Atomic absorption (AA) flame and furnace technology to Inductively coupled plasma – optical emission spectroscopy (ICP-OES), which facilitated the determination of multi-elements in a single analytical run. However, due to the ever-changing sensitivity requirements, inductively coupled plasma mass spectrometry (ICP-MS), is the instrument of choice of analytical chemists all over the world. It functions based on the principle of atomizing the sample, resulting in the creation of atomic and small polyatomic ions, which are then detected. Another major advantage of ICP-MS lies in the fact that it can even detect different isotopes of the same element leading to isotopic labeling as well.

Genetically modified organisms (GMOs) are another significant parameter gaining global importance in a lot of food commodities, both processed

Table 14.4 Comparison of targeted and non-targeted analyses.

Subject	Targeted analysis	Non-targeted analysis
Approach	Bottom-up	Top-down
Analytes	Targeted compounds	Fingerprint
Sensitivity	High	Low
Throughput	Low	High
Sample preparation	Selective	Non-selective
Data analysis	Univariate	Multivariate
Control limits	Publicly available	Specific reference databases
Ease	Simple	Complex

and unprocessed. Whether or not to allow GM foods has been a topic of scientific debate and discussion, although no conclusions have been reached. Various countries have formed technical committees to evaluate the safety of consuming such foods but there has been no conclusive evidence to either support or rebuff either of the claims. Currently, individual countries have their policies ranging from not allowing GM foods to permitting a percentage. However, in the majority of the countries where such foods are allowed, there is a mandatory requirement of labeling them appropriately. Thus, the detection of target compounds is two-pronged - Targeted and Non-targeted, which can be summarized as below in Table 14.4.

14.1.1.3 Food Quality Standards and its Requirements

Food quality standards have been in existence since ancient times and were part of all major civilizations. One of the earliest known food quality standards, of course in very primitive form, is mentioned in *Arthashastra*, an ancient Indian treatise; Chanakya in 375 BC writes about food adulteration and punishments for not maintaining food quality. There are references to the addition of undesirable substances to food and extraction of valuable nutrients from food leading to decreased food quality. Early testaments allowed the consumption of meat of only those animals that were slaughtered. Early records in Athens indicate control of wine and beer to ensure their purity. Rome also had laws protecting consumers against food fraud. Arab scientist, Al Chazini, constructed a high-sensitive balance to prevent fraud in measurements. These ancient documents indicate the prevalence of food adulteration from times immemorial, resulting in the implementation of food quality laws, though in primitive form. With the advent of modern technologies, especially in chemistry, it has become possible to detect such adulterants with precision. However, the same technologies are being exploited for committing food fraud.

Currently, almost all countries, with varying degrees of strictness, have food laws to protect consumers from food fraud. Countries within the European Union and certain developed countries like the USA, Canada, and Israel have very strict laws on food quality, resulting in a better global ranking in terms of food quality and safety. Table 14.5 lists the performance of 113 countries out of the world's 195 countries based on their food security score during 2019.

The overall Global Food Security Index (GFSI) was developed by considering various parameters. It is important to note that even if a country ranks high in GFSI, it may not necessarily be higher in terms of food quality

Table 14.5 Food security score of select countries (Source: Economist Intelligence Unit, Economist Group).

	Overall		Quality and safety		Affordability		Availability	
Country	**Rank**	**Score**	**Rank**	**Score**	**Rank**	**Score**	**Rank**	**Score**
Singapore	1	87.4	25	79.4	2	95.4	2	83.0
Ireland	2	84.0	7	87.7	3	90.5	11	76.8
United States	3	83.7	4	89.1	6	87.4	8	78.3
Switzerland	4	83.1	27	78.2	=17	83.8	1	84.3
Finland	=5	82.9	1	91.8	16	84.1	=6	78.6
Norway	=5	82.9	2	90.5	=26	81.9	3	81.0
Sweden	7	82.7	3	89.4	12	85.0	9	78.1
Canada	8	82.4	10	86.7	20	83.3	4	80.0
Netherlands	9	82.0	5	88.9	9	85.6	=12	76.2
Austria	10	81.7	17	81.1	=10	85.4	=6	78.6
Germany	11	81.5	21	79.8	13	84.9	5	79.1
Australia	12	81.4	20	79.9	7	86.6	10	77.1
Qatar	13	81.2	13	84.1	1	98.9	38	64.0

(Continued)

Table 14.5 Food security score of select countries (Source: Economist Intelligence Unit, Economist Group). (*Continued*)

	Overall		Quality and safety		Affordability		Availability	
Country	**Rank**	**Score**	**Rank**	**Score**	**Rank**	**Score**	**Rank**	**Score**
Denmark	14	81.0	8	87.2	=10	85.4	=15	74.8
Belgium	15	80.7	15	83.9	15	84.4	=12	76.2
France	16	80.4	9	87.1	=17	83.8	=15	74.8
United Kingdom	17	79.1	18	80.9	19	83.6	17	74.4
Israel	18	79.0	16	83.8	21	83.0	18	73.6
New Zealand	19	78.8	=35	73.5	14	84.6	14	75.5
Portugal	20	77.8	6	88	29	81.3	22	70.9
Japan	=21	76.5	28	76.7	24	82.4	21	71.0
United Arab Emirates	=21	76.5	26	78.5	4	89.8	39	63.7
Italy	23	75.8	22	79.7	23	82.5	25	68.3
Poland	24	75.6	=23	79.5	30	81.1	24	69.3
Chile	=25	75.5	33	74.7	32	80.5	19	71.3

(*Continued*)

Table 14.5 Food security score of select countries (Source: Economist Intelligence Unit, Economist Group). (*Continued*)

	Overall		**Quality and safety**		**Affordability**		**Availability**	
Country	**Rank**	**Score**	**Rank**	**Score**	**Rank**	**Score**	**Rank**	**Score**
Spain	=25	75.5	12	84.7	25	82.3	31	65.9
Kuwait	27	74.8	29	75.9	5	88.1	=43	62.3
Malaysia	28	73.8	42	70.6	28	81.7	26	67.7
South Korea	29	73.6	32	74.9	45	75.8	20	71.2
Saudi Arabia	30	73.5	=35	73.5	8	86.3	46	61.8
Greece	31	73.4	11	86	=39	77.8	33	64.9
Czech Republic	32	73.1	46	68.1	22	82.6	29	66.3
Uruguay	33	72.8	37	73.3	=34	79.3	28	66.7
Hungary	34	72.7	43	70.5	31	80.8	30	66.1
China	35	71.0	38	72.6	50	74.8	27	66.9
Belarus	36	70.9	19	80.2	44	76.0	42	62.9
Argentina	37	70.8	=23	79.5	37	78.9	51	60.2

(*Continued*)

Table 14.5 Food security score of select countries (Source: Economist Intelligence Unit, Economist Group). (*Continued*)

	Overall		Quality and safety		Affordability		Availability	
Country	**Rank**	**Score**	**Rank**	**Score**	**Rank**	**Score**	**Rank**	**Score**
Romania	38	70.2	52	64.1	=34	79.3	36	64.3
Brazil	=39	70.1	14	84	43	77.0	58	58.8
Costa Rica	=39	70.1	30	75.6	46	75.6	=40	63.1
Turkey	41	69.8	40	71.1	51	74.7	34	64.8
Russia	42	69.7	41	70.9	33	79.8	=52	60.1
Colombia	=43	69.4	44	69.3	54	73.7	32	65.6
Mexico	=43	69.4	31	75.2	49	74.9	=43	62.3
Panama	45	68.8	39	71.8	53	73.8	=40	63.1
Oman	46	68.4	34	74.4	=39	77.8	=67	57.6
Slovakia	47	68.3	61	59.4	38	78.6	45	62.1
Kazakhstan	=48	67.3	45	68.3	41	77.5	=65	57.7
South Africa	=48	67.3	49	66.2	56	70.8	35	64.5

(*Continued*)

Table 14.5 Food security score of select countries (Source: Economist Intelligence Unit, Economist Group). *(Continued)*

	Overall		Quality and safety		Affordability		Availability	
Country	**Rank**	**Score**	**Rank**	**Score**	**Rank**	**Score**	**Rank**	**Score**
Bahrain	50	66.6	67	56.9	=26	81.9	70	56.3
Bulgaria	51	66.2	48	66.8	36	79.0	81	54.2
Thailand	52	65.1	75	52.6	42	77.1	59	58.7
Azerbaijan	53	64.8	71	54	47	75.3	56	59.2
Vietnam	54	64.6	77	51.7	48	75.1	55	59.7
Egypt	55	64.5	50	65.9	81	57.6	23	70.2
Dominican Republic	56	64.2	53	62.3	63	68.4	50	61.0
Botswana	57	63.8	68	56.6	59	69.5	=48	61.3
Peru	58	63.3	58	60.4	61	69.1	57	59.0
Ghana	=59	62.8	66	57.1	65	66.3	47	61.7
Morocco	=59	62.8	55	61.9	=75	61.5	37	64.2
Serbia	=59	62.8	56	61.8	52	73.9	83	53.0

(Continued)

Table 14.5 Food security score of select countries (Source: Economist Intelligence Unit, Economist Group). *(Continued)*

	Overall		Quality and safety		Affordability		Availability	
Country	**Rank**	**Score**	**Rank**	**Score**	**Rank**	**Score**	**Rank**	**Score**
Indonesia	62	62.6	84	47.1	58	70.4	=48	61.3
Ecuador	63	61.8	63	58.4	60	69.4	=71	56.1
Jordan	=64	61.0	70	54.2	57	70.5	=79	54.8
Philippines	=64	61.0	80	50.3	62	68.9	=65	57.7
Sri Lanka	66	60.8	76	52.4	69	65.0	54	60.0
El Salvador	67	60.7	62	58.9	72	63.8	60	58.6
Guatemala	68	60.6	65	57.5	68	65.3	=67	57.6
Tunisia	69	60.1	54	62.2	=75	61.5	63	58.0
Algeria	70	59.8	74	53	64	66.9	74	55.8
Uzbekistan	71	59.0	73	53.4	67	65.6	77	55.1
India	72	58.9	=85	47	70	64.2	61	58.4
Honduras	73	58.0	57	60.6	82	57.2	64	57.8

(Continued)

Table 14.5 Food security score of select countries (Source: Economist Intelligence Unit, Economist Group). (*Continued*)

	Overall		**Quality and safety**		**Affordability**		**Availability**	
Country	**Rank**	**Score**	**Rank**	**Score**	**Rank**	**Score**	**Rank**	**Score**
Paraguay	74	57.9	51	65.4	55	72.0	103	42.4
Bolivia	75	57.7	64	58.3	66	65.8	=89	50.0
Ukraine	76	57.1	60	59.6	71	63.9	=89	50.0
Myanmar	77	57.0	78	51.3	78	59.1	69	57.2
Pakistan	78	56.8	93	43.6	74	63.2	75	55.7
Nepal	79	56.4	72	53.7	80	58.5	76	55.4
Mali	80	54.4	59	59.9	98	45.9	=52	60.1
Senegal	81	54.3	69	56.1	88	51.6	=71	56.1
Nicaragua	82	54.2	83	48.2	73	63.5	=94	47.9
Bangladesh	83	53.2	=107	30.6	77	60.4	=79	54.8
Cote d'Ivoire	84	52.3	=104	33.1	87	53.5	62	58.1
Benin	85	51.0	=88	46.4	94	48.6	78	54.9

(*Continued*)

Table 14.5 Food security score of select countries (Source: Economist Intelligence Unit, Economist Group). (*Continued*)

	Overall		Quality and safety		Affordability		Availability	
Country	**Rank**	**Score**	**Rank**	**Score**	**Rank**	**Score**	**Rank**	**Score**
Kenya	86	50.7	94	43.2	=83	56.7	93	48.0
Burkina Faso	87	50.1	95	41.6	97	47.0	73	55.9
Cameroon	88	49.9	=85	47	86	53.7	=96	47.6
Niger	89	49.6	=97	37.4	92	50.2	82	53.6
Cambodia	90	49.4	100	34.6	=83	56.7	92	48.1
Ethiopia	91	49.2	96	39	93	49.7	84	52.6
Laos	92	49.1	=97	37.4	85	55.5	=96	47.6
Tajikistan	93	49.0	87	46.6	79	58.8	104	41.1
Nigeria	94	48.4	79	50.7	90	50.4	99	45.8
Rwanda	95	48.2	82	48.5	103	43.8	86	52.0
Tanzania	96	47.6	91	45.9	102	45.1	88	50.4
Guinea	97	46.7	110	29	95	47.4	85	52.4

(*Continued*)

Table 14.5 Food security score of select countries (Source: Economist Intelligence Unit, Economist Group). (*Continued*)

	Overall		Quality and safety		Affordability		Availability	
Country	**Rank**	**Score**	**Rank**	**Score**	**Rank**	**Score**	**Rank**	**Score**
Uganda	98	46.2	81	49.1	99	45.8	101	45.5
Sudan	99	45.7	90	46	96	47.1	102	44.4
Angola	100	45.5	92	44.9	89	51.3	105	40.5
Zambia	101	44.4	102	33.6	105	41.8	87	50.7
Togo	102	44.0	106	31	100	45.6	98	47.2
Haiti	103	43.3	99	35.9	91	50.3	108	39.6
Malawi	104	42.5	=104	33.1	108	39.4	91	48.6
Mozambique	105	41.4	112	20.6	104	42.5	=94	47.9
Sierra Leone	106	39.0	=107	30.6	106	40.8	106	40.3
Syria	107	38.4	=88	46.4	112	34.6	109	38.9
Madagascar	108	37.9	111	22.1	111	35.7	100	45.7
Chad	109	36.9	103	33.5	107	40.3	110	34.9

(*Continued*)

Table 14.5 Food security score of select countries (Source: Economist Intelligence Unit, Economist Group). (*Continued*)

	Overall		**Quality and safety**		**Affordability**		**Availability**	
Country	**Rank**	**Score**	**Rank**	**Score**	**Rank**	**Score**	**Rank**	**Score**
Congo (Dem. Rep.)	110	35.7	113	19.8	109	37.3	107	40.0
Yemen	111	35.6	109	30.2	101	45.5	113	28.6
Burundi	112	34.3	101	34.5	110	36.6	=111	32.2
Venezuela	113	31.2	47	66.9	113	15.8	=111	32.2

Rank: 1 = Best; '=' before the rank indicates a tie. Score: 0 – 100 where 100 = Best.

and safety. In general, it can be said that economically developed countries ranked better on food quality and safety. This might be due to increased consumer awareness, demand, and affordability of such quality products, resulting in the stricter implementation of the food laws, leading to better food quality. However, economically weaker countries have different priorities, the most important being to feed their population, resulting in less emphasis on food quality and safety. Ensuring better food quality requires the development of good infrastructure, allocation of resources towards continuous monitoring and implementation, putting strain on available financial resources. Per capita GDP of 2019 (Source: World Bank) has been plotted against food quality and safety rankings (Figure 14.2). It indicates that developed countries with strong economies had a better ranking while economically poorer nations showed poor ranking for food quality and safety; this shows the significance of financial resources required for achieving food safety and quality. In fact, these two parameters showed a correlation of 0.92, indicating a strong association between them.

Even today, there are many countries especially in Sub-Saharan Africa and Asia which either do not have strong rules or are unable to enforce existing rules to ensure safe food for their citizens, risking the lives of millions of people. Apart from this, having such regulations is vital to trade since most of the importing countries have regulatory laws and import only safe and quality foods. When a consignment is tested by an importing nation, should there be a lack of compliance with their regulations, the whole consignment will not be allowed into the country, leading to

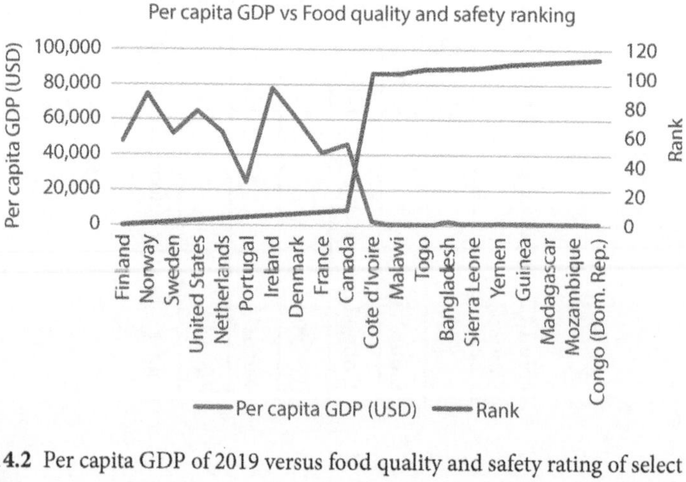

Figure 14.2 Per capita GDP of 2019 versus food quality and safety rating of select countries.

Table 14.6 Codex timelines (Source: FAO).

Year	Event	Brief description
1903	International Dairy Federation	IDF, which later was to be an important catalyst in the conception of the Codex Alimentarius Commission, develops international standards for milk and milk products.
1945	Foundation of the Food and Agriculture Organization of the United Nations	In 1943 44 governments, meeting in Hot Springs, Virginia, United States, commit themselves to found a permanent organization for food and agriculture. In 1945 the first session of the FAO Conference, Quebec City, Canada, establishes FAO as a specialized United Nations agency.
1947	The United Nations Economic Commission for Europe	UNECE sets out norms, standards, and conventions to facilitate international cooperation within and outside the region.
1947	International Standards Organization	ISO, an independent, non-governmental international organization with a membership of 161 national standards bodies. Through its members, it brings together experts to share knowledge and develop voluntary, consensus-based, market-relevant international standards that support innovation and provide solutions to global challenges.
1948	Foundation of the World Health Organization	WHO's constitution came into force on 7 April 1948. The constitution was adopted by the International Health Conference held in New York from 19 June to 22 July 1946, signed on 22 July 1946 by the representatives of 61 states, and entered into force on 7 April 1948. Later, amendments are incorporated into this text.

(Continued)

Table 14.6 Codex timelines (Source: FAO). (*Continued*)

Year	Event	Brief description
1949	Código Latino americano de Alimentos	Argentina proposes a regional Latin American food code, Código Latino americano de Alimentos.
1950	Joint FAO/WHO expert meetings	Joint FAO/WHO expert meetings begin on nutrition, food additives, and related areas.
1953	World Health Assembly	WHO's highest governing body, the World Health Assembly, at its sixth session expresses the view that the increasing use of various chemical substances in the food industry has, in recent decades, presented a new public health problem which might be usefully investigated.
1954	Codex Alimentarius Europaeus	Austria actively pursues the creation of a regional food code, the Codex Alimentarius Europaeus, or European Codex Alimentarius. Dr Hans Frenzel was Minster of Public Alimentation in Austria after the Second World War and he developed the idea of creating a European Codex Alimentarius Commission. With the consent of the European states, the European Council of the Codex Alimentarius Europaeus was established in 1958.
1960	FAO Regional Conference for Europe	The first FAO regional conference for Europe endorses the desirability of international – as distinct from regional – agreement on minimum food standards and invites the Organization's Director-General to submit proposals for a joint FAO/WHO program on food standards to the FAO Conference.

(*Continued*)

Table 14.6 Codex timelines (Source: FAO). (*Continued*)

Year	Event	Brief description
1961	11th FAO Conference decides to establish a Codex Alimentarius Commission	The Council of Codex Alimentarius Europaeus adopts a resolution proposing that its work on food standards be taken over by FAO and WHO. The 11th FAO Conference establishes the Codex Alimentarius Commission and requests early endorsement by WHO of a Joint FAO/WHO Food Standards Programme.
1962	Codex Alimentarius Commission is requested to create the Codex Alimentarius	The joint FAO/WHO Food Standards Conference requests that the Codex Alimentarius Commission implement a joint FAO/WHO food standards program and create the Codex Alimentarius.
1963	Inaugural meeting of the Codex Alimentarius Commission held in Rome	Recognizing the importance of WHO's role in all health aspects of food and considering its mandate to establish food standards, the World Health Assembly approves the establishment of the joint FAO/WHO Food Standards Programme and adopts the Status of the Codex Alimentarius Commission.
1985	UN General Assembly	UN General Assembly stated that where possible governments should adopt Codex Alimentarius standards.
1991	FAO/WHO Conference on Food Standards	FAO/WHO Conference on Food Standards recognized the importance of providing evaluations based on sound science and risk assessment principles.
1992	FAO/WHO International Conference on Nutrition	FAO/WHO International Conference on Nutrition recognized that food regulations should take into account the recommended international standards of the Codex Alimentarius Commission.

(*Continued*)

Table 14.6 Codex timelines (Source: FAO). (*Continued*)

Year	Event	Brief description
1995	Agreement on the Application of SPS and Agreement on Technical Barriers to Trade	Agreement on the Application of SPS and Agreement on Technical Barriers to Trade formally recognized International Standards, guidelines, and recommendations, including the Codex Alimentarius, as reference points for facilitating international trade and resolving trade disputes in international law.
1996	FAO World Food Summit	FAO World Food Summit committed to applying measures, in conformity with the Agreement on the Application of Sanitary and Phytosanitary Measures and other relevant international agreements, that ensure the quality and safety of food supply.
2000	53rd World Health Assembly	53rd World Health Assembly recognized the importance of the standards, guidelines, and other recommendations of the Codex Alimentarius Commission for protecting the health of the consumers and assuring fair trading practices and urged member states to participates actively in activities in the emerging area of food safety risk analysis.
2002	World Food Summit five years later	"World Food Summit five years later" reaffirmed the important role of effective, science-based, internationally accepted standards of food safety.
2004	Second FAO/WHO Global Forum of Food Safety Regulators	Second FAO/WHO Global Forum of Food Safety Regulators affirmed that developing countries would benefit from greater use of basic texts when building their food control systems.

(*Continued*)

Table 14.6 Codex timelines (Source: FAO). (*Continued*)

Year	Event	Brief description
2009	World Summit on Food Security	World Summit on Food Security reaffirmed the right of everyone to have access to safe and nutritious food.
2014	Second International Conference on Nutrition	Second International Conference on Nutrition, Framework for Action. Recommendation 54: Actively take part in the work of the Codex Alimentarius Commission on nutrition and food safety, and implement, as appropriate, internationally adopted standards at the national level.
2015	The 2030 Agenda for Sustainable Development	The 2030 Agenda for Sustainable Development. Target 2.1: By 2030, end hunger and ensure access by all people, in particular the poor, and people in vulnerable situations, including infants, to safe, nutritious and sufficient food all year round.

losses to the exporters. This will be a double blow for global trade since both imports and exports will be affected. Hence, in the interest of global trade and for ensuring the safety of all citizens, it is vital that all the countries draft and implement strict food laws governing food quality and safety. Although it may have initial financial implications, addressing food safety and quality-related effects on human health, apart from improving the health of the citizens, will result in higher economic returns through reduced burden on the nation's exchequer. Following is a brief description of selected global standards:

Codex Alimentarius, a globally accepted standard, is a collection of internationally adopted food standards aimed at protecting consumers' health and creating fair practices in food trade. With its modest beginning in 1945, its timelines are shown in Table 14.6. With 189 members made up of 188 member countries and a member organization (the European Union), it is one of the most recognized and accepted food standards across the globe. Codex includes standards for all principal foods – raw, semi-processed, and processed, for consumer consumption. Also included in the Codex are provisions for food hygiene, food additives, residues of pesticides and veterinary drugs, contaminants, labeling, methods of analysis and sampling, import-export inspection, and certification. Though these standards are not binding on any country, the majority of the countries either use them in the existing form or develop their own based on these standards. Also, the majority of the countries use them for global trade, thereby according to a global status to Codes Alimentarius.

14.1.1.4 *Role of Technology in Building Food Quality Within the Industry*

Traditionally, food quality is determined through various techniques of chemistry, micro, and molecular biology. However, there has been improvement in these techniques leading to the development of sensitive analytical methods. Unfortunately, most of these methods are expensive, laborious, and time-consuming, and need highly skilled personnel. The majority of those techniques are difficult to implement in the field or at the production site. Alternately, newer IoT-based technologies such as Artificial Intelligence (AI), Machine Learning (ML), Sensors, Hyperspectral Imaging, Blockchain, etc., are currently being exploited for use in the food industry. Although these technologies are not developed for the food and agricultural sector, they have found their way into these sectors also and have revolutionized the industry over the past few years. These have found applications in almost all activities of the food industry – grading, packing, quality and safety monitoring, transportation, traceability, etc. A few of these technologies are discussed below.

Hyperspectral Imaging: It is a technique that analyses a wide spectrum of light instead of just assigning primary colors (Red, Green, and Blue) to each pixel. In simple terms, the light striking each pixel is further broken down into many different spectral bands such that more detailed information on the image can be obtained [5]. Hyperspectral spectroscopy, a non-destructive technique, can be of immense utility in determining the raw material quality. The technology can help in identifying different materials from each other and tell the difference between two samples that look identical but have, for example, different chemical compositions. The added advantage of this technique is its non-invasive nature that allows characterizing surface color or texture as well as many chemical compositions. Figure 14.3 shows the hyperspectral images of rice mixed with plastic.

Due to differences in their chemical composition, each material will have a unique signature which will help in differentiating the two samples.

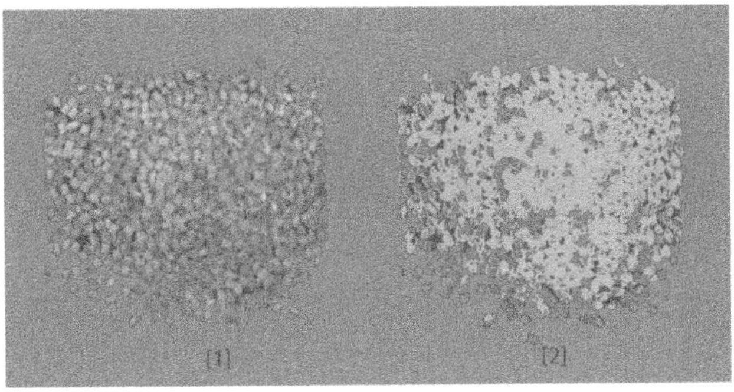

Figure 14.3 Rice mixed with plastic [(1) As seen by naked eye (2) As captured by hyperspectral camera].

This technique also has the potential to differentiate varietal mixtures and will be very useful in determining the quality of high-value raw materials such as spices, which have a high risk of being adulterated. For example, it will be very difficult to identify when papaya seeds are mixed with black pepper or when cheap rice varieties are mixed with expensive basmati rice varieties. Hai *et al.* [6] inspected six rice varieties for their purity using hyperspectral imaging technology. Since the determination of off-types with the naked eye is a tedious and time-consuming process, the HSI system was relied upon to provide spatial and textural information along with high-resolution spectral signatures. The Near-Infrared HSI system was used for capturing data in the wavelength ranges of 950.73 to 1759.4 using the push broom data acquisition technique. High accuracy was achieved in varietal identification by combining the HSI system on spectral and spatial features. This technique can easily be applied for screening export shipments which require quick turnaround times with high accuracy. Similarly, sensors have been developed for determining various chemical parameters and their refinement is still in progress. They can also be used for monitoring the quality and safety of raw materials before processing. It will help in cutting down the cost of producing safe foods since analysis using classical methods involves high costs.

Blockchain: Another recent technology that is finding a growing foothold in traceability is Blockchain [7]. It is a growing list of records called blocks, which are linked using cryptography. Each block contains a cryptographic hash of the previous block, a timestamp, and transaction data. By design, a blockchain is resistant to modification of the data. It is "an open, distributed ledger that can record transactions between two parties efficiently and in a verifiable and permanent way". It is typically managed by a peer-to-peer network collectively adhering to a protocol for inter-node communication and validating new blocks. Once recorded, the data in any given block cannot be altered retroactively without the alteration of all subsequent blocks, which requires consensus of the network majority, thereby resulting in unbroken traceability. Although it is developed primarily to serve as a public transaction ledger of the cryptocurrency bitcoin, like other applications, it found its way into the food and agriculture sector too. It is forecasted that the global blockchain technology in agriculture will grow at an annual rate of 56.4% in the period 2018–22 [8]. Blockchain has found its significant application in supply chain management, which is essential for the food industry to ensure traceability. Agricultural supply chains are plagued by numerous transparency and efficiency issues apart from being risky in nature. With the increasing demand for traceability, it is essential to have tamper-proof systems such as blockchain that are transparent, reliable, and efficient. A few of the major food companies - Dole, Driscoll, Golden State Foods, Kroger, McCormick and Company, Nestlé, Tyson Foods, and Walmart are already utilizing IBM's blockchain technologies to make their food supply chains more transparent and traceable and to streamline payments [9]. Walmart is one of the foremost companies to utilize blockchain technology in tracking its products. It started with the hypothesis that blockchain technology might be a good fit for the decentralized food supply ecosystem for which a good traceability system was created based on Hyperledger fabric. It has initiated two proof of concept projects with IBM. One was about tracking mangoes sold in its US stores while the other was to trace pork sold in its China stores. Its utilization of blockchain technology reduced the tracking time of mangoes from seven days to just 2.2 seconds, whereas the technology allowed uploading of authenticity certificates, bringing critically needed reliability for pork authenticity [10]. Walmart now traces over 25 products from five different suppliers using IBM blockchain which is built atop Hyperledger Fabric. The products include produce such as mangoes, strawberries, and leafy greens; meat and poultry such as chicken and pork; dairy such as yogurt and almond milk; and even multi-ingredient products such as packaged salads and baby foods.

Sensors in Food quality: Sensors are one of the most economical yet reliable means for determining food quality. A wide range of sensors having application in the food industry has been developed. A major area that is exploiting sensors technology is the packaging sector [11, 18, 19] in which there are certain sensors such as those used for monitoring temperature, humidity, gases, pressure, etc. Temperature and humidity are the most important parameters that can influence product shelf life. Increased temperature and humidity levels accelerate product spoilage. Hence, monitoring these parameters using mobile technologies is an easier option for product monitoring. Likewise, spoilage of products releases certain gases and creates pressure inside the package. Sensors for monitoring pressure and gases including methane which will be emitted during food spoilage will indicate product quality. Manufacturers can deploy these sensors in packaging to monitor their products throughout the product life cycle [20]. Integration of food quality and safety limits into the database along with making available sensor data can be accessed by stakeholders in evaluating the product quality. Such wireless sensor networks will be useful not only for evaluating food quality but also will assist the manufacturers in recalling products when safety violations are noticed. Such a system will also help in preventing food fraud. Also, sensors along with instant alerts can be used for monitoring premises of the food processing industry to bring in operational efficiency and improve food quality and compliance. The food industry found immense usefulness when sensors are combined with a camera creating an optical image. Such images are used for efficient sorting and grading of products which otherwise would have been done manually, leading to inconsistent results.

Artificial intelligence and machine learning: Artificial intelligence and machine learning are being touted to play a big role by utilizing big data in the food industry. Like in various industries which are benefiting from this technology, the food industry is also poised to capitalize on its benefits. They can be utilized for bringing in efficiency in plant operations, food market analysis, predictive equipment maintenance to minimize downtime, improved hygiene, supply chain optimization, etc. Examples of its applications include the development of Self-Optimising Clean-in-Place (SOCIP) [12], a pioneering cleaning system that will use artificial intelligence to autonomously optimize the cleaning process for food manufacturing equipment. Current CIP processes are designed for the worst-case scenario resulting in wastage of resources and increased downtime when such extensive cleaning is not required. However, SOCIP will use artificially intelligent, multi-sensor technology to monitor and assess the amount of food and microbial debris that is present inside food manufacturing equipment during the detergent phase of the cleaning process. When fouling is removed SOCIP's intelligent software will halt the detergent phase so the rinse phase can be initiated as soon as the equipment is clean enough - maintaining hygiene standards and improving efficiency, as the equipment is only cleaned for as long as is necessary. Therefore, it will enable food manufacturers to make significant time and cost savings, increasing their productivity and profit margins whilst reducing the use of

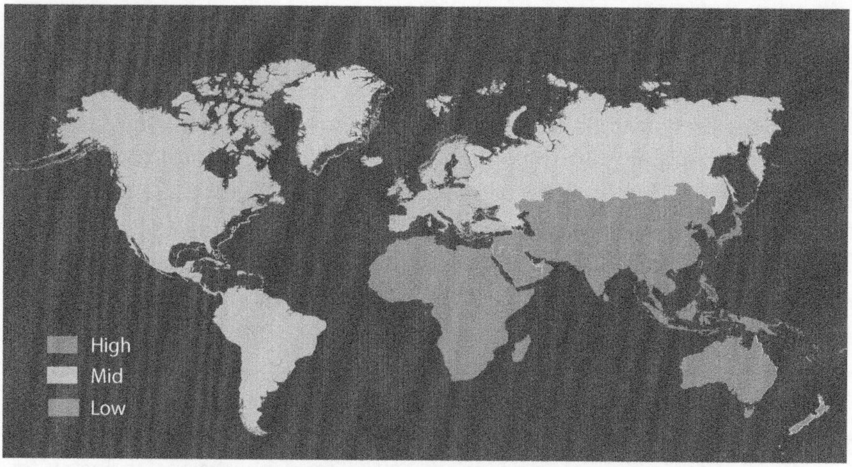

Figure 14.4 Expected growth of AI in food and beverages market during 2019–24. (Source: Mordor Intelligence).

energy, water, and chemicals. In summary, such developments will automate the processes leading to cost savings and reduced dependency on human intervention, resulting in a reduction in errors. The food and beverage market is expected to grow at a CAGR of over 65% during the period 2019–24 [13]. In 2017, the USA had a 29% market share of the AI food industry market – the highest in the world. Over the next few years, Asia-Pacific countries are expected to adopt these technologies the most, as seen in Figure 14.4. The lower growth rate in North American countries such as the USA and Canada is low possibly due to already high usage in these countries.

14.1.1.5 Regulations and Their Requirements

Food safety and quality regulations shall be designed with the primary objective of providing safe and high-quality food for citizens across the globe. They should encompass systems for strengthening surveillance of and response systems since these are essential to control foodborne diseases, facilitate global trade, and ensure health security. More importantly, the international health regulations require that the countries notify the World Health Organization (WHO) about public health events that are of global concern including those caused by contaminated food [14]. As described earlier in this chapter, such regulations have been promulgated since ancient times and have adapted to the changing times. Compiling global regulations is a herculean task. However, the Food, Agriculture and Renewable Natural Resources Legislation Database of FAO compiles such regulations on a countrywide basis, and it is the most comprehensive database.

The Food, Agriculture and Renewable Natural Resources Legislation Database (FAOLEX) [15] is a database of national legislation, policies, and bilateral agreements on food, agriculture, and natural resources management drawn from over 200 countries, territories, and regional integration organizations. It is administered by the Development Law Service of the FAO Legal office. As per the database, there has been a tremendous increase in the number of regulations from 2000 to 2013, possibly due to most of the developing nations joining the developed nations in implementing food safety and quality regulations. One of the primary reasons for such adoption by developing nations is the enforcement of safety and quality standards by the importing countries, which involves rejecting imports when quality standards are not met. According to FAO, global trade in agricultural commodities stands at over USD 1 trillion. The trade-in food and beverages segment stood at about USD 13 billion in 2018 with Germany, USA, the Netherlands, and France the top four countries both in terms of exports and imports. This has created an opportunity for countries to do trade provided they have high-quality and safe food for export. Thus, the significant increase in trade is very important for countries to boost their economies by complying with the quality and safety requirements of the importing nations.

14.1.2 Food Safety

14.1.2.1 Primary and Secondary Food Production

Scientists believe that agriculture was established first in the fertile crescent of the Middle East about ten or eleven thousand years B.C. The region was home to a variety of edible and easily cultivated crops – wheat and barley among cereals; lentils, peas, and chickpeas among vegetables. Also, the region was endowed with wild goats, sheep, pigs, and cattle, all of which were domesticated and became important sources of food. Science and technology played an increasingly important role in food production during the eighteenth and nineteenth centuries. In today's era, primary food production includes growing, cultivation, picking, harvesting, collection, or catching of food, and includes transportation or delivery of food on, from, or between the premises on which it was grown, cultivated, picked, harvested, collected or caught; the packing, treating (for example, washing) or storing of food on the premises on which it was grown, cultivated, picked, harvested, collected or caught. However, primary food production does not include any process involving the substantial transformation of food (for example, manufacturing or canning), regardless of whether the process is carried out on the premises in which the food was grown, cultivated, picked, harvested, collected or caught, or the sale or service of food directly to the consumer. Secondary production means the process of using ingredients produced through primary food processing to create ready-to-cook and/or ready-to-eat foods. An example of this is using flour to make dough followed by baking it to create bread. Other examples include fermenting grape juice with wine yeast to create wine and using ground meat to make sausages, etc. In the recent past, secondary production has grown by leaps and bounds due to various reasons such as changes in lifestyle, food habits, convenience, etc.

14.1.2.2 Historical Trends in Food Safety

Historical accounts of foodborne illness date back to antiquity. The first suggested documented case of a known foodborne illness dates back to 323 B.C. According to doctors at the University of Maryland who studied historical accounts of Alexander the Great's symptoms and death, the ancient ruler is believed to have died from typhoid fever, which was caused by *Salmonella typhi* [16]. The Assize of Bread and Ale was the first known food law; it was proclaimed by King John of England in 1202 and prohibited adulteration of bread with peas or beans. Subsequently, over 400 years later, American colonists enacted a similar resolution in 1646, which was later passed as the Massachusetts Act in 1785 against selling unwholesome provisions. This is believed to the first food safety law of the USA. Later, USDA and FDA were formed to shape US food safety laws which formed a basis for other countries to draft and promulgate their respective regulations. It is ironic to note that, even after 800 years of proclaiming the first known food safety law, the issue not only persists but is of major global concern.

Like with any segment, food safety has witnessed changes with time and evolved, keeping in view global requirements of customer demands and country regulations. Even in the prehistoric era, food safety was ensured by simple methods of adding salt to enhance shelf life by preventing the growth of microorganisms. Our ancestors used to store food products mostly in glass containers to ensure the safety of the product. These simple techniques were implementable at that time since most of the production and consumption used to take place at a local level and there was hardly any trade. But with the advent of modern trade, things have changed drastically. Food safety started to focus on trade and has adapted itself to be trade-centric. Such is the effect of trade that all aspects – manufacturing, processing, packing, transportation, labeling, storing, etc., that ensure food safety have undergone a multitude of changes. There have been significant technological improvements in all these areas with an emphasis on retaining nutrients, enhancing shelf life, and ensuring safety. Improvements made in the processing technologies have helped to retain the nutritional value of the product during storage. Product shelf life has been increased by using various techniques with modified atmospheric packaging (MAP), which effectively modifies the atmosphere (gases) around the product in a way that delays product spoilage, leading to increased shelf life. One of the major changes food safety has brought is the development of global food safety standards.

14.1.2.3 Food Safety Standards and its Requirements

It can be safely said that global food safety standards originated due to increased global food trade and the need to ensure that there is a mechanism on which trading partners can rely to ensure the traded product is produced under hygienic conditions and the required documentation related to the product safety and quality are accurate to warrant product safety. Since trade occurs across the globe and it is practically impossible for an importer to ensure the safety of the supplier's product, which is being produced miles away, a system had to be developed that monitors all aspects of the product. With this background, various global food safety standards have been developed and are implemented by various stakeholders as appropriate to them across the food value chain. Food safety standards are primarily classified based on the type of industry – some of the standards were developed for the primary industry sector like industries involved in the farming of animals, farming of fish, farming of plants, farming of grains and pulses, animal conversion, pre-processing handling of plant products, etc., whereas other standards and their requirements were developed considering the secondary food processing industries like processing of animal perishable products, processing of animal and plant perishable products (mixed products), processing of ambient stable products, production of feed, production of (Bio) chemicals, etc. Examples of the most common global food safety standards across the world are listed in Table 14.7. Implementation requirements described in all these standards are focused primarily on two types of approaches: a) process approach, b) product approach, both of which focus on one aspect, i.e., risk-based

Table 14.7 List of food safety standards implemented across the global food industry.

Food safety and quality standard	**Industry product sector type**
ISO 22000:2018 and FSSC 22000	Animal conversion; Pre-Processing and handling of plant Products; Processing of animal perishable products; Processing of plant perishable products; Processing of animal and plant perishable products (Mixed products); Processing of ambient stable products; Production of feed; Production of (Bio) chemicals; Production of food packaging; Provision of storage and distribution services
BRCGS Food	Animal conversion; Pre-processing and handling of plant products; Processing of animal perishable products; Processing of plant perishable products; Processing of animal and plant perishable products (Mixed products); Processing of ambient stable products; Production of feed; Production of (Bio) chemicals
IFS Food	Animal conversion; Pre-processing and handling of plant products; Processing of animal perishable products; Processing of plant perishable products; Processing of animal and plant perishable products (Mixed products); Processing of ambient stable products; Production of (Bio) chemicals; Provision of storage and distribution services; Production of food packaging
Safe Quality Food (SQF)	Farming of Animals; Farming of plants; Animal conversion; Preprocessing and handling of plant products; Processing of animal perishable products; Processing of plant perishable products; Processing of animal and plant perishable products (Mixed products); Processing of ambient stable products; Production of feed; Production of (Bio) chemicals; Production of food packaging; Provision of storage and distribution services

(*Continued*)

Table 14.7 List of food safety standards implemented across the global food industry. (*Continued*)

Food safety and quality standard	**Industry product sector type**
Global GAP	Farming of fish; Farming of plants; Pre-process handling of plant products
Freshcare and Canada GAP	Farming of plants; Pre-processing and handling of plant products
Global Red Meat Standard GRMS	Animal conversion; Processing of animal perishable products
Asia GAP	Farming of plants; Farming of grains and pulses; Pre-processing and handling of plant products

thinking. HACCP – a seven principal-based system along with prerequisite programs developed over the period of time are the backbone of all these standards. As per International Organization for Standardization (ISO), an international standard-setting body composed of representatives from various national standards organizations, PRPs are basic conditions and activities that are necessary within the organization and throughout the food chain to maintain food safety. Key prerequisite programs for primary and secondary food industries include good agricultural practices (GAP), good veterinary practices (GVP), good manufacturing practices (GMP), construction and layout of buildings, layout of premises and workspace, utilities like air, water, energy, waste disposal, and management, equipment suitability, cleaning and maintenance, management of purchased materials, measures for prevention of cross-contamination, cleaning and sanitation programs, pest control, personnel hygiene, and employee facilities, rework, product recall/withdrawal program, good warehousing requirements, food defense, biovigilance, and bioterrorism control programs, good production practices (GPP), good distribution practices (GDP) and good trading practices (GTP), etc.

14.1.2.4 Role of Technology in Building Food Safety Within Industry

According to a market research agency, Markets and Markets, the global food safety testing market is estimated to grow at a CAGR of 7.7% to USD 24 billion by 2023 from USD 17 billion in 2018. The growth will be driven by:

i. Growth in demand for convenience and packaged foods
ii. Increase in outbreaks of chemical contamination in food processing industries
iii. Rise in consumer awareness on food safety
iv. Increase in global outbreaks of foodborne illnesses
v. Globalization in food trade

As with any industry, technological advancements can be exploited for ensuring food safety well. One such tool, developed by Veritide, a New Zealand-based company, is the use of point-of-care devices which can detect contamination on a real-time basis in meat samples along with processing lines. Such tools are helpful not only in maintaining food safety but also reduce the cost of production due to their real-time detection and help in avoiding product recalls. Another tool that helps in monitoring food safety is the use of on-pack indicators coupled with data carriers such as RFID tags for temperature monitoring of temperature-sensitive products such as dairy, meat, etc.; production of gases such as oxygen, carbon dioxide which will be helpful in monitoring of products such as fruits and vegetables, meat, fermented foods, seafood, etc. One such example of freshness indicator is a sensor label produced by Food Quality Sensor International Inc., Lexington, MA, USA, for detecting biogenic amines which indicate bacterial growth. Similarly, there are tools available to monitor toxin levels. Although such tools are currently limited to very few applications, there is a good prospect for their use in food safety since the sector is expected a grow at a considerable pace. As discussed in earlier sections, blockchain is a powerful technology to monitor traceability throughout the value chain along with preventing food fraud. Beverage giant Coca-Cola has used Artificial Intelligence in developing new products. The company has installed several thousand self-serving soda fountains across the USA and studied customer preferences by collecting data on how they mixed a variety of their products. After analyzing the data, they developed a new product, Cherry Sprite. Such a type of understanding customer tastes and developing new products usually will cost a lot of manpower to conduct surveys and is cost-intensive. Next-Generation Sequencing (NGS), a rapid screening technique for the determination of food adulteration and foodborne pathogens at the molecular level is now available for use in the food industry. Though the technology is well developed, its applications in the food industry are still in the nascent stage. One example is the use of virtual reality in training food safety handlers in improving knowledge on food safety and mock drills in addressing such issues. Another aspect where technology can be exploited is the use of predictive modeling in food preparation areas by mapping high-risk areas and utilizing the information in implementing preventive actions to eliminate contamination. However, currently, certain technologies are cost-intensive. But with time, as with other technologies, it will become more affordable and will be one of the game-changing technologies for the food industry.

Though a wide variety of such tools and technologies are available to a certain extent at a commercial level, their adoption has been very limited. The primary reasons for such a low level of adoption are a limited number of applications and associated higher cost. Of course, with time, more choices will be available, which will also bring down the cost resulting in higher adoption of the advanced tools and techniques. These are expected to shape the food industry in terms of safety and quality in the years to come.

14.2 Future Trends in Quality and Food Safety

Technological advances made in the recent past have had a significant impact on food quality and safety. The future of food safety will be greatly influenced by advanced fields such as AI and ML due to its deep penetration into various areas including food safety. It will find wider applications due to its ease of use and being able to provide almost real-time data to the users. Though it is originally not intended for use in agriculture and food safety, its applications have found their way into all walks of life. The use of sensors is expected to penetrate the food quality segment due to its ability for yielding rapid and non-destructive results. Sensor-based technologies will provide almost instant results which will be in huge demand in the coming years. A lot of research has taken place in this segment and the rapid developments are an indicator of its wider applications in the food industry. The advent of such technologies, in the near future, may make current methodologies obsolete. Many people also believe that technology is taking over the world, and to some degree, they might be right. Although sometimes we do not realize it, technology impacts our every move, even swaying our food choices.

GMOs: While this may seem like a new technology, GMOs have been around for many years. The first genetically modified organism was made in 1994. Additionally, products from GMOs have been in stores since 1997. Essentially, scientists alter certain crops, making them resistant to pests as well as herbicides. Scientists also can increase the nutrition of the crop itself. So, when you bite into a carrot, you're receiving far more nutrition than before. Now, GMO crops are also being modified to grow in habitats that differ from their origin, which can provide a resolution to food crises in various countries.

Precision Agriculture: Precision agriculture, also known as satellite farming, is a form of farming that includes using GPS tracking systems as well as satellite imagery to keep an eye on weather patterns. The process allows farmers to adjust their crop yields and soil levels promoting efficiency. With the rise of food shortages and overpopulation, it is essential to make the most of an area and ensure that the food yields flourish. This technology provides the farmer the information they need to plant only on the parts of land that will yield the most food.

Data Analysis: It is part of the food industry, too! With tons of data coming in about certain foods, analyzing all of that data and determining relationships and trends allows authorities to determine which foods promote human nutrition. It had impacted regulatory agencies such as the FDA, which has revised a few of their protocols about certain foods as new data reveals their impact on humans. In the United States, a lot of food goes to waste. Whether consumers don't buy ripe fruits or throw their leftovers away, food usually ends up in the trash. To combat this waste, a few companies have started to take food waste and turn it into cosmetics. These cosmetics are then sold, giving the food that would otherwise have just been rotting at the bottom of the dumpster a purpose instead.

3D Printed Food: It may seem strange to imagine edible food exiting an electronic printer, but it's a legitimate operation. One company has already managed to make candies using pure sugar and a 3D printer. They have teamed up with Hershey's and 3D-printed chocolate. This new development could be used to create meatless meat, satisfying the protein need for protein-energy malnutrition whilst cutting down on the number of slaughtered cows and the amount of carbon emissions that are created in the meat industry. One product, Foodini, is already selling to homes. Now it is even possible to create 3D-printed pizzas, desserts, and burgers! Of course, as a first step, ingredients must be mixed in a blender. Perhaps in another year, there will be a feature that will eliminate manual intervention of performing blending too [17].

14.3 Conclusion

Food quality and safety had their modest beginnings years ago with the simple objective of preventing fraud. Although the objective has remained more or less the same, rapid advances have been made in detecting fraud. With global communities trying to protect their citizens from food fraud and to ensure food safety, stringent requirements have evolved leading to the formation of various global standards addressing food quality and safety. Such regulations became necessary since food became a tradable commodity. Due to scientific advancements, such adulterants have been developed which are difficult to detect using traditional methods. Such a requirement has led to the development of modern techniques that are sensitive and selective in detecting such adulterants. The future of food quality and safety will be greatly influenced by the most reliable, easy-to-use, and rapid technologies such as artificial intelligence and machine learning in the future which will redefine the way food safety and quality will be monitored in years to come.

References

1. World Health Organization. (2015) WHO estimates of the global burden of foodborne diseases: foodborne disease burden epidemiology reference group 2007-2015.
2. Jaffee, S., Henson, S., Unnevehr, L., Grace, D., and Cassou, E. (2019). The Safe Food Imperative: Accelerating Progress in Low- and Middle-Income Countries. Agriculture and Food Series; Washington, DC: World Bank.
3. ISO 9000:2015. Quality management systems — Fundamentals and vocabulary.
4. Good laboratory practice training manual for the trainer: a tool for training and promoting good laboratory practice (GLP) concepts in disease endemic countries - 2nd ed. (2008). World Health Organization.
5. Armin, S. and Hubertus F. (2017). Diagnostic Procedures. Armin, S. and Hubertus, F. (Eds.) *Biomedical Engineering in Gastrointestinal Surgery*. Academic Press, 87-220.
6. Hai, V., Christos, T., Paul, M., David, H., Trung, K.D., Thi, L.L., Ivan, A., and Stephen, M. (2016). Rice seed varietal purity inspection using hyperspectral imaging. Hyperspectral Imaging and Applications Conference, Coventry, United Kingdom.
7. Prashant Bagade. (2020). Post-harvest research in India. Sanjay, K. (Ed.). *India Commodity Yearbook 2020*. National Collateral Management Services Private Limited.

8. https://www.agriculture.com/technology/data/blockchain-technology-will-improve-farmers-ability-to-market-and-sell-crops. Last accessed on October 9, 2020.
9. Tripoli, M. and Schmidhuber, J. (2018). Emerging Opportunities for the Application of Blockchain in the Agri-food Industry. FAO and ICTSD: Rome and Geneva. Licence: CC BY-NC-SA 3.0 IGO.
10. Reshma, K. (2018) Food Traceability on Blockchain: Walmart's Pork and Mango Pilots with IBM. *Journal of the British Blockchain Association*, 1, 1, 1–12.
11. Alexandru, P., Mihaela, H., Mirel, P., Oana, G., Jude, D.H., Daniel, D., Le, H.S., and Simona, G. (2019). An Intelligent IoT-Based Food Quality Monitoring Approach Using Low-Cost Sensors. *Symmetry*, 11, 374.
12. http://www.marplug-technologies.com/SOCIP. Last accessed on October 12, 2020.
13. https://spd.group/machine-learning/machine-learning-and-ai-in-food-industry/#:~:text=Technology%20in%20the%20Food%20Industry,-When%20talking%20about&text=Artificial%20Intelligence%20and%20Machine%20Learning,well%20as%20in%20food%20manufacturing. Last accessed on October 12, 2020.
14. International Health Regulations. (2005). World Health Organization, 2nd Edition.
15. http://www.fao.org/faolex/en/ Last accessed on October 6, 2020.
16. University Of Maryland Medical Center. "Intestinal Bug Likely Killed Alexander The Great." *ScienceDaily*, 22 June 1998.
17. https://socialnomics.net/2018/03/04/5-ways-new-technologies-are-affecting-the-food-industry/. Last accessed on February 28, 2021.
18. Kumar, N., Kaur, P., Devgan, K., & Attkan, A. K. (2020). Shelf life prolongation of cherry tomato using magnesium hydroxide reinforced bionanocomposite and conventional plastic films. *Journal of Food Processing and Preservation*, 44(4), e14379.
19. Devgan, K., Kaur, P., Kumar, N., & Mahal, A. K. (2020). Development of smart film labels for storage quality assessment of yellow bell pepper (Capsicum annuum L.) under active packaging. *Arch Lebensmittelhyg*, 71, 138-145.
20. Malathi, A. N., Kumar, N., Nidoni, U., & Hiregoudar, S. (2017). Development of soy protein isolate films reinforced with titanium dioxide nanoparticles. *International Journal of Agriculture, Environment and Biotechnology*, 10(1), 141.

Index

Also of Interest

Other Books in the series, "Bioprocessing in Food Science"

NOVEL TECHNOLOGIES IN FOOD SCIENCE, Edited by Navnidhi Chhikara, Anil Panghal, and Gaurav Chaudhary, ISBN: 9781119775584. The third book in the series, "Bioprocessing in Food Science," covers the latest and greatest new technologies in food science. *DUE OUT IN 2022!*

FUNCTIONAL FOODS, Edited by Navnidhi Chhikara, Anil Panghal, and Gaurav Chaudhary, ISBN: 9781119775560. Presenting cutting-edge information on new and emerging food engineering processes, *Functional Foods,* the second volume in the groundbreaking new series, "Bioprocessing in Food Science," is an essential reference on the modeling, quality, safety, and technologies associated with food processing operations today. *DUE OUT IN EARLY 2022!*

MICROBES IN THE FOOD INDUSTRY, Edited by Navnidhi Chhikara, Anil Panghal, and Gaurav Chaudhary, ISBN: 9781119775584. *DUE OUT IN SUMMER 2022!*

Other related titles from Scrivener Publishing

FUNCTIONAL POLYMERS IN FOOD SCIENCE, Part 1: Food Packaging, Edited by Giuseppe Cirillo, Umile Gianfranco Spizzirri, and Francesca Iemma, ISBN: 9781118594896. Provides an extensive overview of the link between polymer properties and food quality. *NOW AVAILABLE!*

FUNCTIONAL POLYMERS IN FOOD SCIENCE, Part 2: Food Processing, Edited by Giuseppe Cirillo, Umile Gianfranco Spizzirri, and Francesca Iemma, ISBN: 9781118595183. Provides an extensive overview of the link between polymer properties and food quality. *NOW AVAILABLE!*

ADVANCES IN FOOD SCIENCE AND TECHNOLOGY Volume 1, Edited by Visakh, P.M., Sabu Thomas, Laura B. Iturriaga, and Pablo Daniel Ribotta,

ISBN: 9781118121023. Written in a systematic and comprehensive manner the book reports recent advances in the development of food science and food technology areas. *NOW AVAILABLE!*

ADVANCES IN FOOD SCIENCE AND NUTRITION Volume 2, Edited by Visakh P.M., Laura B. Iturriaga, and Pablo Daniel Ribotta, ISBN: 9781118137093. This important book comprehensively reviews research on new developments in all areas of food chemistry/science and nutrition. *NOW AVAILABLE!*

Applied Water Science Volume 1: Fundamentals and Applications, Edited by Inamuddin, Mohd Imran Ahamed, Rajender Boddula and Tauseef Ahmad Rangreez, ISBN 9781119724766. Edited by one of the most well-respected and prolific engineers in the world and his team, this is the first volume in a two-volume set that is the most thorough, up-to-date, and comprehensive volume on applied water science available today. *COMING IN SUMMER 2021*

Sustainable Water Purification, by Safiur Rahman and M. R. Islam, ISBN 9781119650997. This is the only book that takes a zero-waste approach to propose 100% sustainable water purification techniques. *NOW AVAILABLE!*